Automation in Proteomics and Genomics

Automation in Proteomics and Genomics

An Engineering Case-Based Approach

Editors

DR GIL ALTEROVITZ

Harvard/MIT Health Science and Technology Division, Massachusetts Institute of Technology, Cambridge, MA, USA

MS ROSEANN BENSON

DR MARCO RAMONI

Harvard/MIT Health Science and Technology Division, Massachusetts Institute of Technology, Cambridge, MA, USA

WILEY

A John Wiley and Sons, Ltd., Publication

This edition first published 2009

Registered Office
John Wiley & Sons Ltd, The Atrium, Southern Gate, Chichester, West Sussex, PO19 8SQ, United Kingdom

For details of our global editorial offices, for customer services and for information about how to apply for permission to reuse the copyright material in this book please see our website at www.wiley.com.

Library of Congress Cataloging-in-Publication Data

Automation in proteomics and genomics : an engineering case-based approach /
[edited by] Gil Alterovitz, Roseann Benson, Marco Ramoni.
p. ; cm.
Includes bibliographical references and index.
ISBN 978-0-470-72723-2 (cloth)
1. Proteomics–Automation. 2. Genomics–Automation. 3. Proteomics–Data processing. 4. Genomics–Data processing.
I. Alterovitz, Gil. II. Benson, Roseann. III. Ramoni, Marco F.
[DNLM: 1. Genomics–methods. 2. Automation–methods. 3. Proteomics–methods. QU 58.5 A939 2008]
QP551.A98 2008
572'.6–dc22

2008038622

A catalogue record for this book is available from the British Library.

ISBN 978-0-470-72723-2

Set in 10/12pt Times by Aptara Inc., New Delhi, India
Printed and bound in Singapore by Fabulous Printers Pte Ltd

Contents

Preface

Over the 10-year period between 1995 and 2005, DNA sequencing costs decreased fifty-fold, primarily as a result of increasing throughput due to incremental advances in tools, technologies and process improvements. As impressive as that accomplishment is the fact that the current outlay for sequencing the three billion base pairs of the DNA found in human or mammal genomes is $10 million dollars (sources: The National Human Genome Research Institute (NHGRI), The National Institute of Health, and genomics.energy.gov). When new high-throughput techniques and the computational machinery needed to analyze data generated are created and implemented, further cost reductions are expected.

Increasing the capacity to capture and analyze proteomic data is paramount on the research scene. To the casual observer, it would appear that proteomic researchers could simply adopt automation technology developed for genomics research. However, technology transfer between genomics and proteomics would need to be modified to account for post-transitional modifications (PTM), which are specific to proteomics. Also, proteomic datasets are expanding exponentially while genomic datasets are saturated. Managing a much larger, and continually growing database presents its own set of organizational opportunities.

The inspiration for *'Automation in Genomics and Proteomics: An Engineering Case-Based Approach'*, was two Massachusetts Institute of Technology (MIT) and Harvard interdisciplinary special studies courses, *Bioinformatics and Proteomics: An Engineering-Based Problem Solving Approach* and numbered 6.092/HST.480, respectively, and the article 'Automation, parallelism, and robotics for proteomics' created by Alterovitz, Liu, Chow and Ramoni. *Bioinformatics and Proteomics: An Engineering-Based Problem* Solving Approach focused on bioinformatics and proteomics with engineering-based approaches. While the 'Automation, parallelism and robotics for proteomics' [1] article discussed various technologies and methods being applied within the proteomics field that facilitate automation to achieve cost- and time-saving benefits and link proteomics-based information with germane research areas.

The book, *'Automation in Genomics and Proteomics: An Engineering Case-Based Approach'*, is the product of an international effort that spanned six countries and involved 18 public, private and academic institutions. The textbook addresses automation technology currently in the areas of analysis, design and integration and describes current technological limitations. The underlying biology concepts are also delineated. By disseminating knowledge of leading experts in the field, the textbook underscores that a collaborative effort across many disciplines is required to advance automation. This book's distinctive engineering-oriented coverage makes the material more intuitive for a technical audience, and is intended for an upper-level undergraduate elective or as a graduate-level foundation class. It can also be used as an industry reference tool. This book is an excellent stand-alone text for an introductory/motivational seminar or course on the subject, or it can serve as a complementary text to traditional texts.

'Automation in Genomics and Proteomics: An Engineering Case-Based Approach', is divided into four sections. In Section 1, the fundamental biology is introduced from an engineering perspective. The first chapter, *The Central Dogma: from DNA to RNA, and to Protein*, presents the needed molecular and cellular biology background and can be treated within a review session, if an introductory biology course is a prerequisite. Molecules and bioprocesses that are related to protein biosynthesis are the focus of this chapter, where DNA is the source of genetic information and amino acids are the raw materials. In the second chapter, *Genomes to Proteomes*, the book moves from the genomics to proteomics by discussing the automatic annotation and the generation of high-quality gene models, the set-up and execution of quantitative and statistically rigorous global proteomic experiments, and proteomics in a biological context.

Sections 2 and 3 focus on design and analysis via automation:

- Chapter 3, *High-Throughput DNA Sequencing*, reviews the current state of traditional dideoxy sequencing workflows and the development of next-generation technologies that are poised to revolutionize genomics.
- Chapter 4, *Modeling a Regulatory Network using Temporal Gene Expression Data: Why and How?*, introduces methodologies to classify genes according to their expression measurements across a set of conditions, and presents sophisticated procedures to infer regulatory networks from gene expression data.
- Chapter 5, *Automated Prediction of Protein Attributes and Its Impact on Biomedicine and Drug Discovery*, systematically explores recent progress in computational methods and, for those prediction methods with web-servers currently available, a step-by-step instruction is presented.
- Chapter 6, *Molecular Interaction Networks: Topological and Functional Characterizations*, introduces both basic concepts and current research trends in network biology.
- Chapter 7, *DNA Synthesis*, evaluates the standard methods, chemistry and instrumentation of DNA and gene synthesis and the applications, strategies and the trends of applying high-throughput gene synthesis in synthetic biology to design and engineer biological systems.
- Chapter 8, *Computational and Experimental RNA Nanoparticle* Design, discusses some of the recent advances in the ability to computationally design RNA nanoparticles, drawing upon a wide array of known structural motifs as well as the issues that are involved in experimentally constructing and testing their validity.

- Chapter 9, *New Paradigms in Droplet-Based Microfluidics and DNA Amplification*, addresses the confluence of new genetic and microfluidic technologies that will be used to automate *in vitro* evolution, genomics and molecular and cellular screening on a scale that was previously impractical.
- Chapter 10, *Synthetic Networks*, demonstrates several interesting synthetic networks and provides valuable engineering tools to study motifs, modularity and robustness of cellular networks. Also in this chapter, the current understanding of cellular networks, synthetic network construction and remaining challenges towards automating biochemical processes using synthetic circuitry are reviewed.

Chapters 11 and 12 are included in Integration, the fourth and final section of *Automation in Genomics and Proteomics: An Engineering Case-Based Approach*. In this section, Chapter 11, *Molecular Modeling of CYP Proteins and Its Implication for Personal Drug Design*, addresses the existing computational methods for modeling 3-D protein structures with Shanghai Molecular Modeling (SAMM), the molecular modeling software developed at Shanghai Jiaotong University, while Chapter 12, *Recent Progress of Bioinformatics in Membrane Protein Structural Studies*, examines the recent progress of computational work and automation process in membrane protein structural studies.

In Chapter 13, *Trends in Automation for Genomics and Proteomics*, the book concludes with a field summary and an exploration of future avenues of research. For those interested in additional resources, source code and related materials: the book's internet site can be accessed at: http://bcl.med.harvard.edu/proj/automation.

As an international effort, there are many people whose contributions were critical to the publication of this work. The editors would like to thank the contributing authors to the text, including: Masahiko Sisido and Takashi Ohtsuki (Chapter 1); Scott E. Baker, Ellen A. Panisko, Igor Grigoriev, Don S. Daly and Bobbie-Jo Webb-Robertson (Chapter 2); Tarjei S. Mikkelson (Chapter 3); Sophie Lebre and Gaelle Lelandais (Chapter 4); Kou-Chen Chou (Chapters 5, 11 and 12); Jake Chen and Xiaogang Wu (Chapter 6); Jingdong Tian (Chapter 7); Isil Severcan, Cody Geary, Luc Jaeger, Eckart Bindewald, Wojciech Kasprzak and Bruce A. Shapiro (Chapter 8); Jonathan Rothberg, Michael Samuels, John Leamon, Ronald Godiska, Thomas Schoenfeld and David Mead (Chapter 9); Jongmin Kim (Chapter 10); Cheng-Cheng Zhang, Jing-Yi Yan, Jing-Fang Wang and Dong-Qing Wei (Chapter 11); Hong-Bin Shen, Jun-Feng Wang, Li-Xiu Yao and Jie Yang (Chapter 12); and Dmitriy Sonkin (Chapter 13). The editors would especially like to thank the Wiley commissioning editor, Paul Deards, who invited us to write this book.

Thank you to William H. Down and Jonathan Dreyfuss for reviewing and editing the manuscript. A special thanks also to the anonymous book proposal and book draft reviewers.

Gil Alterovitz, PhD
Roseann Benson
Marco F. Ramoni, PhD

Reference

1. G. Alterovitz, J. Liu, J. Chow, and M.F. Ramoni. (2006) Automation, parallelism, and robotics for proteomics. *Proteomics*, **6**(14), 4016–22.

List of Contributors

Gil Alterovitz Harvard/MIT Health Science and Technology Division, Massachusetts Institute of Technology, Cambridge, MA, USA.

Scott E. Baker Pacific Northwest National Laboratory, Richland, Washington, USA.

Roseann Benson Harvard Partners Center for Genetics and Genomics, Harvard Medical School, Boston, MA, USA.

Eckart Bindewald SAIC-Frederick, Inc., Basic Research Program, NCI-Frederick, Frederick, MD, USA.

Jake Chen Indiana University School of Informatics/Purdue University School of Science, Indianapolis, IN, USA.

Kuo-Chen Chou Gordon Life Science Institute, San Diego, California, USA.

Don S. Daly Pacific Northwest National Laboratory, Richland, Washington, USA.

Cody Geary Department of Chemistry and Biochemistry, Biomolecular Science and Engineering program, University of California at Santa Barbara, Santa Barbara, CA, USA.

Ronald Godiska Lucigen Corporation, Middleton, WI, USA.

Igor Grigoriev US DOE Joint Genome Institute, Walnut Creek, California, USA.

Luc Jaeger Department of Chemistry and Biochemistry, Biomolecular Science and Engineering program, University of California at Santa Barbara, Santa Barbara, CA, USA.

Wojciech Kasprzak SAIC-Frederick, Inc., Basic Research Program, NCI-Frederick, Frederick, MD, USA.

Jongmin Kim Biotechnology Team, *CbsBioscience Inc.* Daejeon, Korea.

John Leamon RainDance Technologies, Guilford, CT, USA.

Sophie Lèbre Centre for Bioinformatics, Division of Molecular Biosciences, Imperial College London, London, UK.

Gaëlle Lelandais Equipe de Bioinformatique Génomique et Moléculaire, Université Paris 7, Paris, France.

David Mead Lucigen Corporation, Middleton, WI, USA.

Tarjei S. Mikkelsen Harvard-MIT Division of Health Sciences and Technology, Broad Institute of MIT and Harvard, Cambridge, MA, USA.

Takashi Ohtsuki Department of Bioscience and Biotechnology, Okayama University, Japan.

Ellen A. Panisko Pacific Northwest National Laboratory, Richland, Washington, USA.

Marco Ramoni Harvard/MIT Health Science and Technology Division, Massachusetts Institute of Technology, Cambridge, MA, USA.

Jonathan Rothberg RainDance Technologies, Guilford, CT, USA.

Michael Samuels RainDance Technologies, Guilford, CT, USA.

Isil Severcan Department of Chemistry and Biochemistry, Biomolecular Science and Engineering program, University of California at Santa Barbara, Santa Barbara, CA, USA.

Thomas Schoenfeld Lucigen Corporation, Middleton, WI, USA.

Bruce A. Shapiro Center for Cancer Research Nanobiology Program, National Cancer Institute, Frederick, MD, USA.

Hong-Bin Shen Institute of Image Processing & Pattern Recognition, Shanghai Jiaotong University, Shanghai, China.

Masahiko Sisido Department of Bioscience and Biotechnology, Okayama University, Japan.

Dmitriy Sonkin Harvard Partners Center for Genetics and Genomics, Harvard Medical School, Boston, MA, USA.

Jingdong Tian Department of Biomedical Engineering & Institute for Genome Sciences and Policy, Duke University, Durham, NC, USA.

Jun-Feng Wang Department of Biological Chemistry and Molecular Pharmacology, Harvard Medical School, Boston, MA, USA.

Jing-Fang Wang Department of Bioinformatics and Biostatistics, College of Life Sciences and Technology, Shanghai Jiao Tong University, Shanghai, China.

Bobbie-Jo Webb-Robertson Pacific Northwest National Laboratory, Richland, Washington, USA.

Dong-Qing Wei Department of Bioinformatics and Biostatistics, College of Life Sciences and Technology, Shanghai Jiao Tong University, Shanghai, China.

Xiaogang Wu Indiana University School of Informatics/Purdue University School of Science, Indianapolis, IN, USA.

Jing-Yi Yan Department of Bioinformatics and Biostatistics, College of Life Sciences and Technology, Shanghai Jiao Tong University, Shanghai, China.

Jie Yang Institute of Image Processing & Pattern Recognition, Shanghai Jiaotong University, Shanghai, China.

Li-Xiu Yao Institute of Image Processing & Pattern Recognition, Shanghai Jiaotong University, Shanghai, China.

Cheng-Cheng Zhang Department of Bioinformatics and Biostatistics, College of Life Sciences and Technology, Shanghai Jiao Tong University, Shanghai, China

About the Editors

Gil Alterovitz, PhD

Dr Alterovitz received his PhD in Electrical and Biomedical Engineering at MIT through the Harvard/MIT Division of Health Sciences and Technology. He is a biomedical informatics fellow with the Harvard/MIT Division of Health Sciences and Technology (HST), Children's Hospital Informatics Program, and the Harvard Medical School-Partners Center for Genetics and Genomics. He is currently heading a new class that he initiated at Harvard University, Bio.95hfa: 'Proteomics and Cellular Network Engineering'. He has served on the Harvard/MIT Division of Health Science and Technology MD Curriculum and the Harvard/MIT Division of Health Science and Technology PhD Admission committees. He was a US Fulbright to Canada (University of Toronto) in 1998–1999. Dr Alterovitz has an S.M. degree from the Massachusetts Institute of Technology (MIT) in Electrical Engineering and Computer Science, where he was a NDSEG Fellow. His B.S. is in Electrical and Computer Engineering from Carnegie Mellon University.

Dr Alterovitz has worked at Motorola (where he won the Motorola Intellectual Property Award), at IBM, and as a consultant for several national clients. As an invited contributor, he wrote the 'Proteomics' section for the *Wiley Encyclopedia of Biomedical Engineering*. Dr Alterovitz has appeared or has been cited for achievements in several national media outlets, including three separate editions of USA Today and National Public Radio. He was also featured in the Boston Globe. In 2001, he was selected as one of approximately 20 international delegates to the Canada25 forum (to discuss healthcare/technology) covered by CBC radio, a national TV special and Canada's *Maclean's*.

Roseann Benson

For eleven years, Ms Benson was a chemical engineer, and participated in all phases of automation implementation projects in a variety of industries: nuclear, semiconductor, aluminum, specialty chemical and environmental. Her engineering experience ran the project gamut, from inception to completion, and included framing the technical problems correctly; designing bench tests to establish system specifications; setting and calculating design parameters; selecting, purchasing and installing equipment; and utilizing general equipment to meet particular applications. As part of these engineering projects, Ms Benson acquired extensive experience with documentation. One paper, 'Shock Deionization of the K and L Basins', presented the research and outcome of a successful project where indefinite scientific principles were defined more accurately within a particular context. She won an American Institute of Chemical Engineers' (AIChE) National 'Outstanding Paper' award for the paper that she wrote and presented at an AIChE meeting. Ms Benson's formal credentials include a Clarkson University Chemical Engineering bachelor's degree, a Nova Southeastern executive Masters in Business Administration and a Harvard Certificate in Administration and Management.

Ms Benson writes nonfiction outside of the engineering field, and is a contributor to three Cadogan Guides. Her first book, *101 Puppy-Buying Tips* has recently been published by LifeTips.com. In addition, she serves on the City of Beverly's Library Board of Trustees.

Marco F. Ramoni, PhD

Marco F. Ramoni is Assistant Professor of Pediatrics and Medicine at Harvard Medical School and Assistant Professor of Health Sciences and Technology at the Harvard University and the Massachusetts Institute of Technology Division of Health Sciences and Technology. He is also Associate Director of Bioinformatics at the Harvard Partners Center for Genetics and Genomics and the Director of the National Library of Medicine Training Fellowship in Biomedical Informatics at Children's Hospital Boston. He is also the Director of the course 'Biomedical Informatics' at the Harvard-MIT Division of Health Sciences and Technology, core faculty of the course Genomic Medicine at Harvard Medical School and a member of the curriculum committee of the Cellular and Molecular Medicine track of the Medical Physics and Medical Engineering graduate program at Harvard-MIT Division of Health Sciences and Technology. He is cofounder of Bayesware LLC, a software company developing machine-learning programs based on Bayesian methods. He received a PhD in Biomedical Engineering and a BA in Philosophy (Epistemology) from the University of Pavia (Italy), and completed his postdoctoral training at McGill University, Montreal (Canada). He has held academic and visiting positions at the University of Massachusetts, the University of London (United Kingdom), the Knowledge Media Institute (United Kingdom) and the University of Geneva (Switzerland). He is author of over 90 publications in genetics, biomedical informatics, statistics and artificial intelligence.

Section 1

Fundamentals of Molecular and Cellular Biology

1

The Central Dogma: From DNA to RNA, and to Protein

Takashi Ohtsuki and Masahiko Sisido

Department of Bioscience and Biotechnology, Okayama University, Japan

Within a single cell – the minimum unit of every living organism – many millions of different types of molecule are working to maintain the cell, to promote its replication, or even to cause its suicide. The bioprocesses conducted within the cell are chemical reactions that proceed under the control of a highly organized network of molecular interactions between relevant biomolecules.

Among these biomolecules, three types of biopolymer are crucial, namely nucleic acids, proteins and polysaccharides. Nucleic acids preserve, replicate and transform the genetic information that serves to design a number of different proteins and low-molecular-weight biomolecules. Proteins function at almost all stages of the bioprocesses, from the birth to the death of a cell. Polysaccharides play important roles in communicating molecular network information and in storing chemical energy. Biopolymer concentrations are regulated to optimum levels for each stage of the bioprocess, but decompose when their roles are complete. This chapter will focus on the molecules and bioprocesses that are related to protein biosynthesis, where DNA is the source of genetic information and the amino acids are the raw materials.

1.1 Chemistry of DNA

Deoxyribonucleic acid (DNA) is a biopolymer that is located inside the nucleus of mammalian cells or in the cytosol of bacterial cells. DNA stores the genetic information that will be converted into the amino acid sequences of protein molecules in the cell.

Automation in Proteomics and Genomics: An Engineering Case-Based Approach Edited by Gil Alterovitz, Roseann Benson and Marco Ramoni

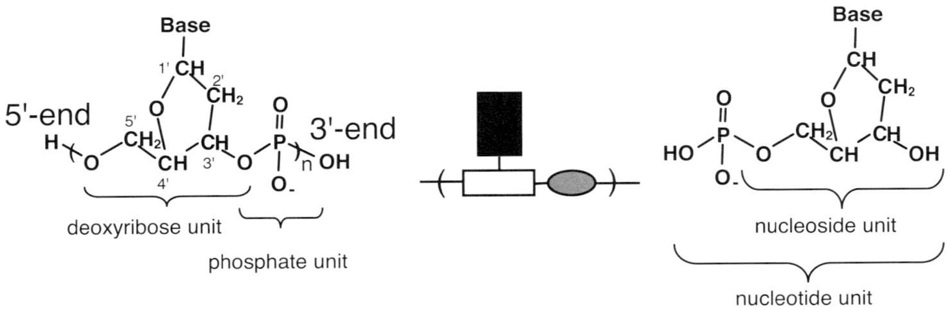

Figure 1.1 *The chemical structure of DNA and its monomer unit*

DNA, as shown in Figure 1.1, is a polyester made through condensations between a deoxyribose as the diol unit and a phosphoric acid as the bivalent acid unit. The negatively charged phosphates make the DNA molecule water-soluble. Due to the asymmetric arrangement of the 5′-OH and 3′-OH groups on the deoxyribose unit, DNA is a directional biopolymer. The chain end with the 5′-OH or 5′-O-phosphate unit is called the 5′-end, while the end with the 3′-OH or 3′-O-phosphate unit is called the 3′-end.

DNAs are characterized by the sequences of base groups that are linked to the deoxyribose units. There are four types of base group: adenine (A), thymine (T), guanine (G) and cytosine (C). Different DNAs carry different sequences of nucleobases that are read from the 5′ end to the 3′ end.

Nucleobases form hydrogen bonds between A and T and between G and C exclusively, as shown in Figure 1.2. With few exceptions, the A-T/G-C pairing is a basic rule common to all organisms. As a result of this exclusive pairing, a DNA strand that has a 5′-A-T-G-C-A-T-G-C-3′ sequence, for instance, forms a stable hybrid only with a DNA strand of a 5′-G-C-A-T-G-C-A-T-3′ sequence. Note that the two DNA strands hybridize in an antiparallel manner, as shown in Figure 1.3. The two DNA strands that carry fully matched sequences are called a complementary pair. Watson and Crick discovered that the complementary DNA strands form a double-helical structure, as shown in Figure 1.4.

As the base sequences are kept safely inside the cylinder of negatively charged, double-helical chains, the genetic information has been stored securely for many generations in

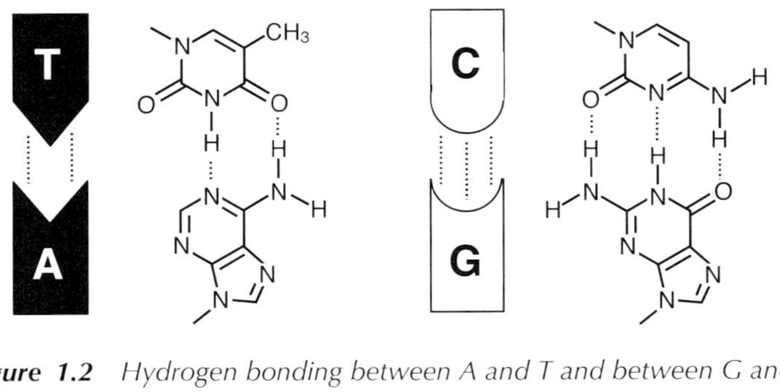

Figure 1.2 *Hydrogen bonding between A and T and between G and C*

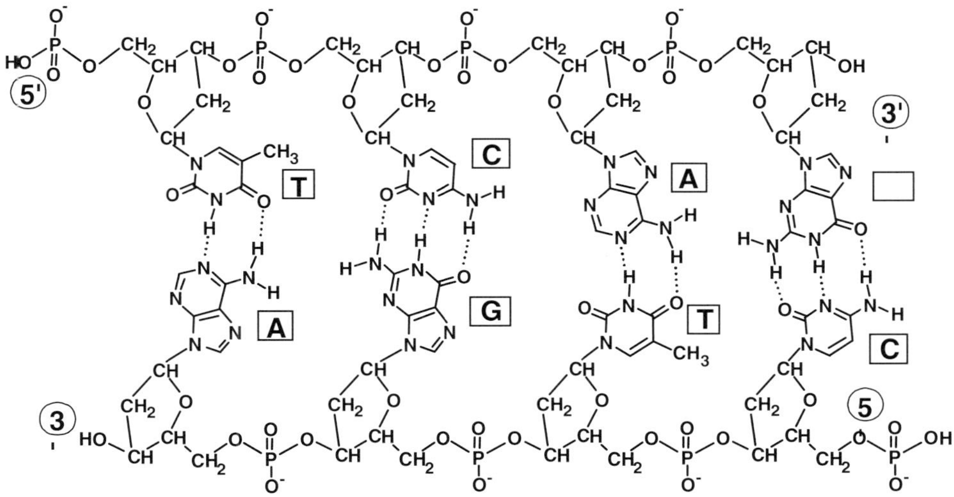

Figure 1.3 Chemical structure of double-stranded DNA

the form of the sequences of nucleobases. The double-helical structure, however, is not absolutely stable, and unfolds at high temperatures or by the action of an enzyme called a helicase.

1.2 Replication of DNA

In order for genetic information to be transferred to the next generation, DNA must first be copied to replicate itself. DNA replication is conducted with an aid of an enzyme called DNA polymerase. The basic chemistry of the replication proceeding inside the enzyme is illustrated in Figure 1.5.

First, the double-helical chain is unfolded and one of the DNA chains is copied to create its complementary chain. The monomer units involved in this polymerization are activated nucleotide units, dATP, dTTP, dGTP and dCTP. The triphosphate unit of the dNTP units is very susceptible to the attack of the 3′-OH group, and forms a diphosphate linkage. Guided

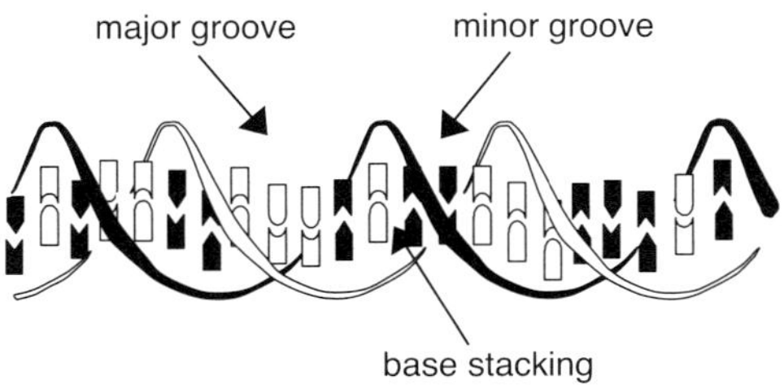

Figure 1.4 Double-helical structure of a complementary pair of two DNA strands

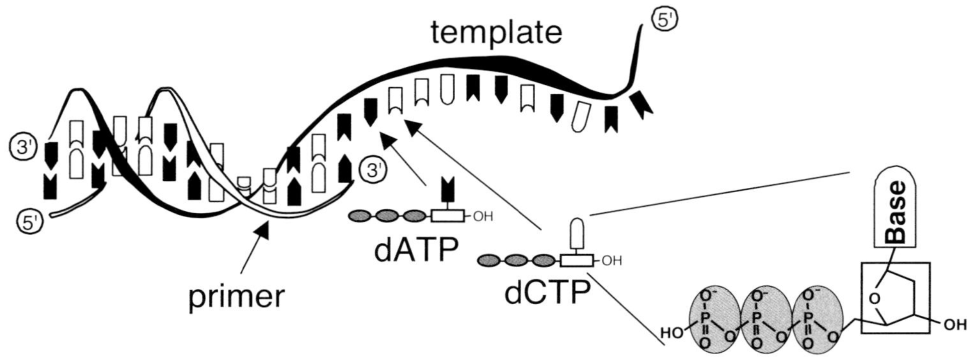

Figure 1.5 Replication of one of the DNA chains to create a complementary chain

by the enzyme, a correct monomer binds to the template DNA chain and reacts with the 3′-OH group of the growing chain. In this way, the new chain grows from the 5′ end to the 3′ end.

1.3 Transcription from DNA to RNA

Although the stable and inflexible DNA double-helical structure is suitable for the storage of genetic information, its large size necessitates that a smaller, more flexible biopolymer, is used to translate the stored genetic code into proteins. To that end, the base sequences are copied into another type of biopolymer nucleic acid, specifically ribonucleic acid (RNA).

RNA is structurally different from DNA in two ways (see the left part of Figure 1.6). The first difference is that an OH group is attached to the 2′C atom of deoxyribose unit; the 2′-OH derivative is called a ribose unit. The second difference is that a methyl group is removed from the thymine unit to make a uracil unit, U. The introduction of a 2′-OH

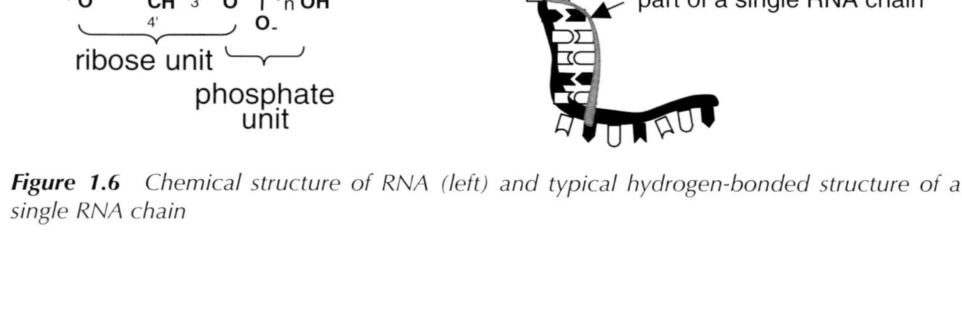

Figure 1.6 Chemical structure of RNA (left) and typical hydrogen-bonded structure of a single RNA chain

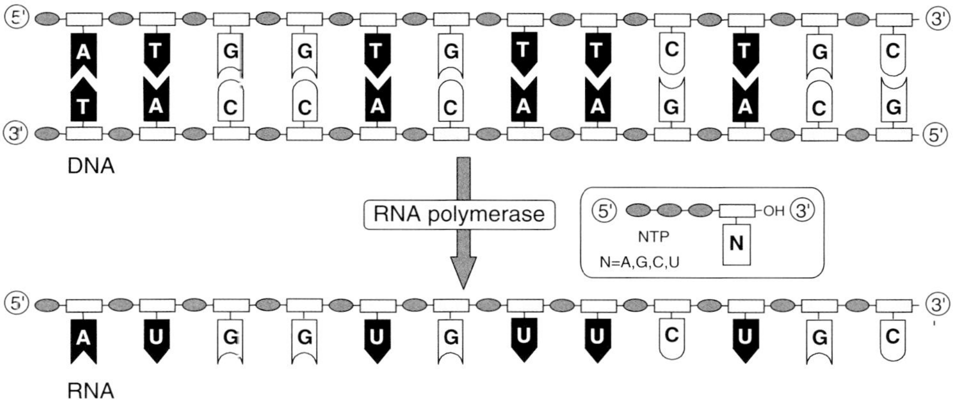

Figure 1.7 *Transcription of one of the double-stranded DNA chains to a complementary RNA chain with the aid of RNA polymerase*

group causes a small conformational change on the ribose unit such that the RNA chain will favor single-stranded conformations. The single-stranded RNAs, however, often assume an intramolecularly hydrogen-bonded structure, such as a stem-loop structure (see Figure 1.6, right).

Similar to DNA replication, one of the double-stranded DNA chains is copied to a single RNA chain of the complementary nucleobase sequence, except for the alteration of T to U, as shown schematically in Figure 1.7. This procedure is known the transcription process, and is conducted with an enzyme called RNA polymerase. The chemistry of transcription is similar to the replication process, and the monomers are ATP, UTP, GTP and CTP.

1.4 Translation of the Nucleobase Sequence of mRNA to the Amino Acid Sequence of Protein

The information stored in the form of a nucleobase sequence along an RNA chain is translated to an amino acid sequence of a protein, as shown schematically in Figure 1.8. RNAs that serve the translation process are called messenger RNAs (mRNAs). In the translation process, three consecutive nucleobases on a mRNA are taken together and converted to a specific amino acid. The set of three nucleobases is called a codon. As four possibilities (A, U, G and C) exist for each nucleobase, there are $4^3 = 64$ different codons.

Adapter molecules bridge the codons and the amino acids. A class of small RNAs, called transfer RNAs (tRNAs), serve as those adapters. The base sequence of a yeast tRNA that bridges between a codon UUC and an amino acid, phenylalanine, is shown in Figure 1.9.

tRNAs commonly have stem-loop structures with three loops and four stems. Among the loops, the anticodon loop contains three consecutive nucleobases that bond specifically to its complementary codon; thus, a tRNA of a specific anticodon binds to a specific codon on an mRNA. If a particular amino acid is linked to a specific tRNA of specific anticodon, the amino acid will be called up by the codon. In this way, the sequence of nucleobases is translated to the sequence of amino acids.

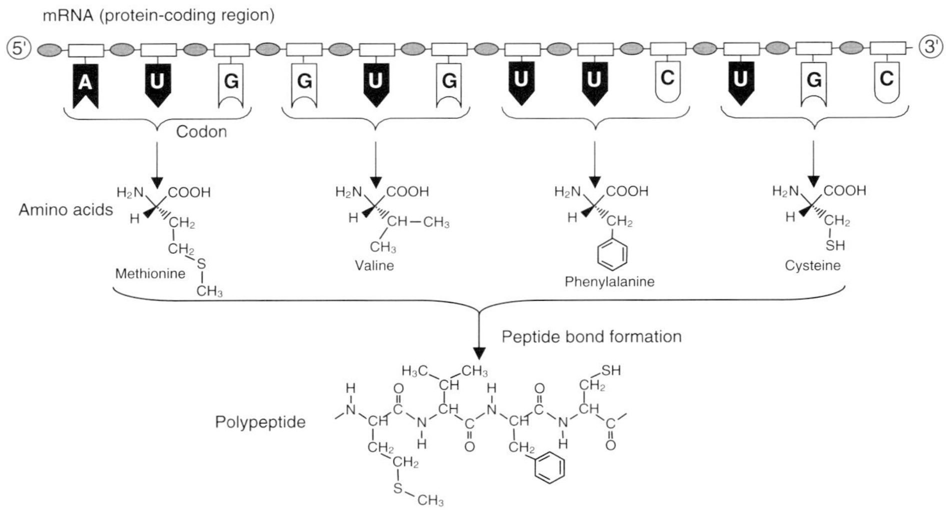

Figure 1.8 *Translation of a nucleobase sequence on an mRNA to an amino acid sequence of a protein. A set of three consecutive nucleobases (codon) corresponds to a specific amino acid. The amino acids will be linked together to produce a polypeptide chain*

1.5 The Codon Table

A list that correlates between the base sequences of codons and the amino acids is called a codon table (see Figure 1.10). The codon table is common to almost all organisms on the

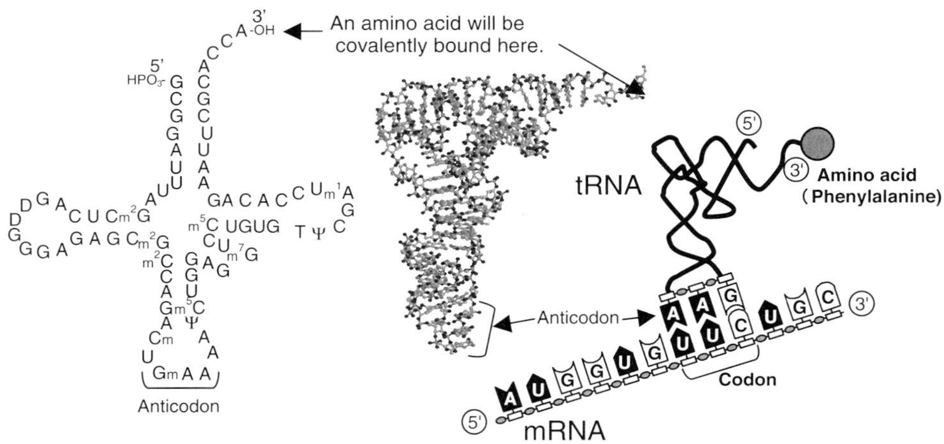

Figure 1.9 *Yeast transfer RNA (tRNA) that bridges between a codon UUC and an amino acid, phenylalanine. Nucleobase sequence (left), crystal structure (center) and a schematic illustration of the codon/anticodon pairing. The tRNA contains modified nucleobases, such as m^2G, m^2C, Cm, Gm, Ψ and D*

1st letter \ 2nd letter	U	C	A	G
U	UUU Phe	UCU Ser	UAU Tyr	UGU Cys
	UUC Phe	UCC Ser	UAC Tyr	UGC Cys
	UUA Leu	UCA Ser	UAA Stop	UGA Stop
	UUG Leu	UCG Ser	UAG Stop	UGG Trp
C	CUU Leu	CCU Pro	CAU His	CGU Arg
	CUC Leu	CCC Pro	CAC His	CGC Arg
	CUA Leu	CCA Pro	CAA Gln	CGA Arg
	CUG Leu	CCG Pro	CAG Gln	CGG Arg
A	AUU Ile	ACU Thr	AAU Asn	AGU Ser
	AUC Ile	ACC Thr	AAC Asn	AGC Ser
	AUA Ile	ACA Thr	AAA Lys	AGA Arg
	AUG Met	ACG Thr	AAG Lys	AGG Arg
G	GUU Val	GCU Ala	GAU Asp	GGU Gly
	GUC Val	GCC Ala	GAC Asp	GGC Gly
	GUA Val	GCA Ala	GAA Glu	GGA Gly
	GUG Val	GCG Ala	GAG Glu	GGG Gly

Figure 1.10 *The codon table*

earth, except for several violations found in mitochondria. The codon table is, therefore, the second basic rule of living organisms.

Because 64 codons correspond to 20 amino acids, there is redundancy in the use of codons. For example, phenylalanine is coded by both UUU and UUC, while leucine is coded by six codons. It must be noted that UUA, UAG and UGA do not correspond to any amino acid, and so are called stop codons. Thus, if they appear on an mRNA, the protein synthesis will cease. Compared with the stop signal, the mechanism of the start of protein synthesis is a little complicated, and is different in bacteria and eukaryotic cells. In prokaryotic bacteria, mRNA has a special region, called the Shine–Dalgano (SD) sequence, and the first AUG codon after the SD sequence works as the start codon. In eukaryotic cells, the first AUG codon from the 5′ terminal of an mRNA is the start codon. In any case, the N-terminal amino acid of a newly synthesized protein will be methionine.

1.6 The Twenty Amino Acids

The types of amino acid that constitute proteins, again, are common to all organisms; the chemical structures of the 20 amino acids are listed in Figure 1.11.

Amino acids are classified into five types, depending on chemical and physical properties of their side groups. The first group (Gly, Ala, Val, Leu, Ile, Met and Pro) has hydrophobic

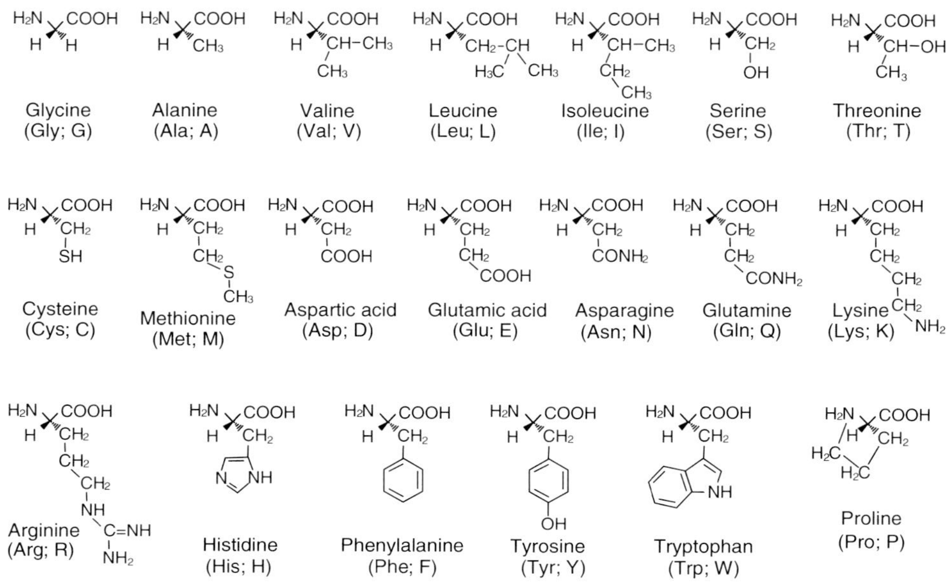

Figure 1.11 *Amino acids that constitute proteins*

side groups, while the second group (Ser, Cys, Thr, Asn and Gln) has nonionic polar side groups. The third group (Asp and Glu) has anionic side groups, and the fourth (Arg and Lys) has cationic groups. The fifth group (His, Phe, Tyr and Trp) has aromatic groups. Protein conformations depend on the combination of the different groups along a polypeptide chain, such as water-soluble globular proteins, membrane-penetrating hydrophobic proteins, and so on.

1.7 Aminoacylation of tRNA

The process of linking a specific amino acid to a specific tRNA is called the aminoacylation of tRNA, and is governed by a single enzyme, aminoacyl tRNA synthetase (ARS) for each amino acid. For example, phenylalanine (Phe) is charged onto a tRNA that has an anticodon UUU or UUC, with an enzyme PheRS. Aminoacylation consists of two stages, as illustrated in Figure 1.12 (top). First, a particular amino acid is bound to its specific ARS and is activated with adenosine triphosphate (ATP) to form an adenylated amino acid. In the second stage, a particular tRNA is bound to its specific ARS that holds the adenylated amino acid. The latter then reacts with the 3′-terminal OH group of the tRNA to form an ester linkage between the amino acid and the tRNA.

ARS is a 'super' enzyme that recognizes three different substrates: ATP, a specific amino acid, and a specific tRNA. In the first stage of aminoacylation, the formation of an adenylated amino acid activates an amino acid. The mixed anhydride of carboxylic acid and phosphoric acid of the adenylated amino acid is very susceptible to water. Inside the enzyme, however, the mixed anhydride is kept safe, until a correct tRNA is bound in

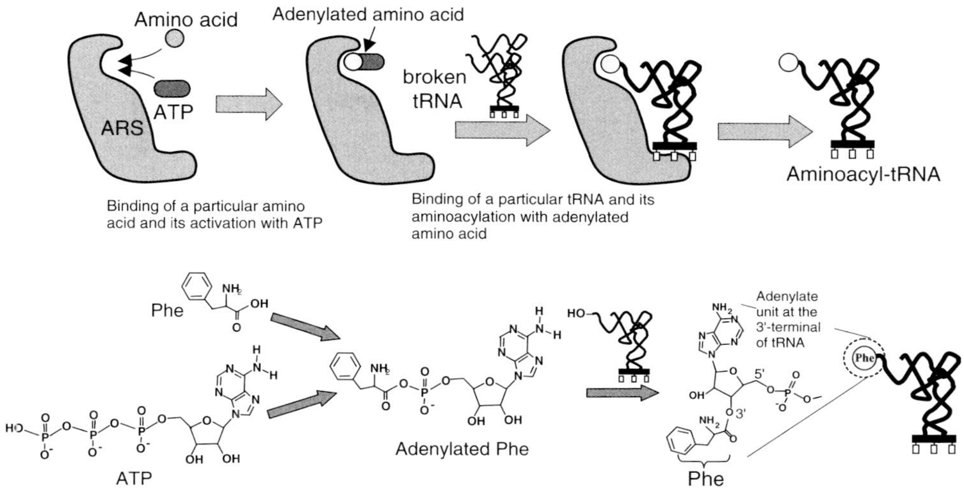

Figure 1.12 *Schematic illustration (top) and chemical processes (bottom) of tRNA aminoacylation*

proximity so as to induce the aminoacylation. The accuracy of the ARS/amino acid/tRNA selection is very high, and the probability of an erroneous aminoacylation is less than 10^{-4}.

1.8 Protein Synthesis in Ribosomes

Codon/anticodon pairing consists of only three base pairs, and is not strong enough to hold the tRNA/mRNA hybrids. Consequently, aminoacyl tRNAs do not bind to mRNA in solution, even if they have correct anticodons against the codons on mRNA. The codon/anticodon pairing takes place only inside a huge molecular assembly, called a ribosome, which is constructed from RNAs and proteins (Figure 1.13). Inside a ribosome, there are two 'rooms' – one for an aminoacyl tRNA (A site) and the other for a tRNA linked with the growing peptide (P site). There are also two 'tunnels' – one for an mRNA and the other for the growing peptide. Protein synthesis proceeds inside a ribosome, as illustrated in Figures 1.13 and 1.14.

After a tRNA has been aminoacylated with the relevant ARS, it is brought into the A site of the ribosome by the aid of an enzyme, elongation factor-Tu (EF-Tu). In the A site, the aminoacyl tRNA is oriented to locate its amino group in close proximity to the C-terminal ester group of the growing peptide on the tRNA in the P site (see Figure 1.14, top, left). The amino group then attacks the ester group, leading to the formation of a new peptide bond. As the result of this peptide bond formation, the growing peptide transfers to the tRNA in the A site (peptidyl transfer; see Figure 1.14, top, right). The A site tRNA, carrying the growing peptide, is then translocated to the P site, leaving the A site vacant (Figure 1.14, bottom). Finally, the next aminoacyl tRNA will be brought into the vacant A site. This polymerization cycle will be repeated until one of stop codons (UAA, UAG and UGA)

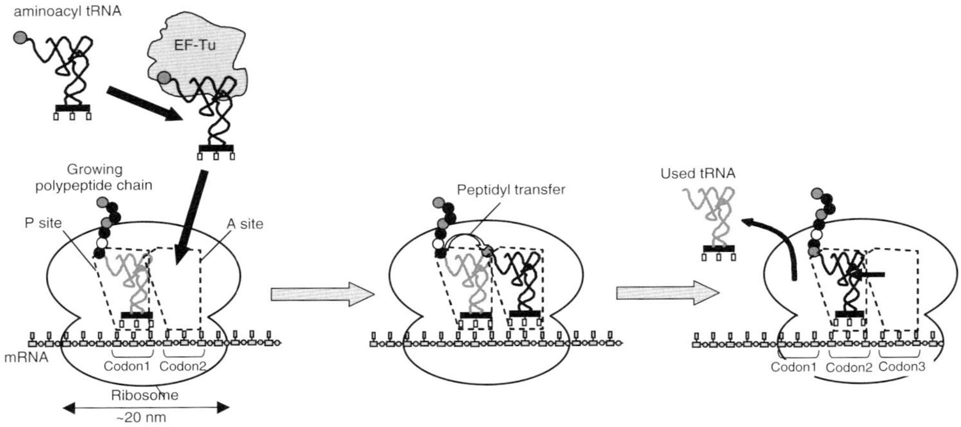

Figure 1.13 Schematic illustration of the course of protein synthesis inside a ribosome

appears on the mRNA. The polypeptide synthesis proceeds at a rate of about two amino acids per second in eukaryotic cells, and about 20 amino acids per second in bacteria.

1.9 The Total Process of Protein Synthesis: 'The Central Dogma'

The entire bioprocess – from DNAs to proteins and from amino acids to polypeptides – is summarized in Figure 1.15. The protein biosynthetic process is essentially the same in all organisms and is referred to as the 'central dogma', although several important differences exist between bacterial and eukaryotic systems.

The central dogma consists of two paths: one path for a flow of information from the nucleobase sequences of DNAs to the amino acid sequences of proteins, and a second path for a flow of materials from amino acids to polypeptides.

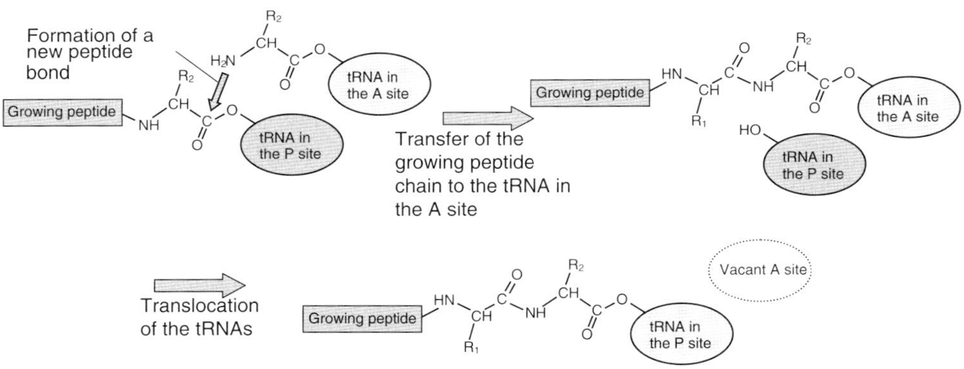

Figure 1.14 Chemistry of protein synthesis occurring inside a ribosome

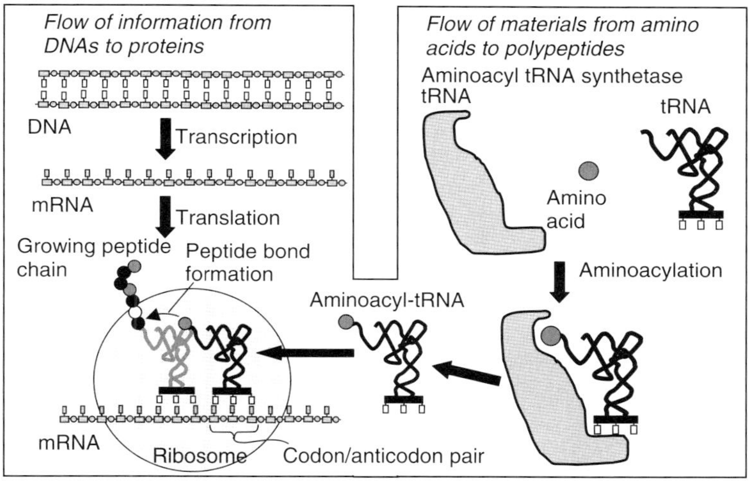

Figure 1.15 *The central dogma: The protein biosynthetic process*

It is counterintuitive that all organisms from bacteria, to plants and to animals, share essentially the same biosynthetic mechanism, due to the obvious differences in their physical appearances. As no organism lives with only 18 types of amino acid or with six types of nucleobase, it can be deduced that all living organisms are descendants from a single common cell that was comprised of 20 types of amino acid, four types of nucleobase and, essentially, the same protein-biosynthesizing system, as shown in Figure 1.15. Currently, a number of chemists are attempting to expand the central dogma and to create a 'new life' that lives with more than 20 types of amino acid or with more than four types of nucleobase.

1.10 Proteins: Polypeptides with a Variety of Specialty Side Groups that are Spatially Arranged to Achieve Biological Functions

Proteins are constructed from polypeptide chains along which a variety of functional groups are rationally arranged to play individual roles. Unlike most synthetic polymers, the polypeptide main chain is relatively stiff; such rigidity is due to amide groups favoring a planar and trans geometry resulting from the partial shift of an electron from nitrogen to oxygen (Figure 1.16).

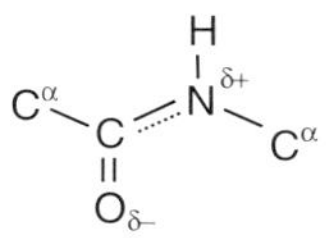

Figure 1.16 *Planar and trans geometry of an amide group*

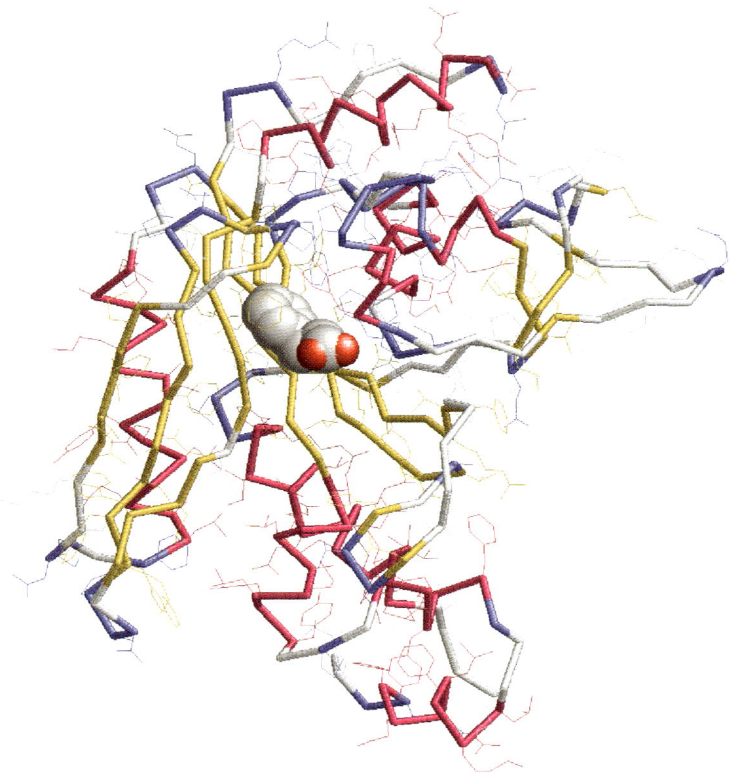

Figure 1.17 *Three-dimensional X-ray crystallographic structure of bacterial phenylalanyl tRNA synthetase (PheRS). PheRS consists of two polypeptide chains. In this figure only the phenylalanine binding part (A chain) is shown. The bound phenylalanine is shown by a space-filling model. The main chain is shown by rods. The dark gray portions are in α-helical conformation, while the light gray portions are in β-sheet forms. The side chains are shown with fine lines*

Moreover, because of the electronic polarization, the N–H group is an excellent proton donor, while the oxygen atom, in turn, is an excellent proton acceptor. As a result, the amide groups in a polypeptide chain are able to build a strong hydrogen bond network with each other. If the hydrogen bonds were formed between amide groups that are separated by every three α-carbon atoms along a single polypeptide chain, then the latter will take a right-handed α-helical conformation. If hydrogen bonds form to assemble several antiparallel-running chains together, then the polypeptide chains will assume a β-sheet structure.

By combining these structural motifs, such as α-helices and β-sheets, proteins may take a variety of main chain conformations. As an example, a main chain structure of a bacterial phenylalanyl tRNA synthetase (PheRS) is shown in Figure 1.17. The α-helical parts of the main chain are shown in dark gray, and the β-sheet parts in light gray. The bound phenylalanine is presented by a space-filling model.

As amino acid side groups appear in every three atoms along a polypeptide chain, severe crowding is expected between them. Thus, the orientations of the side groups are very constrained and, if they were properly arranged, the side groups would form a three-dimensional space for the effective binding of external molecules, or build a functional region for achieving enzymatic reactions. Figure 1.17 shows how a substrate (phenylalanine, shown by the space-filling model) is bound to its binding site that is made of constrained orientations of the side groups located nearby.

1.11 Genetic Engineering

The central dogma tells us that the amino acid sequences of proteins are determined solely by the nucleobase sequences of protein-coding DNA. Therefore, if new DNA can be synthesized, or if some nucleobases can be substituted with other nucleobases, and the new DNA is introduced into the protein-biosynthesizing system, then new or partially mutated proteins will be created. This technique is known as 'genetic engineering', and is widely applied in agricultural, pharmaceutical and medical fields.

In order to introduce new or mutated DNAs into living organisms, for example *Escherichia coli*, a small cyclic double-helical DNA, called a plasmid, is used as the transporter or a vector of the gene (Figure 1.18). The plasmid contains functional units, as depicted in the figure. The new gene is inserted into the protein-coding region by cutting off a portion by restriction enzymes, *EcoR*I and *Hind*III (see Figure 1.18) and pasting a new gene in place of the missing portion by an enzyme called a DNA ligase.

Restriction enzymes cleave double-stranded DNA chains at their specific sites, as typically exemplified for *EcoR*I and *Hind*III in Figure 1.19.

Figure 1.18 *Arrangement of functional components along a plasmid. The plasmid is a cyclic double-helical DNA of several thousand base pairs. The promoter sequence determines the start point of RNA polymerization. The Shine–Delgado (SD) sequence determines the point of ribosome attachment. Protein synthesis starts from the start codon (ATG) to one of the stop codons (TAG, TAA, TGA). The terminator sequence determines the end of transcription. The plasmid contains two restriction sites (*EcoR*I site and *Hind*III site in the above example) for inserting new genes. In the above example, the promoter and terminator sequences are taken from those of T7-phage, because of their high efficiencies*

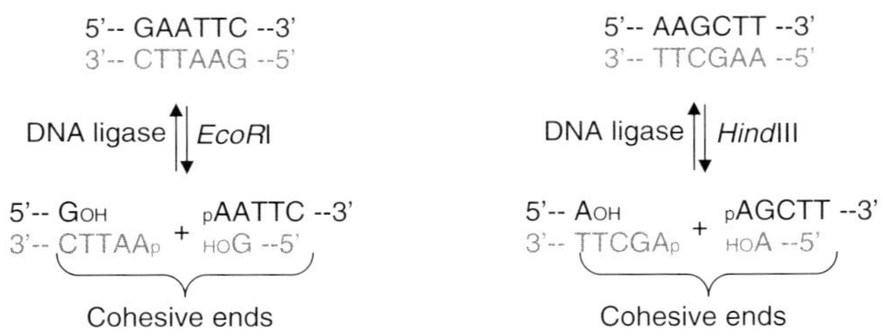

Figure 1.19 *DNA cleavage with restriction enzymes (downward arrows) and ligation with DNA ligase (upward arrows).* EcoR*I cleaves a double-helical DNA at the GAATTC/CTTAAG site.* Hind*III cleaves at the AAGCTT/TTCGAA site, leaving cohesive ends, respectively. The cohesive ends can be ligated again with the DNA ligase*

The cleavage leaves a pair of short complementary chains (cohesive ends) which will be linked again with an enzyme, DNA ligase. Ligation also takes place between the cohesive ends that are produced from different double-stranded DNAs, cleaved by the same type of restriction enzyme. Therefore, if the same set of restriction sites were to exist on a plasmid (Figure 1.20, top) and on a DNA fragment that included the protein-coding region (bottom), the latter would be inserted into the plasmid after cleavage by restriction enzymes, followed by the ligation with DNA ligase (Figure 1.20).

1.12 Large-Scale Production of Engineered Proteins

The complete procedure for obtaining a target protein from the plasmid is illustrated in Figure 1.21. The plasmid inserted with the protein coding region is introduced into

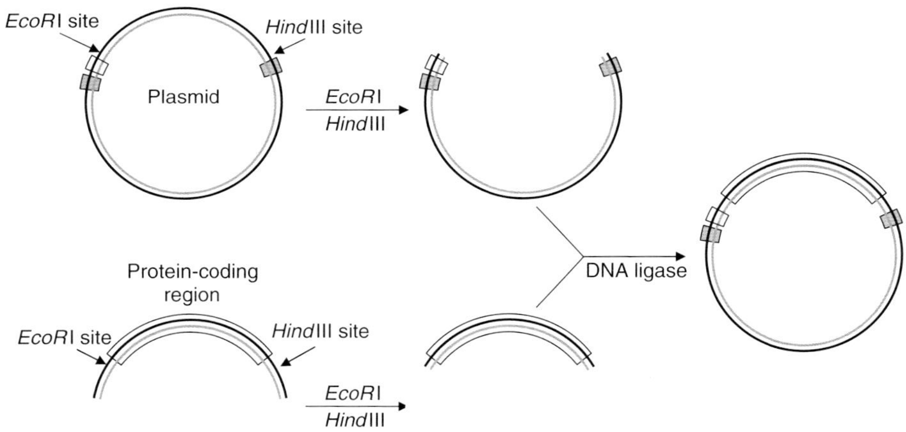

Figure 1.20 *Insertion of a protein-coding region on a fragment of double-stranded DNA into a plasmid by the use of a pair of restriction sites on both the plasmid and the DNA*

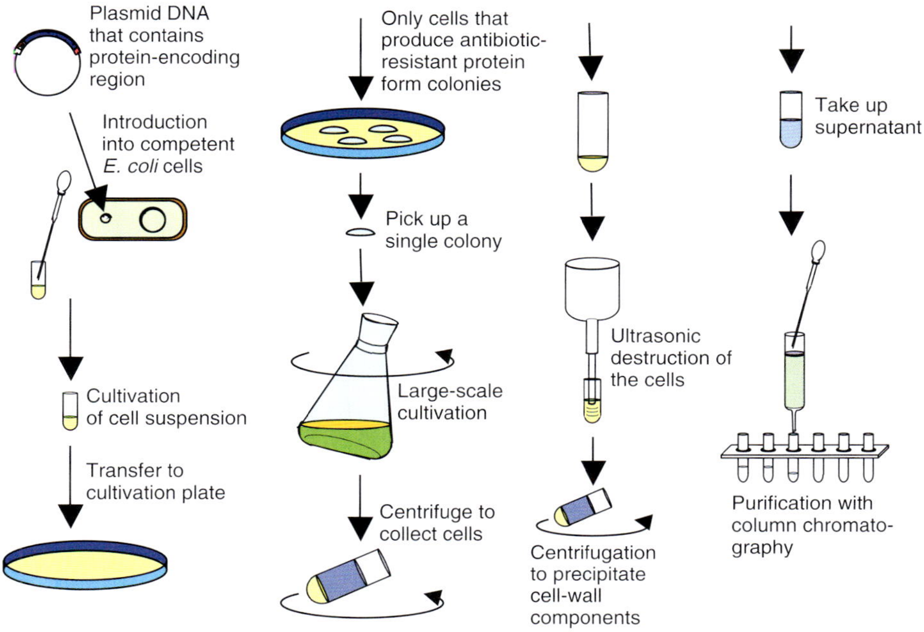

Figure 1.21 *Total procedure for the large-scale production of proteins by using* E. coli *transformed with a plasmid*

E. coli cells, the cell walls of which are temporarily made permeable to DNAs (competent cells). These transformed cells are cultivated first in suspension, and then transferred onto a cultivation plate with an antibiotic (ampicillin). On the cultivation plate, only those cells that are successfully producing the target protein, together with an ampicillin-resistant protein, can survive and grow to form colonies. Next, one of the colonies is picked up and cultivated in large quantity. After harvesting cells by centrifugation, the cells are lysed by ultrasonic agitation and the insoluble components precipitated by centrifugation. The protein in the supernatant is then purified using column chromatography.

1.13 Cell-Free Protein Synthesis and its Automated Process

Protein synthesis using living cells is advantageous for producing a large quantity of any single type of protein, because the transformed cells can be stored and used repeatedly. However, this approach is not appropriate for synthesizing many different types of protein as, usually, it takes a week (or even longer) to obtain a large quantity of transformed cells. Another drawback of the living cell system is that the expressed proteins often form insoluble aggregates (inclusion bodies) inside the host cells, that are not easily resolved. It is also clear that proteins which are toxic to the host cells cannot be synthesized. Nonetheless, these limitations can be avoided if all of the macromolecules that are functioning in the central dogma are extracted from living cells and then assembled in a test tube to conduct

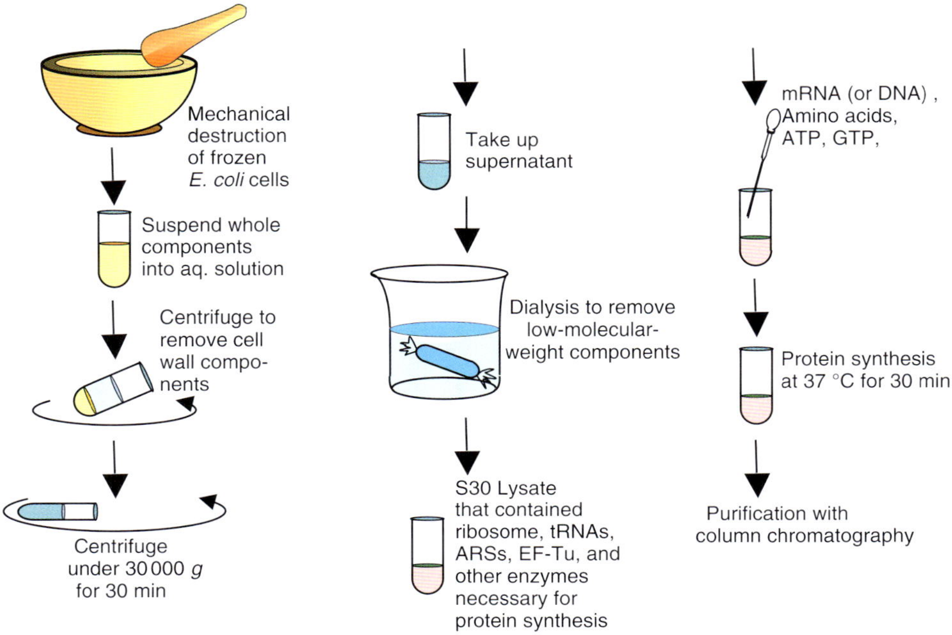

Figure 1.22 *Procedure for preparing* E. coli *S30 lysate and cell-free protein synthesis*

protein synthesis. The technique is referred to as cell-free protein synthesis, and protein-synthesizing mixtures, taken from *E. coli* for example, are now commercially available.

The procedure for preparing a cell-free protein-synthesizing system (*E. coli* S30 lysate) is illustrated in Figure 1.22. The frozen cells are mechanically destroyed and suspended in aqueous solution. After removal of the insoluble components, the soluble portion is centrifuged at 30 000 $\times$ g for 30 min. The supernatant is then removed and dialyzed against phosphate-buffered saline to remove any low-molecular-weight components. The remaining solution contains tRNAs, ARSs, ribosomes and other enzymes that are necessary for protein synthesis. Following centrifugation, this protein-synthesizing mixture is known as an S30 mixture.

By adding DNA or mRNA and an amino acid mixture, together with energy sources (ATP and GTP) to the S30 mixture, protein synthesis starts rapidly such that within 30 min the target protein is obtained in quantities of approximately 1 $\mu g\,ml^{-1}$ lysate.

As the cell-free synthesis will cease when one of amino acids or NTPs is exhausted, the materials must be fed continuously in order to continue the synthesis. In addition, waste materials such as diphosphates, nucleotide diphosphates (NDPs) and nucleotide monophosphates (NMPs) must be removed from the reaction mixture. This can be accomplished by using a reaction chamber equipped with an autofeeder separated with a semipermeable membrane, as illustrated in Figure 1.23.

By using such a continuous reaction system the protein yield can be increased to 10-fold that obtained when using a batch system.

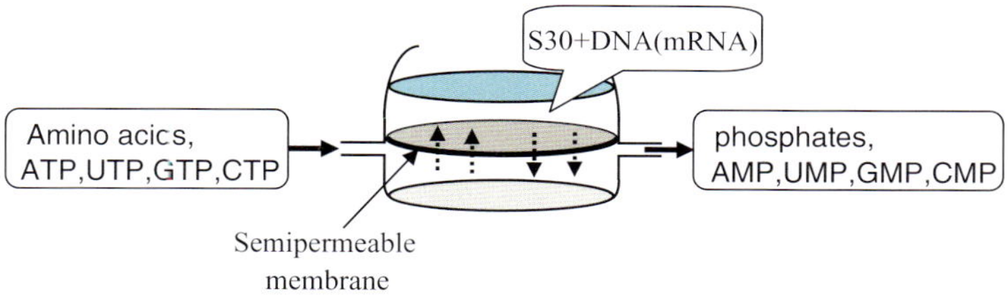

Figure 1.23 *Schematic representation of a continuous protein-synthesizing chamber equipped with an autofeeder for amino acids and NTPs*

One of the limiting factors of widespread cell-free synthesis is the cost of S30 or other cell lysates. In order to prepare the S30 lysate, a large quantity of *E. coli* cells is required, together with processes that are not suited to large-scale production. However, when the cell-free system becomes less cost-prohibitive, it is poised to become a major protein-producing procedure. Due to the flexibility of the system to synthesize a variety of protein types, the cell-free system is more suited for automated processes than for conventional protein synthesis using living cells.

Acknowledgement

This work was supported by the grants from the National 863 bioinformatics projects under the contract No. 2C07AA02Z333, and the Chinese National Science Foundation under the contract No. 20773085, 30870476 and 30770502, as well as the Virtual Laboratory for Computational Chemistry of CNIC, and the Supercomputing Center of CNIC, Chinese Academy of Sciences.

Reference

1. Alberts, B., Johnson, A., Lewis, J. *et al.* (2008) *Molecular Biology of the Cell*, 5th edn, Chapters 5 and 6, Garland Publishing, Inc., New York.

2

Genomes to Proteomes

Ellen A. Panisko[1], Igor Grigoriev[2], Don S. Daly[1], Bobbie-Jo Webb-Robertson[1] and Scott E. Baker[1]

[1] *Pacific Northwest National Laboratory, Richland, Washington, USA*
[2] *US DOE Joint Genome Institute, Walnut Creek, California, USA*

2.1 Introduction

Today, biologists are 'awash' with genomic sequence data, due in large part to the rapid acceleration in the generation of DNA sequences that has occurred during the race between public and private research institutes to sequence the human genome. In parallel with the large Human Genome Project effort, smaller genomes of other important model organisms have been sequenced. Subsequent projects have effectively utilized both the technological advances and the DNA sequencing infrastructure developed for the human and other organism's genome projects. As a result, the genome sequences of many organisms are today available in high-quality draft form.

Although the availability of draft data is a promising treasure trove of information, there are limitations to the biological insights that can be gleaned from DNA sequences alone, as genome sequences offer only a 'bird's eye view' of the biological processes endemic to an organism or community. Fortunately, the genome sequences now being produced at record pace can serve as the foundation for other global experimental platforms such as proteomics.

Proteomic methods offer a 'snapshot' of the proteins present at any one point in time for a given biological sample. Current global proteomic methods combine enzymatic digestion, chromatographic separations, mass spectrometry and database searching for peptide identification. One key aspect of proteomics is the prediction of peptide sequences from mass spectrometry data. 'Global' proteomic analysis uses the computational matching

Automation in Proteomics and Genomics: An Engineering Case-Based Approach Edited by Gil Alterovitz, Roseann Benson and Marco Ramoni

of experimental mass spectra with predicted spectra based on databases of gene models that are often generated computationally. Thus, the quality of gene models predicted from a genome sequence is crucial in the generation of high quality peptide identifications. Once peptides are identified they can be assigned to their parent protein. Proteins identified as expressed in a given experiment are most beneficial when compared to other expressed proteins in a larger biological context or biochemical pathway.

This chapter discusses the automatic annotation and the generation of high-quality gene models, the set-up and execution of quantitative and statistically rigorous global proteomic experiments, and proteomics in a biological context.

2.2 Gene Modeling

Genome sequencing has evolved dramatically during the past few years. The sequence of the first bacterial genome of *Haemophilus influenzae* was published in 1995 [1], shortly followed by the first sequenced eukaryotic genome of *Saccharomyces cerevisiae* [2]. Several large genome sequencing centers were established around the world for large-scale production sequencing, and have an average sequencing capacity of three giga bases per month, or roughly an equivalent of the human genome size. New short-read sequencing technologies promise to make genome sequence affordable for small laboratories and research groups. The substantial amounts of sequence data anticipated require adequate efforts and tools for analysis and interpretation. Genome annotation is one of the first steps in the analysis of a genome sequence, and includes finding genes and then describing their structures and functions. Approaches used for gene prediction in prokaryotes and eukaryotes are different. Finding genes in prokaryotes is a relatively straightforward task because of the simple gene structure (uninterrupted open reading frames; ORFs) and high gene density, with almost the entire DNA used for coding; therefore, automated approaches to predict gene models in prokaryotes genomes is feasible. In contrast, eukaryotic genes have complex exon–intron structures, and a significant fraction of eukaryotic genome sequence corresponds to noncoding DNA (e.g. 'gene deserts' in human [3]).

Despite the significant efforts made by many research groups, there are as yet no completely automated methods to predict gene models in eukaryotic genomes. Most of the eukaryotic gene predictors that have been developed and tuned for human or other higher eukaryote genomes are not applicable to another genome, and show low accuracy even between vertebrate genomes [4]. Eukaryotic gene predictors require training for every organism on a set of known genes from that organism's genome. This information is used to derive genome-specific parameters that then are utilized to predict genes in the whole genome. Several benchmarks have been developed to evaluate current gene predictors for human (EGASP [5]), fruit fly (GASP [6]), maize [7] and other genomes (e.g. NGASP, www.wormbase.org).

2.2.1 Gene Predictors

Eukaryotic gene predictors can be roughly described as *ab initio* (e.g. Fgenesh [8]; Augustus [9]; SNAP [10]; GeneMark [11]), homology-based (GeneWise [12]; Fgenesh+ [8]), expressed sequence tag (EST)-based (GrailEXP [13]; PASA [14]), synteny-based

(Twinscan [15]) and hybrid methods (EuGene [16]; Combiner [17]; TWAIN [18]). They differ in balance between *content-based* (distinguishing exons from introns or intergenic regions by, for example, nucleotide composition) and *signal-based* parameters (defining starts and ends of exons and genes) [19]. The *content* information can come from homology to proteins, ESTs and genome conservation, as well as coding potentials derived from a training set of genes. *Signals*, while mostly conserved, can be refined based on homology gene models and ESTs aligned to genomic sequence. In general, the predicted models will be highly inaccurate if the genome to which the gene-finding algorithm is applied differs in gene structure from the genome on which the algorithm was trained [10].

Given a sufficient number of known genes or full-length cDNAs for a particular genome, gene prediction parameters can be computed and used for genome wide gene prediction. Often, for most newly sequenced genomes, full-length cDNA sequence is not available; however, some characteristics of gene structure in a given genome can be inferred from ESTs. They can be directly mapped to genome assembly or used in EST-based gene predictors such as PASA [14]. Reliable homology-based gene models built with GeneWise [12] or Fgenesh+ [8] offer another source of information for training gene predictors. While these predictions lack untranslated regions (UTRs), close protein homologues often retain very similar exon–intron structures. In addition, genomes of closely related organisms can help to recover content and signal information using synteny-based gene prediction methods. These methods have been used successfully in human, mouse and rat (SLAM [20]), *Caenorhabditis elegans* (TwinScan [21]), *Aspergillus* genomes (TWAIN [18], *Cryptococcus neomorphans* (TwinScan [15]) and *Phytophthoras* [22]. Although these methods predict exons with a reasonable quality, they suffer from chimerism in genome scale applications and so are often used mostly to correct models of orthologous genes.

2.2.2 Annotation Pipelines

As each gene prediction method has its own advantages and drawbacks, combining different methods can improve the overall quality of gene models. Methods to select entire gene models such as Bayesian framework [23], to assemble model fragments into *de novo* models (e.g. EuGene [16]), or to combine multiple sources of information such as gene models and ESTs [17], have been proposed. Annotation pipelines employed at the genome-sequencing centers normally use several gene predictors. In addition to increasing the overall accuracy of annotations, they offer scalable solutions. The ENSEMBL pipeline was used for most vertebrate genomes [24], while the US DOE Joint Genome Institute (JGI) Annotation Pipeline includes Fgenesh [8], GeneWise [12] and Fgenesh+ [8] with a number of in-house developments that use ESTs and select a best representative model for every locus among the number of predicted genes. The Broad Institute used similar set of tools for the annotation of fungal genomes. The Institute for Genome Research (TIGR)/J. Craig Venter Institute (JCVI) annotation team trains several gene predictors, but use a subset of them for final annotations. Genome-sequencing consortia use additional gene predictors [25–27].

When the gene models have been predicted, the corresponding predicted proteins are functionally annotated. Functions can be inferred by sequence similarities to other proteins from, for example, UniProt or GenBank, as determined by protein sequence alignments using Blast [28]. InterProScan [29] combines several domain-search methods to predict domains, including SignalP and TargetP [30] for more specialized analysis. Comparison with the specialized databases (e.g. KEGG [31]) allows one to map the predicted proteins onto metabolic pathways; Gene Ontology [32] and KOG [33] categories provide the user with multiple entry points into the annotation data.

The overall workflow is similar between the different pipelines, and includes the following major steps that are common to all:

1. Repeat masking to exclude transposons from the final set of gene models
2. Mapping ESTs and homologues as seeds for gene predictors
3. Gene prediction using several methods
4. Gene annotation via domain prediction and homology searches.

Additional experimental data available at the time of genome annotation is becoming more often an integral part of validation modules of these pipelines.

2.2.3 Experimental Validation and Annotation of Predicted Gene Models

The accuracy of predicted genes depends on derived parameters, and varies from genome to genome. A significant fraction of predicted genes with no similarity to any protein in GenBank lacks annotation. Experimentally derived data (ESTs, microarrays, proteomics) may not only validate predicted gene structures but also add annotation by describing the conditions under which a particular gene or protein was expressed. Predicted gene models can be validated using gene-expression data. Evidence for predicted transcripts can be collected from ESTs/cDNAs overlapping with a gene model, microarrays with oligonucleotide probes corresponding to the predicted transcripts, and tiling arrays where probes are evenly distributed throughout the genome sequences. In addition, the comparative analysis of ESTs from different libraries or microarray probe hybridization levels under different conditions provides biological insights and annotation information. These resources are stored in genome databases, as well as larger repositories (e.g. ArrayExpress [34]; GEO [35]). In addition to a wide variety of proteomics biomedical studies, other examples include secreted proteins in fungi that degrade biomass [36–38], or have symbiotic relationships with plants [39]. For example, 10,048 genes were predicted for the genome of *Phanerochaette crysosporium* using a 10.6× genome sequence assembly processed with the JGI Annotation Pipeline. The processing of mass spectroscopy data resulted in the identification of 4697 peptides supporting 1489 genes, including 193 peptides supporting splice sites. The genome browser of the JGI Genome Portal illustrates peptide support for predicted gene models (Figure 2.1) [40].

2.2.4 Challenging Genes that Require Validation with Proteomics

While proteomics is valuable in validating the predictions of protein-coding genes, an additional value is derived from its ability to distinguish between protein-coding and noncoding genes, transcripts for both of which can be equally supported by ESTs or microarrays.

2.2.4.1 Pseudogenes

Remnants of genes that are no longer transcriptionally active are called pseudogenes. Based on their origin, they are subdivided into either *processed* (emerged through the

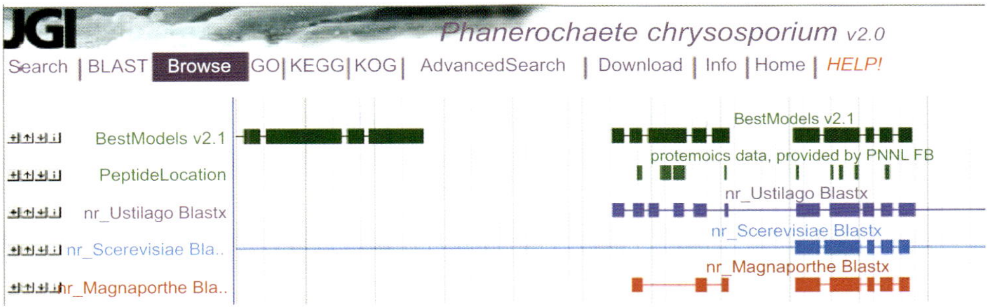

Figure 2.1 *Peptides mapped to genome assembly provide experimental support for predicted gene models in the* Phanerochaete chrysosporium *genome*

retrotransposition of processed transcripts back into genomic sequence) or *nonprocessed* (duplicated, not active and therefore mutated genes). Pseudogenes often have features that make them appear to be genes and, at times, are expressed based on EST or microarray evidence. Increased rates of mutation can introduce either stop codons or frameshifts; the frameshifts can be the result of either sequencing error or genomic mutation, especially for nonexpressed genes, and possibly be resolved with proteomics.

2.2.4.2 *Seleno Proteins*

Selenocysteine (Sec) is a rare amino acid that significantly increases the enzymatic activity of a protein. A nucleotide triplet, UAG, which normally is interpreted as a stop codon, codes for Sec. In the presence of a *cis*-acting mRNA structure, called the selenocysteine insertion sequence (SECIS) element, this codon is recognized by selenocysteinyl tRNA, which integrates a Sec amino acid into the protein sequence. The presence of a stop codon in the middle of a predicted gene/transcript sequence makes it a viable pseudogene candidate. However, as some pseudogenes are expressed in the form of RNA, only protein expression can support this type of protein.

2.2.4.3 *Noncoding Genes*

Often, clusters of ESTs suggest missing genes in places where no gene model was predicted. The lack of a gene model indicates a lack of significant coding potential, and/or homology in that locus, which could be due to incorrect training or specific genes. The identification of an ORF does not necessarily mean that coding genes are present, as a long ORF can be found even in noncoding RNAs.

2.2.4.4 *Polycystronic Genes*

Proteomics data can resolve conflict between the different ORFs found in the same genes, either in genes with low GC and without any stops, or in polycystronic genes, where several genes are expressed as the same transcript to be processed before translation.

2.2.4.5 Caveat

The resolution of mass spectra requires a database of protein sequences derived from predicted gene models. In order to support predicted gene models, the same gene models are used to resolve mass spectra. One option is to use all ORFs in a six-frame translation derived directly from a genomic sequence, although the exon–intron structure of eukaryotic genes makes this difficult. Only peptides that align entirely within a single exon can be resolved in this way, and peptides aligned across a splice site will be lost.

2.3 Proteomics: Experimental Design

In the majority of proteome studies, investigators are interested in comparing the proteins expressed in a cell or tissue under one condition versus another (i.e. normal versus diseased). Originally, proteome studies were conducted using two-dimensional polyacrylamide gel electrophoresis (2D-PAGE [41]), where the proteins are separated first by isoelectric point and then by size. After running the sample, the gel is stained with a protein-binding dye, and an image analysis is then performed to compare each stained gel from the two (or more) conditions. A 'spot' of interest can be excised from the gel and the protein identified using mass spectrometry [42]. Today, the technology has been developed to allow two samples to be analyzed on a single gel by labeling each sample with a different fluorescent dye [43]. However, the field of proteomics is currently dominated by relatively high-throughput mass spectrometry based approaches.

In these methods, high pressure liquid chromatography (HPLC) is coupled directly with mass spectrometry (MS). As the peptides are eluted from the chromatographic column they are converted to the gas phase by electrospray ionization (ESI) [44,45] and drawn into the inlet of the mass spectrometer. Currently, different types of MS and HPLC platforms are used across proteomics laboratories. For the basic 'shotgun' approach to proteomics, peptides isolated from cells are digested with a protease (typically trypsin) having defined cleavage sites and the resultant peptides are then analyzed using HPLC/MS techniques. An example of the complexity of the resulting sample can be found in the worm *C. elegans*, the genome of which encodes approximately 20,000 ORFs that, in theory, can produce close to one million tryptic peptides [46]. Given that the tryptic digest samples are too complex to resolve each individual peptide in time by HPLC, an identical sample run through a HPLC/MS system will have a limited number of overlap of peptide identifications (this is often referred to as 'under-sampling'; see Figure 2.2).

Determining differences in protein expression between samples is less intuitive than with the 2D-PAGE method, and several such techniques are currently in use in research projects (for reviews, see Refs [47, 48]). Perhaps the most straightforward approach is that of spectral counting, which entails counting the number of spectra that a peptide (or peptides) from a protein produce during a full HPLC/MS analysis [49]. The counts from two different sample types are compared to identify proteins that are differentially expressed.

For model systems with defined growth media, stable isotope-labeling strategies are often utilized [50]. Here, one cell condition (normal) may be grown in a baseline medium, while the other cell is grown in a medium in which stable, heavy isotope-labeled amino acids are substituted. Equivalent amounts of cells or protein extracts from the two cells are

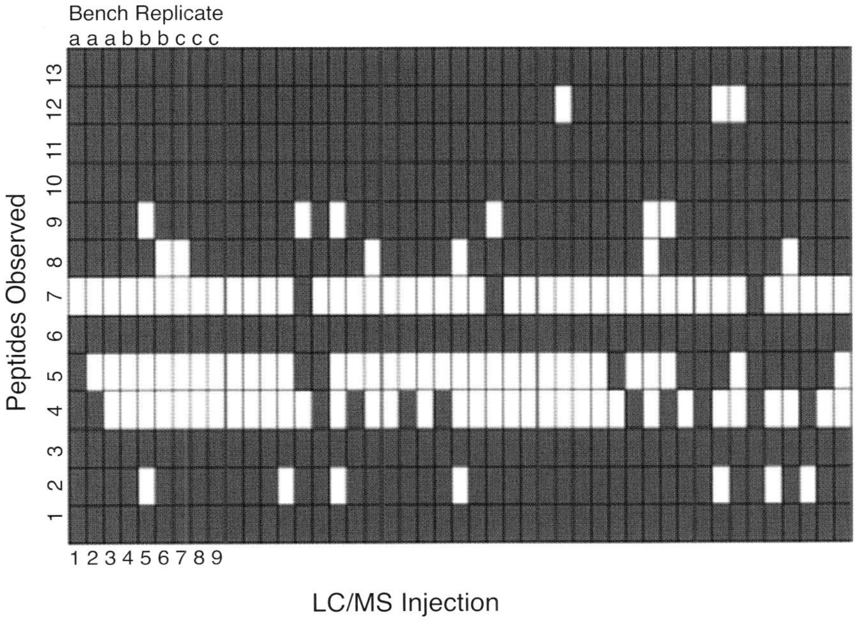

Figure 2.2 *Peptides observed from NADP-dependent isocitrate dehydrogenase in total soluble proteome analyses of* Trichoderma reesei. *Each column represents an individual LC/MS injection of sample, while each row represents a peptide from isocitrate dehydrogenase that has been observed at some time in previous experiments. The white blocks indicate that the peptide was observed in the sample; gray blocks indicate that the peptide was not observed. The blue lines separate the five different samples examined. Each sample has three bench-top replicates, each of those replicates had three LC/MS technical replicates, resulting in nine injections for each of the five samples*

combined and processed for analysis, after which the ratio of the mass spectral intensities between the heavy and light isotopically labeled peptides are used for relative quantitation. For cells with undefined growth media, such as human tissue samples, a similar strategy can be used with affinity labels (e.g. ICAT [51], iTRAQ [52]). The affinity tag is produced in two versions – heavy and light. The protein extract from one condition is treated with the heavy reagent, while the extract from the other condition of interest is treated with the light reagent. Equivalent amounts of labeled extract are then combined for processing. One advantage of affinity labeling methods is that they can isolate specific peptides (with ICAT, only cysteine-containing peptides), thereby reducing the overall sample complexity.

Software such as MASIC [53] is used to determine the mass spectral intensities of peptides. The process begins with the parent ion (mass-to-charge ratio) that was identified for a peptide, and extracts the elution profile (extracted ion chromatogram) of that ion from the mass spectra collected for that HPLC/MS injection. Essentially, this is a plot of parent ion intensity over time, after which the peak area and maximum peak intensity can be calculated and used for quantitation.

The same method can be applied to nonisotopically labeled samples for relative quantitation. The number of heavy isotopically labeled amino acids or affinity tags available does not limit nonlabeled experiments. Regardless of the method utilized, all experiments benefit from a strong experimental design. However, caution must be taken to reduce sample preparation variability and to prevent any experimental processing from biasing the data. Experimental bias can result from preparing all 'like' samples together and separated from the remaining conditions of an experiment, or including only one replicate per sample.

2.4 Proteomics Sample Processing

A myriad of approaches is available for sample processing. Often, an investigator will focus on isolating a specific type of protein from a sample. For example, if they are interested in isolating only phosphorylated proteins, they can choose from an affinity labeling technique (phosphoprotein isotope-coded affinity tag; PhIAT [54]) or a chromatographic method (immobilized metal affinity chromatography; IMAC [55]). Although the sample processing methods are too numerous to discuss at this point, they are excellently reviewed elsewhere [56, 57]. However, a basic procedure for total soluble protein proteome sample processing is outlined as follows.

After harvesting the cells or tissues, the samples are typically stored frozen until all biological replicates can be processed in parallel (or according to the experimental design). Depending on the sampling techniques used, experimental (bench) replicates can be initiated either before or after cell lysis. For instance, if the cell number can be easily determined, then placing equal numbers of cells into separate tubes can produce replicate samples, with each tube being processed separately throughout the entire method. The cells may be lysed either chemically or mechanically, depending on the model system employed. Lysis is often performed in the presence of a high-molarity chaotrophic salt, such as urea or guanidine, so that the proteins are denatured as soon as the cell contents are released. Protease inhibitors may also be added to the lysis buffer. The cell debris is then removed from the sample by centrifugation and the supernatant reserved for further processing. The cell lysate is subsequently assayed to determine protein concentration, usually using the bicinchoninic acid (BCA) [58] method, due to its tolerance of high salt concentrations. Experimental replicates may also be introduced at this stage for those systems where the cell number is not easily assayed, simply by aliquoting equivalent amounts of protein to separate tubes. Many investigators also denature the protein sample by incubating with tris(2-carboxyethyl) phosphine) (TCEP), after which the cysteines are chemically modified by incubation with iodoacetamide to prevent disulfide bond formation. The samples are now ready for tryptic digestion.

For effective digestion, the samples must be diluted with buffer to reduce the concentration of salt to a level which is tolerated by the enzyme, and to ensure that the sample is at the appropriate pH. Trypsin is added to the sample at a ratio of anywhere from one part trypsin to between 20 and 100 parts of the sample protein. The trypsin is prepared according to the manufacturer's instructions, added to the sample, and incubated at 37 °C for 4 hours upto overnight.

Before processing by HPLC/MS, peptides from the sample are separated from salts and concentrated using solid-phase extraction. This procedure uses a matrix of silica beads to which are attached chains of hydrocarbons that are 18 carbons in length (reverse-phase). Solid-phase extraction cartridges are typically made to attach to a vacuum manifold that allows the liquid to be pulled through the column and provide a place to collect the final eluate of peptides. Adding two to four column volumes of methanol first activates the resin (after activation, the column must not be allowed to run dry). Water (2–4 column volumes) is then added to equilibrate the resin to aqueous conditions. The sample is then added to the column and washed with six to eight column volumes of water or volatile buffer (ammonium bicarbonate). The peptides are eluted with two column volumes of organic solvent (often 80–100% acetonitrile). The samples are dried using a centrifugal vacuum concentrator, and can be stored frozen until HPLC/MS analysis.

It is difficult to describe a 'typical' HPLC/MS experiment, as the development of techniques to improve the chromatographic separation of peptides and detection by MS currently form a very active area of research. Usually, reverse-phase chromatography (essentially a separation of peptides by hydrophobicity) is used, and the eluate from the HPLC column is injected directly into the mass spectrometer, in real time. Multidimensional separations can also be performed where the sample is separated in one dimension by strong cation exchange into fractions that are then subjected to reverse-phase separation. It is possible to perform this 'on-line', with a single biphasic column [59].

The columns used are fused silica with inner diameters of 75–150 μm that are packed with reverse-phase particles of 3–5 μm in size. When the column has been equilibrated in an aqueous form with a dilute volatile acid (this ensures that the peptides will be positively charged), the samples can be injected onto the column. The peptide samples isolated by solid-phase extraction are resuspended in the same solvent used to equilibrate the column. The peptides are eluted from the column by adding increasing amounts of organic solvent containing the same concentration of acid. Gradient profiles used vary across investigators; an example is shown in Figure 2.3.

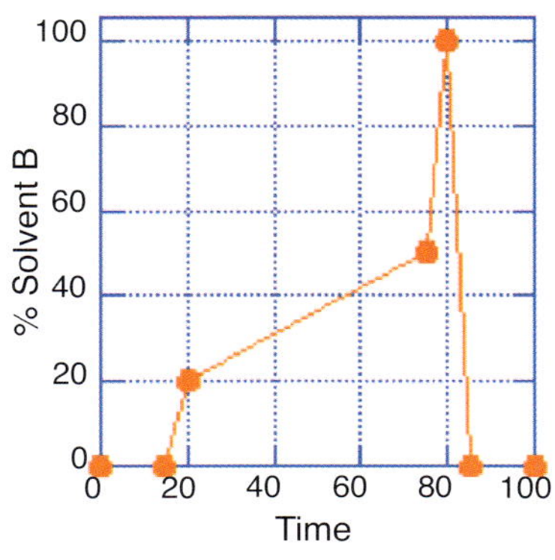

Figure 2.3 *An example of a solvent gradient used for eluting peptides from a HPLC column. For 15 min after sample injection the column remained at 100% solvent A (0.1% formic acid). There followed a linear gradient, from 100% to 20% solvent B (90% acetonitrile, 0.1% formic acid) over 5 min, followed by another linear gradient from 20% solvent B to 50% solvent B over 55 min, and finally a linear gradient from 50% solvent B to 95% solvent B over 5 min. Note that the column must be re-equilibrated with solvent A before the next sample is injected*

Mass spectrometers are typically operated in a data-dependent mode, choosing the most intense ions observed in a survey (MS) scan and isolating those ions for fragmentation (MS/MS scan) in subsequent scans. Thousands of scans can be collected for a single HPLC/MS injection. As the manner of peptide fragmentation is predictable, the mass-to-charge ratio from the survey scans and the ions produced by fragmentation are used for identification. Software programs such as Sequest [60] and X!tandem [61] perform probability-based identification by utilizing the data from the mass spectrometer and comparing it to the expected peptide fragmentation for all tryptic peptides from an *in silico* tryptic digest of all proteins from a defined protein database (i.e. SwissPro or the predicted proteins from an organism the genome of which has been sequenced).

Data analysis is an extremely important aspect of proteome studies, which deserves more than the cursory mention included here. Often, an experiment will involve dozens (if not hundreds) of HPLC/MS injections, and consequently the data files are very large. Indeed, thousands of peptides may be identified in a single analysis, such that laboratories which specialize in proteomics often have their own data management systems [62].

2.5 Statistical Modeling of Proteomics Data

A biological study of peptides by HPLC coupled with MS produces a large and complex but somewhat sparse dataset due to the design of the study, the HPLC/MS queuing plan for study samples, and the (often incomplete) observation of numerous peptides across the study's sample collection. As an example, consider a study of cells grown with exposure to five different concentrations of pesticide, with each condition having multiple samples with replicate HPLC/MS injections. The realized design has an intricate structure spanning thousands of peptide measurements perforated with missing observations. To ensure valid and objective biological conclusions, a statistical method for a HPLC/MS-based biological study should formulate a design-complementing queuing plan that complements the experimental design to ensure that those data suitable for the appropriate statistical modeling are collected. A matching statistical analysis can then be performed.

Statistical modeling relates to the defining, fitting and interpreting of a probability model. The simplest statistical algorithm is an exercise in statistical modeling if the intention is to make inferences about problems behind the data. A (potentially invalid) probability model implicitly looms under each application, such that the validity of this exercise depends upon an understanding and application of basic statistical concepts that underpin the designing, fitting and interpreting of probability models. Numerous Internet resources offer quick, outstanding refreshers about important basic statistical concepts; these include Wikipedia [63], NIST SEMATECH e-Handbook of Statistical Methods [64], Electronic Statistics Textbook [65], EBook [66] and MathWorld Probability and Statistics [67].

2.5.1 Mixed-Effects Modeling

Mixed-effects linear statistical modeling [68] is an established statistical methodology for the analysis of comparative, screening and time-course experiments. A mixed-effects model includes terms for both fixed effects such as researcher-set treatments, and random effects due to subject response, instrument variability or other nuisance factors. The mixed-effects modeling approach is uniquely suited to producing a HPLC/MS sample queuing plan and statistical analysis complementary to the often complex realized design of a biological study. Whilst a detailed discussion and example is described in Ref. [69], we offer here a brief overview of the topic.

The basic steps are to:

1. Identify the HPLC/MS nuisance factors
2. Evaluate the design of the biological study
3. Formulate the HPLC/MS queuing plan
4. Explore the HPLC/MS dataset
5. Define and fit protein-level mixed-effects models
6. Group proteins based on estimates of biological parameters
7. Draw biological conclusions about individual proteins and protein groups
8. Alternatively, to draw conclusions about the quality and performance of the HPLC/MS process.

The mixed-effects statistical model is characterized by three important elements:

- The model describes a HPLC/MS abundance measurement as a multiplicative function of study and processing factors. To facilitate modeling fitting, this multiplicative model is log-transformed so that log(peptide abundance) is expressed as an additive model of study and processing factors. The model generates estimates of model goodness-of-fit, treatment and peptide effects, standard errors and confidence intervals. Pertinent results are then transformed back to the original scale for biological interpretation.
- The model has two disparate sets of terms – one set represents the biological design, while a second set represents the HPLC/MS sample processing plan.
- The relative difference in HPLC/MS measurability between peptides is represented by a component measurability factor.

A biologically induced difference between two conditions is often inferred from the ratio of a peptide's HPLC/MS abundance estimates (i.e. a component's relative abundance). The acceptance of forming this ratio to eliminate or to significantly minimize the systematic effects of HPLC/MS processing, coupled with the common assumption that any measurement error is relative (i.e. MS measurement errors increase with measurement values), suggests that any variation in MS abundances may be explained adequately with a multiplicative error probability model. Further, sample effects due to dilution/titration, fractionation and so on, are often multiplicative in nature. Consequently, an additive statistical model may effectively describe log-transformed MS abundances (i.e. in matrix notation, model terms and coefficients are separable).

Restricted maximum likelihood estimation (REML) is the method used for model fitting. REML was developed and refined to estimate more accurately variance components in random and mixed-effects models [70–72]. REML correctly tabulates the degrees of freedom for unbalanced data, improving error estimates and inferences, and is better suited to fitting linear models to the often-incomplete HPLC/MS datasets than other techniques, such as ordinary least-squares analysis or analysis of variance.

2.5.2 Data Quality Issues

Variance in HPLC/MS analysis represents a significant challenge. Ideally, each protein processed would be extracted, digested, purified, separated by HPLC and observed by MS with equal efficiency. Proteins that are equimolar in a sample would have comparable MS abundances proportional to their concentration. In particular, abundance peptides from their parent protein would be replicate measurements of the parent protein's abundance. In reality, however, some peptides are more easily measured (i.e. identified and quantified) by HPLC/MS than others [73]. Whether caused by peptide digestion efficacy, hydrophobicity or ionization potential, these nuisance factors directly affect the quantification of a component's abundance. This HPLC/MS peptide measurability effect varies across peptides due to the cumulative – yet differential – effects of nuisance factors. Relative HPLC/MS peptide measurability, however, is very reliable across samples measured under similar conditions on the same HPLC/MS platform; that is, unique peptides of a given protein most often display similar MS abundance profiles randomly perturbed by measurement error across a biological study. Differences in the HPLC/MS peptide measurability can be estimated and removed by mixed-effects modeling to eliminate this source of variability and allow pooling of data from peptides of the same parent protein [69]. The mixed-effects modeling produces one model for each fitted protein. A single study may result in hundreds to a few thousand acceptable individual protein models.

HPLC/MS processing introduces many nuisance factors that are unrelated to the biological factors of greatest interest, such as variability in instrument performance ('instrument drift'), the use of different HPLC columns and electrospray emitters. Often, one group in one location at one time executes a biological study, while an independent group in a separate location at a later time analyzes the resultant samples using HPLC/MS. The study designers are advised to include various quality control samples and to use a complementary HPLC/MS sample queuing plan to guard the validity and objectivity of their study. Here, the important statistical principles are randomization, replication and blocking, where blocking is key.

A block is a set of samples spanning the interesting factors over which the nuisance factors are assumed to have a constant effect (although the nuisance effect may vary from block to block). The nuisance factor combinations determine the block size, or number of samples in a block. Consider a study investigating protein expression in diabetic tissue, where age, gender and body mass index (BMI) could be nuisance factors. A block in the diabetes design would be a sample from the diabetic tissue of interest with one combination of all nuisance factors – one tissue sample each from a nondiabetic, prediabetic and diabetic subject matched on age, gender and BMI. In its simplest form, a block is one replicate of the full biological design, or a complete mini-experiment containing one sample from each treatment combination. Blocking is quite common in biological studies, and an experiment's blocks are the natural blocks for HPLC/MS processing. If the study design does not feature blocks, then study blocks solely for queuing HPLC/MS samples may be formed. It is necessary to select one sample at random from each treatment combination in order to fill a block (Figure 2.4).

The general stability of an HPLC/MS processing line, HPLC column or MS instrument may be assessed with a controlled experiment featuring the sequential processing of numerous replicates of the same quality control sample across one or more processing lines. Here, the objective is to identify the longest run of injections, or block size, over

	HPLC/MS Injection	Sample Description
Block 1	1	Control
	2	+ Concentration B
	3	+ Concentration C
	4	+ Concentration A
	5	+ Concentration D
Block 2	6	+ Concentration A
	7	+ Concentration C
	8	Control
	9	+ Concentration B
	10	+ Concentration D
Block 3	11	+ Concentration A
	12	+ Concentration B
	13	+ Concentration D
	14	Control
	15	+ Concentration C

Figure 2.4 *A simple example of a HPLC/MS queue. Shown is a partial HPLC/MS queue for an experiment where a cell line is exposed to four different concentrations of pesticide. Here, a block contains five HPLC/MS injections, one from each sample type-control, and those exposed to concentrations B, C and D. Note that within each block the five different samples of peptides are in random order within each block*

which the nuisance effects of a HPLC/MS processing line are relatively constant. Suppose the HPLC/MS block size is larger than the study block size, then the study blocks effectively become the HPLC/MS blocks. That said, the samples within each existing study block should be randomly ordered, and then these blocks should be randomly queued for HPLC/MS processing. The aim of the HPLC/MS sample queuing plan is to control the confounding of nuisance HPLC/MS processing factors with the biological factors of interest (Figure 2.5). Although specific HPLC/MS nuisance factors are many in number, most can be sufficiently controlled by grouping under the major categorical variables, namely sample preparation set, HPLC column internal diameter and HPLC/MS data acquisition start time.

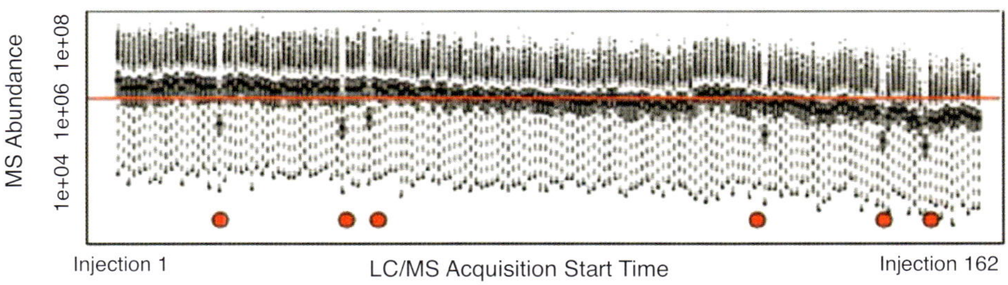

Figure 2.5 *Advantages of proper HPLC/MS queuing. A boxplot of MS peptide abundance (y-axis) for 162 HPLC/MS injections of 27 separate total soluble digests of* Trichoderma reesei. *The red trend line is the median peptide abundance across all samples. The red dots lie below six HPLC/MS injections that were well below the median observed for the remaining samples. All six were HPLC/MS technical replicates (injections) from a single bench-top replicate, suggesting that there was a problem in the processing of this sample*

The mixed-effects model may also include terms that reflect a more complex design of the biological experiment, and terms that break out other HPLC/MS processing effects, such as differences in sample preparations and time of MS acquisition. Those terms which are not supported by measurements, such as peptides only observed in one HPLC/MS injection, are excluded. The effectiveness of this modeling is limited by the amount and pattern of missing observations. In effect, only the information in the observed abundances is retained, while the information in missing observations is discarded. However, this need not be the case; if the abundance data are converted to observation presence/absence (or binary 1/0) data, then the differences in the probability of a peptide observation across treatments may be modeled using additional statistical methods.

Overall, the goal is to draw valid, objective, statistically defensible conclusions. As in the first examination of the data, visual and tabular summaries are very effective. Here, however, the strength of the evidence need not be anecdotal, because an appropriate analysis produces valid estimates of standard errors and confidence intervals, although careful interpretation is required. It is important that the interpreter knows the statistically valid interpretation of standard errors and confidence intervals (each parameter estimate, or contrast of parameter estimates, has its own standard error and confidence interval).

2.6 Integrating Proteomic Data with Other High-Throughput Data

In recent years, technological advances in high-throughput technologies have fueled a revolution in biology, enabling the analysis of entire systems on a global scale (e.g. whole cells, tumors or environmental communities). Thus far, the discussion has focused on the global profiling of proteins using high-throughput MS (e.g. normalization approaches). In the context of systems biology, however, this proteome information must be integrated with a plethora of additional information, both from other high-throughput ‘omic’ technologies (e.g. transcriptomics and metabolomics), and supplementary information (e.g. functional annotations, cellular location predictions, regulatory elements). This task of data integration, with an eye on systems biology, requires multiple layers of computational tasks, including linking to data management systems, bioinformatics tools and statistical and visualization methods.

2.6.1 Data Management and Connectivity to Bioinformatics Tools

There are many challenges with managing data from heterogeneous data sources, ranging from simple access to the data to performing complex queries and workflows on the data to answer targeted questions of interest. In practice, the need to integrate and perform complex analyses on heterogeneous data sources results in ad hoc connections between databases and software tools by writing small scripts, cutting and pasting queries, and basic manual labor. As a result, the recent years have seen a surge in the development of new software tools which focus on automating and simplifying these tasks. These bioinformatics resource tools tend to fall into four categories: semantic mapping; interoperation of heterogeneous bioinformatics databases; automated workflow analyses; and programs that integrate the data with bioinformatics software. Semantic mapping approaches, such as ToolBus [74] and Taverna [75], focus on defining translation engines which ensure that entities across environments are appropriately related. Alternatively, other approaches focus on the capability to access heterogeneous databases and merge directly on the data sources, such as BRIDGE [76]. These investigations have led to subsequent tools, such as BioWarehouse [77] and GenFlow [78], that offer a combination of semantic mapping and database access capabilities. Alternatively, one can focus on the goal of integration and define specialized workflows [79,80]. In some cases, these methods are linked to statistical and visualization tools [81,82].

Some current approaches, such as BRM [83], Gaggle [84] and FACT [85], focus on facilitating all of these capabilities (object mapping, database access and generic workflows) into a single environment. As both of these systems biology environments are built in JAVA, they can be easily installed and run by biologists and bioinformatics experts on publicly available web sites: BRM (http://www.sysbio.org/dataresources/brm.stm) and Gaggle (http://gaggle.systemsbiology.net). These two integration and analysis tools have commonalities, and the underlying programming languages allow them to work together. The BRM working environment (Figure 2.6) is given as an example of the multilayer analyses allowed by these multicapability software programs. At the top left is the project browser, which allows the user to manage multiple heterogeneous datasets in a single space that provides information on each source, such as the number of rows and columns. The

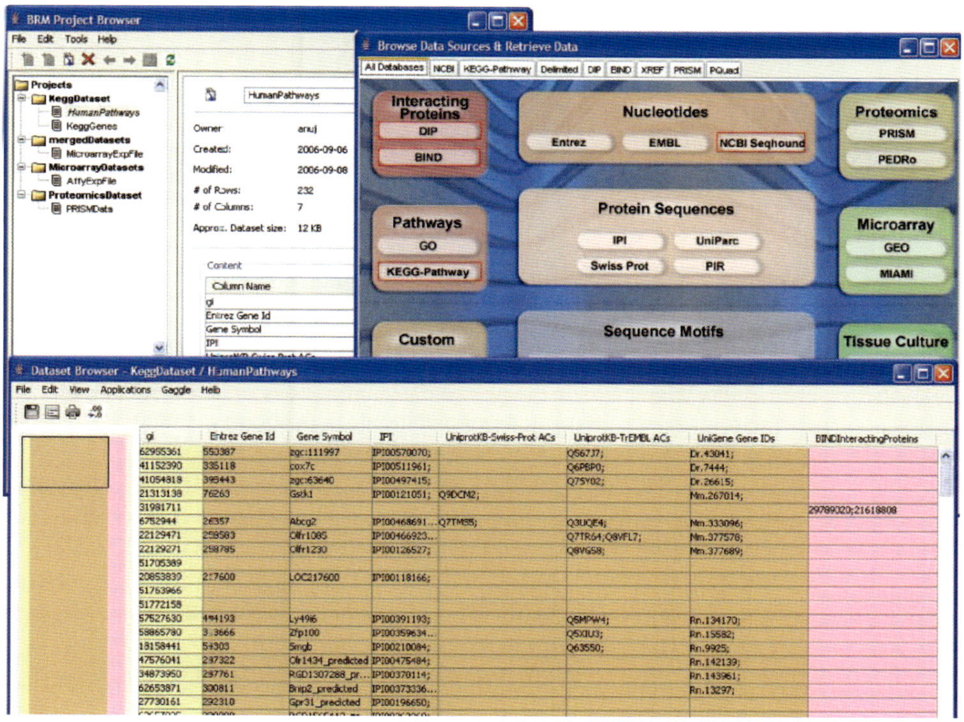

Figure 2.6 A collage of the Bioinformatics Resource Manager (BRM) working environment, including the project browser (top left), dataset browser (bottom) and data retrieval panel (top right) capabilities

dataset browser at the bottom allows the user to evaluate multiple data types in one view, and codes each data source by color. To retrieve additional information associated with one or more data source, the data retrieval panel (top right) allows direct access to bioinformatics resources, such as protein interactions [86, 87], pathways [88] and annotation data [89–91]. In addition, from this retrieval data panel, visualization software [92, 93] associated with different types of data can be launched directly, without any additional installations from the user.

2.6.2 Statistical Integration

There are many levels of integration that can be performed when evaluating multiple omics data sources, as well as ancillary information (e.g. gene ontologies). The task is often complicated due to the heterogeneity of the data; for example, quantitative variables on multiple scale and categorical information. Although many reviews have also been completed on the integration of omics' data focused towards systems biology [94–96], these tend to spotlight a generalized need and not the specifics of the statistical methods that may be employed. A comprehensive discussion of the guiding principles of integration, such as balancing sensitivity and false discovery rates, global versus query specific analyses, supervised versus unsupervised methods and sequential versus concurrent methods, is provided elsewhere [97].

In general, statistical methods tend to fall into two categories: unsupervised (exploratory) analyses; or supervised learning. *Unsupervised methods*, such as principal component analysis (PCA) [98], optimize some features of the data, such as variance, which may reveal clustering tendencies of the data in a lower dimensionality. These methods are for exploration purposes and try to identify underlying structure in the data. Alternatively, supervised learning assumes that the response is known, or has been measured, and the goal is to find a correlative model between the set of features and the response, such as with regression [99]. These methods are predictive in the sense that, if one attains the set of features for a new observation, then the response can be predicted from the model.

In respect to statistical data integration, which is irrelevant to the actual statistical model employed, there are generally three basic approaches to merge the data for statistical analysis; these are highly dependent on the type of data being considered. The first approach is that of feature integration, where the individual datasets are merged into a global dataset and then evaluated using either supervised or unsupervised learning. The second approach is to evaluate each dataset individually with methods, such as clustering, and then statistically to merge the results. The final method is to transform each dataset into an alternate representation, such as a network or kernel, and to merge the data in this new dimensional space: these methods are normally used in conjunction with supervised learning. Here, these three strategies are described briefly, as well as the benefits and caveats of each approach.

2.6.3 Feature Integration

One of the most common approaches in data integration is simply feature integration. If dataset A consists of m features, $D_A = [f_1, f_2, \ldots, f_m]$ and dataset B consists of n

features, $D_B = [g_1, g_2, \ldots, g_n]$, then the integrated dataset is simply:

$$D_{AB} = [f_1, f_2, \ldots, f_m, g_1, g_2, \ldots, g_n]. \quad (2.1)$$

This can be achieved by merging the two datasets such that each observation in one dataset matches that in another. For example, each protein corresponds to a gene, or by attempting to separate specific events, such as the toxicity of a compound so that the biological samples are the observations and the biomolecular molecules are the features. The task of merging data for the goal of integrating microarray and proteomic data has been reviewed [100], and can be accomplished using tools such as those described previously. Additionally, Cox *et al.* [101] review clustering and correlation-based approaches for merged datasets. This approach is only feasible when the variables are of a common type (e.g. qualitative), as normalization is typically a necessity to place each variable on the same scale. However, given the appropriate scaled dataset, most multivariate statistical methods (such as clustering or regression) could be employed to analyze trends or relationships in the data. In the field of proteomics, this approach is most commonly used to merge ancillary information with peptide identification results to improve the quality of the results – that is, to improve sensitivity [102, 103]. These approaches describe a peptide as a set of disparate features associated with identification metrics, such as the cross-correlation score from SEQUEST [60] and fraction of matched peaks, and use the supervised learning algorithm support vector machine (SVM) [104, 105] to determine correct from incorrect identifications.

The primary caveats of feature integration are the need for a one-to-one correspondence between objects in each dataset, and that variable types (such as categorical information) are difficult to merge. In addition, the contribution of each feature is often not evident but is of interest. Thus, feature selection methods typically follow the initial analyses. Overall, with feature integration, care must be taken to assure that the data are appropriate to merge and properly normalized.

2.6.4 Individual Analyses Followed by Integration

An alternative approach to integration is to evaluate each dataset individually and then to merge the results of each analysis. An unsupervised approach to this task is to cluster each dataset into some number of clusters and then to merge the clusterings; this is commonly referred to as metaclustering [106, 107]. These methods have been shown to be of use in biology [108–110]. If each observation is treated as a probability of being associated with a specific cluster, then a simple Bayesian approach can be taken to merge these results. Two primary benefits of this approach are the capability to integrate multiple data formats (e.g. qualitative and quantitative), and that the low dimensionality of the results is conducive to visualization [111]. Figure 2.7 provides an example of three types of experimental data (Powerblot, FTICR proteomics and microarray) over a time-course. For each dataset at each time point the datasets are clustered into three classes (upregulated, downregulated, neither). The top and bottom axes demonstrate which colored lines belong to each data type, and their respective results at each time point. As seen in the figure, common trends among the three datasets can be easily observed. In addition, the bottom metaclustering is a merged result over the entire time-course to highlight statistical trends among the data.

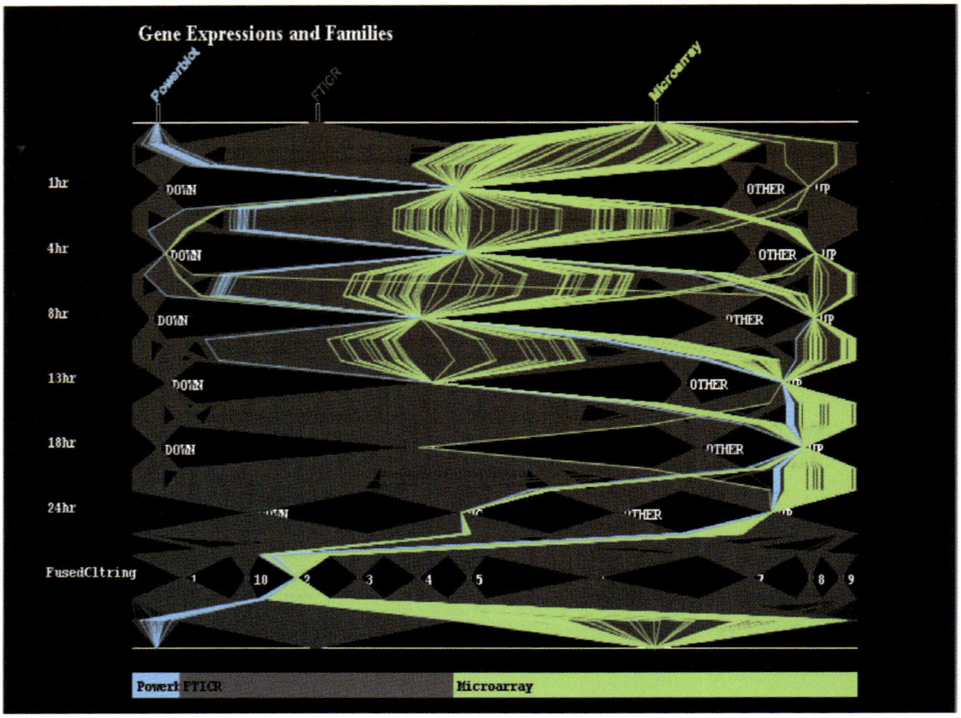

Figure 2.7 The Juxter visualization tool demonstrates the capability to integrate individual datasets (in this case, Powerblot, FTICR and Microarray data), each at an individual time point. The top and bottom tiers represent the data type that each layer in the visualization shows when genes or proteins fall into the categories of up-, down- or nonregulated at each time point. The last layer in the visualization gives the statistically merged results over the entire time-course

As an alternative to classifying or clustering the data (as seen in Figure 2.7), in proteomics and biology the end goal is often to find a set of biomolecules that are relevant to the question of interest – for example, which proteins are associated with a virulent versus nonvirulent pathogen. In this case, the most common approach is to use statistical tests of significance and to assign p-values, or normalized false discovery rates, to each biomolecule. Thus, all of the datasets can be reduced down to a set of p-values, which can then be merged into a level of significance associated with related entities. In recent years these methods have become much more robust by accounting for biological nuances and using multiple statistics to evaluate the significance of individual biolomolecular species (e.g. genes, proteins, metabolites). In Ref. [112] each dataset is first evaluated using PCA to visualize relationships in each data source and reduce the dimensionality of the integration task. A coinertia analysis is then used to evaluate correlations across the datasets, which better accounts for any biological issues that arise in direct correlation analyses from post-transcriptional and post-translational regulations. POINTILLIST [113] uses a weighted version of several statistical metrics of significance to derive a network model where the integrated p-value measure indicates the degree of confidence in a node or edge, being a true component of the system of interest where a node represents a biomolecular species.

The benefit of these methods – both integration of clusters and statistical levels of significance – is that they can better handle datasets of vastly different sizes and types. Additionally, a normalization need only be performed within each dataset. Even further, as seen in Figure 2.7, there is no need for one-to-one mapping between datasets under the condition that clusterings are the end goal. Although some of the statistical significance integration approaches can account for missing data (a common problem in proteomics), the caveat is that, in many cases, there may not be a one-to-one mapping between the datasets, and the interpretation may be both difficult and time-consuming.

2.6.5 Integration in Feature Space via Data Transformation

In supervised learning it is often the case that the data are transformed into an alterative representation, such as a relationship or a kernel matrix. In biology, data are often represented as the relationship between biomolecular entities, for example correlations between genes that might relate to a common regulation, or links between proteins that represent possible interactions. These relationship matrices can be merged into a more accurate view of the system by using methods such as Bayesian networks, where the relationship matrices are the input [114–116]. This approach is slightly different from that described above as the relationship matrix itself is not typically the final result for an individual dataset, but an intermediate representation used for the task of learning. Seeing as the data are merged at an intermediate form, a major benefit of this approach is that the data do not have to have a one-to-one mapping. The largest caveat is that these methods are often computationally intensive in learning the parameters of the model.

A more abstract approach to the integration of transformed data is that of kernel fusion. A kernel function is a transformed projection of the data that, in principle, enhances linear separability. Kernel functions are especially powerful for datasets that are not linearly separable by mapping the data into a space that can be linearly separated by a SVM. Individual kernel functions for each dataset can be merged into an integrated kernel [117],

$$K_{\mathrm{Int}} = m_1 K_1 + m_2 K_2 + \cdots + m_n K_n, \tag{2.2}$$

where K_i is the kernel associated with the i-th dataset. K_{Int} can be used to build a supervised model in the same manner as for a single data source. Although this is a very powerful statistical approach, it possesses the same limitations as the feature integration method, namely a one-to-one mapping between biomolecular entities. However, with this approach it is much easier to integrate information from other computational tools, such as similarity between entities by protein domains or sequence similarity.

2.7 Summary

In 1958, Francis Crick laid out the 'Central Dogma' of biology:

> *...once 'information' has passed into protein* it cannot get out again. *In more detail, the transfer of information from nucleic acid to nucleic acid, or from nucleic acid to protein may*

> *be possible, but transfer from protein to protein, or from protein to nucleic acid is impossible. Information means here the* precise *determination of sequence, either of bases in the nucleic acid or of amino acid residues in the protein. [118]*

In other words, proteins are an end-point for the information encoded within the genome, and thus they should – and do – merit a strong research focus. The structure of proteins, however, makes their study at a global scale more challenging than that of nucleic acids, for which high-throughput sequencing and hybridization approaches already exist. The study of all proteins encoded by the genome, or the 'proteome' (this term was coined by the Australian biochemist, Marc Wilkins, in 1994), relies on protein or peptide separation followed by MS. In the past, this has proven to be the most efficient method for identifying protein sequences *en masse*; however, given the complex mixtures and the nature of the approach, there are several caveats associated with modern proteomics.

This chapter has explored the many limitations associated with current proteomics methods. Gene models are crucial to proteomics, because they serve as the basis for database searching algorithms used to match mass spectra generated by global proteomics. Indeed, incorrect models lead to both false-positive and false-negative peptide sequence information. The processing of samples for proteomic analysis is also important as, while the development of a method for protein isolation can be generalized, sample handling can vary with the breadth of organisms being studied. Also crucial to processing is the development of a statistically rigorous experimental design, this being integral to downstream analysis and to the identification of samples that have failed due to processing or instrument errors. The correct design of an experiment also translates to an ability to apply statistical modeling approaches, which moves proteomics from a qualitative to a quantitative method. Finally, the tools and methods for the integration of proteomic and other high-throughput global analyses, such as microarrays and metabolomics, are needed because the proteome is only one tool of several required to build hypotheses and models of biological systems.

Today, whilst the field of proteomics continues to advance rapidly, further advances in gene modeling, MS, rigorous statistical approaches and bioinformatics tools for proteomics are needed to secure the robustness of those methods currently in use for genome sequencing and transcriptome analysis. Although, undoubtedly, improvements and refinements will be made as we move forwards, there is still much to be gained from the currently available proteome analysis approaches.

References

1. Fleischmann, R.D., Adams, M.D., White, O. *et al.* (1995) Whole-genome random sequencing and assembly of *Haemophilus influenzae* Rd. *Science*, **269**, 496–512.
2. Goffeau, A., Barrell, B.G., Bussey, H. *et al.* (1996) Life with 6000 genes. *Science*, **274**, 546, 63–47.
3. Taylor, J. (2005) Clues to function in gene deserts. *Trends in Biotechnology*, **23**, 269–71.
4. Burset, M. and Guigo, R. (1996) Evaluation of gene structure prediction programs. *Genomics*, **34**, 353–67.
5. Guigo, R., Flicek, P., Abril, J.F. *et al.* (2006) EGASP: the human ENCODE genome annotation assessment project. *Genome Biology*, **7**(Suppl.1), S21–31.

6. Reese, M.G., Hartzell, G., Harris, N.L. *et al.* (2000) Genome annotation assessment in *Drosophila melanogaster*. *Genome Research*, **10**, 483–501.
7. Yao, H., Guo, L. and Fu, Y. *et al.* (2005) Evaluation of five ab initio gene prediction programs for the discovery of maize genes. *Plant Molecular Biology*, **57**, 445–60.
8. Salamov, A.A. and Solovyev, V.V. (2000) Ab initio gene finding in *Drosophila* genomic DNA. *Genome Research*, **10**, 516–22.
9. Stanke, M. and Waack, S. (2003) Gene prediction with a hidden Markov model and a new intron submodel. *Bioinformatics*, **19**(Suppl. 2), ii215–25.
10. Korf, I. (2004) Gene finding in novel genomes. *BMC Bioinformatics*, **5**, 59.
11. Lukashin, A.V. and Borodovsky, M. (1998) GeneMark.hmm: new solutions for gene finding. *Nucleic Acids Research*, **26**, 1107–15.
12. Birney, E., Clamp, M. and Durbin, R. (2004) GeneWise and Genomewise. *Genome Research*, **14**, 988–95.
13. Xu, Y., Mural, R.J. and Uberbacher, E.C. (1997) Inferring gene structures in genomic sequences using pattern recognition and expressed sequence tags. *Proceedings International Conference on Intelligent Systems for Molecular Biology*, **5**, 344–53.
14. Haas, B.J., Delcher, A.L., Mount, S.M. *et al.* (2003) Improving the *Arabidopsis* genome annotation using maximal transcript alignment assemblies. *Nucleic Acids Research*, **31**, 5654–66.
15. Tenney, A.E., Brown, R.H., Vaske, C. *et al.* (2004) Gene prediction and verification in a compact genome with numerous small introns. *Genome Research*, **14**, 2330–5.
16. Schiex, A., Moisan, A. and Rouze, P. (2001) EuGene: An eukaryotic gene finder that combines several sources of evidence, in Computational Biology: First International Conference on Biology, Informatics, and Mathematics, JOBIM 2000 Montpellier, France, May 3–5, 2000 Selected Papers (eds O. Gascuel and M.-F. Sagot), Springer, Heidelberg.
17. Allen, J.E., Pertea, M. and Salzberg, S.L. (2004) Computational gene prediction using multiple sources of evidence. *Genome Research*, **14**, 142–8.
18. Majoros, W.H., Pertea, M. and Salzberg, S.L. (2005) Efficient implementation of a generalized pair hidden Markov model for comparative gene finding. *Bioinformatics*, **21**, 1782–8.
19. Solovyev, V.V. (2002) Structure, properties and computer identification of eukaryotic genes, in *Bioinformatics – from Genomes to Drugs* (ed. T. Lengauer), Wiley-VCH Verlag GmbH, Weinheim.
20. Dewey, C., Wu, J.Q., Cawley, S. *et al.* (2004) Accurate identification of novel human genes through simultaneous gene prediction in human, mouse, and rat. *Genome Research*, **14**, 661–4.
21. Wei, C., Lamesch, P., Arumugam, M. *et al.* (2005) Closing in on the *C. elegans* ORFeome by cloning TWINSCAN predictions. *Genome Research*, **15**, 577–82.
22. Tyler, B.M., Tripathy, S., Zhang, X. *et al.* (2006) Phytophthora genome sequences uncover evolutionary origins and mechanisms of pathogenesis. *Science*, **313**, 1261–6.
23. Pavlovic, V., Garg, A. and Kasif, S. (2002) A Bayesian framework for combining gene predictions. *Bioinformatics*, **18**, 19–27.
24. Potter, S.C., Clarke, L., Curwen, V. *et al.* (2004) The Ensembl analysis pipeline. *Genome Research*, **14**, 934–41.
25. Braun, B.R., van Het Hoog, M., d'Enfert, C. *et al.* (2005) A human-curated annotation of the *Candida albicans* genome. *PLoS Genetics*, **1**, 36–57.
26. Dujon, B., Sherman, D., Fischer, G. *et al.* (2004) Genome evolution in yeasts. *Nature*, **430**, 35–44.
27. Jaillon, O., Aury, J.M., Noel, B. *et al.* (2007) The grapevine genome sequence suggests ancestral hexaploidization in major angiosperm phyla. *Nature*, **449**, 463–7.
28. Altschul, S.F. and Koonin, E.V. (1998) Iterated profile searches with PSI-BLAST – a tool for discovery in protein databases. *Trends in Biochemical Science*, **23**, 444–7.

29. Zdobnov, E.M. and Apweiler, R. (2001) InterProScan–an integration platform for the signature-recognition methods in InterPro. *Bioinformatics*, **17**, 847–8.
30. Emanuelsson, O., Brunak, S., von Heijne, G. and Nielsen, H. (2007) Locating proteins in the cell using TargetP, SignalP and related tools. *Nature Protocols*, **2**, 953–71.
31. Kanehisa, M., Goto, S., Kawashima, S. *et al.* (2004) The KEGG resource for deciphering the genome. *Nucleic Acids Research*, **32**, D277–80.
32. Ashburner, M., Ball, C.A., Blake, J.A. *et al.* (2000) Gene ontology: tool for the unification of biology. The gene ontology consortium. *Nature Genetics*, **25**, 25–9.
33. Tatusov, R.L., Fedorova, N.D., Jackson, J.D. *et al.* (2003) The COG database: an updated version includes eukaryotes. *BMC Bioinformatics*, **4**, 41.
34. Parkinson, H., Sarkans, U., Shojatalab, M. *et al.* (2005) ArrayExpress – a public repository for microarray gene expression data at the EBI. *Nucleic Acids Research*, **33**, D553–5.
35. Barrett, T., Suzek, T.O., Troup, D.B. *et al.* (2005) NCBI GEO: mining millions of expression profiles – database and tools. *Nucleic Acids Research*, **33**, D562–6.
36. Medina, M.L., Haynes, P.A., Breci, L. and Francisco, W.A. (2005) Analysis of secreted proteins from *Aspergillus flavus*. *Proteomics*, **5**, 3153–61.
37. Medina, M.L., Kiernan, U.A. and Francisco, W.A. (2004) Proteomic analysis of rutin-induced secreted proteins from *Aspergillus flavus*. *Fungal Genetics and Biology*, **41**, 327–35.
38. Wymelenberg, A.V., Sabat, G., Martinez, D. *et al.* (2005) The *Phanerochaete chrysosporium* secretome: database predictions and initial mass spectrometry peptide identifications in cellulose-grown medium. *Journal of Biotechnology*, **118**, 17–34.
39. Bestel-Corre, G., Dumas-Gaudot, E. and Gianinazzi, S. (2004) Proteomics as a tool to monitor plant-microbe endosymbioses in the rhizosphere. *Mycorrhiza*, **14**, 1–10.
40. Zhou, K., Panisko, E.A., Magnuson, J.K. *et al.* (in press) Proteomics for validation of automated gene model predictions, in *Mass Spectrometry of Proteins and Peptides* (eds M. Lipton and L. Pasa-Tolic), Humana Press.
41. Righetti, P.G., Castagna, A., Antonucci, F. *et al.* (2004) Critical survey of quantitative proteomics in two-dimensional electrophoretic approaches. *Journal of Chromatography A*, **1051**, 3–17.
42. Gevaert, K. and Vandekerckhove, J. (2000) Protein identification methods in proteomics. *Electrophoresis*, **21**, 1145–54.
43. Marouga, R., David, S. and Hawkins, E. (2005) The development of the DIGE system: 2D fluorescence difference gel analysis technology. *Analytical and Bioanalytical Chemistry*, **382**, 669–78.
44. Cole, R.B. (2000) Some tenets pertaining to electrospray ionization mass spectrometry. *Journal of Mass Spectrometry*, **35**, 763–72.
45. Griffiths, W.J., Jonsson, A.P., Liu, S. *et al.* (2001) Electrospray and tandem mass spectrometry in biochemistry. *Biochemical Journal*, **355**, 545–61.
46. Conrads, T.P., Anderson, G.A., Veenstra, T.D. *et al.* (2000) Utility of accurate mass tags for proteome-wide protein identification. *Analytical Chemistry*, **72**, 3349–54.
47. Bantscheff, M., Schirle, M., Sweetman, G. *et al.* (2007) Quantitative mass spectrometry in proteomics: a critical review. *Analytical and Bioanalytical Chemistry*, **389**, 1017–31.
48. Nesvizhskii, A.I., Vitek, O. and Aebersold, R. (2007) Analysis and validation of proteomic data generated by tandem mass spectrometry. *Nature Methods*, **4**, 787–97.
49. Mann, M. and Wilm, M. (1994) Error-tolerant identification of peptides in sequence databases by peptide sequence tags. *Analytical Chemistry*, **66**, 4390–9.
50. Ong, S.E., Blagoev, B., Kratchmarova, I. *et al.* (2002) Stable isotope labeling by amino acids in cell culture, SILAC, as a simple and accurate approach to expression proteomics. *Molecular and Cellular Proteomics*, **1**, 376–86.

51. Gygi, S.P., Rist, B., Gerber, S.A. *et al.* (1999) Quantitative analysis of complex protein mixtures using isotope-coded affinity tags. *Nature Biotechnology*, **17**, 994–9.
52. Ross, P.L., Huang, Y.N., Marchese, J.N. *et al.* (2004) Multiplexed protein quantitation in *Saccharomyces cerevisiae* using amine-reactive isobaric tagging reagents. *Molecular and Cellular Proteomics*, **3**, 1154–69.
53. Monroe, M.E., Shaw, J.L., Daly, D.S. *et al.* (2008) MASIC: A software program for fast quantitation and flexible visualization of chromatographic profiles from detected LC-MS(/MS) features. *Computational Biology and Chemistry*, **32**, 215–17.
54. Goshe, M.B., Conrads, T.P., Panisko, E.A. *et al.* (2001) Phosphoprotein isotope-coded affinity tag approach for isolating and quantitating phosphopeptides in proteome-wide analyses. *Analytical Chemistry*, **73**, 2578–86.
55. Ficarro, S.B., McCleland, M.L., Stukenberg, P.T. *et al.* (2002) Phosphoproteome analysis by mass spectrometry and its application to *Saccharomyces cerevisiae*. *Nature Biotechnology*, **20**, 301–5.
56. Canas, B., Pineiro, C., Calvo, E. *et al.* (2007) Trends in sample preparation for classical and second generation proteomics. *Journal of Chromatography A*, **1153**, 235–58.
57. Bodzon-Kulakowska, A., Bierczynska-Krzysik, A., Dylag, T. *et al.* (2007) Methods for samples preparation in proteomic research. *Journal of Chromatography B: Analytical Technologies in the Biomedical and Life Sciences*, **849**, 1–31.
58. Smith, P.K., Krohn, R.I., Hermanson, G.T. *et al.* (1985) Measurement of protein using bicinchoninic acid. *Analytical Biochemistry*, **150**, 76–85.
59. Washburn, M.P., Wolters, D. and Yates, J.R. 3rd (2001) Large-scale analysis of the yeast proteome by multidimensional protein identification technology. *Nature Biotechnology*, **19**, 242–7.
60. Eng, J.K., McCormack, A.L. and Yates, J.R. III (1994) An approach to correlate tandem mass spectral data of peptides with amino acid sequences in a protein database. *Journal of the American Society of Mass Spectrometry*, **5**, 976–89.
61. Craig, R. and Beavis, R.C. (2003) A method for reducing the time required to match protein sequences with tandem mass spectra. *Rapid Communications in Mass Spectrometry*, **17**, 2310–16.
62. Kiebel, G.R., Auberry, K.J., Jaitly, N. *et al.* (2006) PRISM: a data management system for high-throughput proteomics. *Proteomics*, **6**, 1783–90.
63. Wikipedia (2008).
64. NIST/SEMATECH (2008) *e-Handbook of Statistical Methods*. http://www.itl.nist.gov/div898/handbook
65. StatSoft, I. (2007) *Electronic Statistics Textbook*, StatSoft.
66. UCLADepartment of Statistics (2008) *EBook*, UCLA.
67. Wolfram, I. (2008) *Mathworld Probability and Statistics*.
68. Pinheiro, J.C. and Bates, D.M. (2000) *Mixed-Effects Models in S and S-PLUS*, Springer, New York.
69. Daly, D.S., Anderson, K.K., Panisko, E.A. *et al.* (2008) Mixed-effects statistical model for comparative LC-MS proteomics studies. *Journal of Proteome Research*, **7**, 1209–17.
70. Laird, N.M. and Ware, J.H. (1982) Random-effects models for longitudinal data. *Biometrics*, **38**, 963–74.
71. Patterson, H.D. and Thompson, R. (1971) Recovery of interblock information when block sizes are unequal. *Biometrika*, **58**, 545–54.
72. Searle, S.R., Casella, G. and McCulloch, C.E. (1992) *Variance Components*, John Wiley & Sons, Inc., New York.
73. Purvine, S., Picone, A.F. and Kolker, E. (2004) Standard mixtures for proteome studies. *Omics*, **8**, 79–92.

74. Eckart, J.D. and Sobral, B.W. (2003) A life scientist's gateway to distributed data management and computing: the PathPort/ToolBus framework. *Omics*, **7**, 79–88.
75. Oinn, T., Addis, M., Ferris, J. *et al.* (2004) Taverna: a tool for the composition and enactment of bioinformatics workflows. *Bioinformatics*, **20**, 3045–54.
76. Goesmann, A., Linke, B., Rupp, O. *et al.* (2003) Building a BRIDGE for the integration of heterogeneous data from functional genomics into a platform for systems biology. *Journal of Biotechnology*, **106**, 157–67.
77. Lee, T.J., Pouliot, Y., Wagner, V. *et al.* (2006) BioWarehouse: a bioinformatics database warehouse toolkit. *BMC Bioinformatics*, **7**, 170.
78. Oikawa, M.K., Broinizi, M.E.B., Dermargos, A. *et al.* (2004) GenFlow: Generic flow for integration, management and analysis of molecular biology data. *Genetic and Molecular Biology*, **27**, 691–5.
79. Lu, Q., Hao, P., Curcin, V. *et al.* (2006) KDE Bioscience: platform for bioinformatics analysis workflows. *Journal of Biomedical Information*, **39**, 440–50.
80. Peleg, M., Yeh, I. and Altman, R.B. (2002) Modeling biological processes using workflow and Petri Net models. *Bioinformatics*, **18**, 825–37.
81. Facius, A., Englbrecht, C., Birzele, F. *et al.* (2005) PRIME: a graphical interface for integrating genomic/proteomic databases. *Proteomics*, **5**, 76–80.
82. Watson, M. (2005) ProGenExpress: visualization of quantitative data on prokaryotic genomes. *BMC Bioinformatics*, **6**, 98.
83. Shah, A.R., Singhal, M., Klicker, K.R. *et al.* (2007) Enabling high-throughput data management for systems biology: the bioinformatics resource manager. *Bioinformatics*, **23**, 906–9.
84. Shannon, P.T., Reiss, D.J., Bonneau, R. and Baliga, N.S. (2006) The Gaggle: an open-source software system for integrating bioinformatics software and data sources. *BMC Bioinformatics*, **7**, 176.
85. Kokocinski, F., Delhomme, N., Wrobel, G. *et al.* (2005) FACT–a framework for the functional interpretation of high-throughput experiments. *BMC Bioinformatics*, **6**, 161.
86. Bader, G.D., Betel, D. and Hogue, C.W. (2003) BIND: the biomolecular interaction network database. *Nucleic Acids Research*, **31**, 248–50.
87. Xenarios, I., Rice, D.W., Salwinski, L. *et al.* (2000) DIP: the database of interacting proteins. *Nucleic Acids Research*, **28**, 289–91.
88. Kanehisa, M. and Goto, S. (2000) KEGG: Kyoto encyclopedia of genes and genomes. *Nucleic Acids Research*, **28**, 27–30.
89. Bairoch, A., Apweiler, R., Wu, C.H. *et al.* (2005) The universal protein resource (UniProt). *Nucleic Acids Research*, **33**, D154–9.
90. Kersey, P.J., Duarte, J., Williams, A. *et al.* (2004) The international protein index: an integrated database for proteomics experiments. *Proteomics*, **4**, 1985–8.
91. Wheeler, D.L., Chappey, C., Lash, A.E. *et al.* (2000) Database resources of the national center for biotechnology information. *Nucleic Acids Research*, **28**, 10–14.
92. Shannon, P., Markiel, A., Ozier, O. *et al.* (2003) Cytoscape: a software environment for integrated models of biomolecular interaction networks. *Genome Research*, **13**, 2498–504.
93. Webb-Robertson, B.J., Peterson, E.S., Singhal, M. *et al.* (2007) PQuad–a visual analysis platform for proteomic data exploration of microbial organisms. *Bioinformatics*, **23**, 1705–7.
94. Aggarwal, K. and Lee, K.H. (2003) Functional genomics and proteomics as a foundation for systems biology. *Briefings in Functional Genomics and Proteomics*, **2**, 175–84.
95. Nie, L., Wu, G., Culley, D.E. *et al.* (2007) Integrative analysis of transcriptomic and proteomic data: challenges, solutions and applications. *Critical Reviews in Biotechnology*, **27**, 63–75.
96. Reif, D.M., White, B.C. and Moore, J.H. (2004) Integrated analysis of genetic, genomic and proteomic data. *Expert Reviews in Proteomics*, **1**, 67–75.

97. De Keersmaecker, S.C., Thijs, I.M., Vanderleyden, J. and Marchal, K. (2006) Integration of omics data: how well does it work for bacteria? *Molecular Microbiology*, **62**, 1239–50.
98. Johnson, R.A. and Wichern, D.W. (1992) *Applied Multivariate Statistical Analysis*, Prentice Hall, Englewood Cliffs, N.J.
99. Neter, J. (1996) *Applied Linear Regression Models*, Irwin, Chicago, Ill.
100. Waters, K.M., Pounds, J.G. and Thrall, B.D. (2006) Data merging for integrated microarray and proteomic analysis. *Briefings in Functional Genomics and Proteomics*, **5**, 261–72.
101. Cox, B., Kislinger, T. and Emili, A. (2005) Integrating gene and protein expression data: pattern analysis and profile mining. *Methods*, **35**, 303–14.
102. Anderson, D.C., Li, W., Payan, D.G. and Noble, W.S. (2003) A new algorithm for the evaluation of shotgun peptide sequencing in proteomics: support vector machine classification of peptide MS/MS spectra and SEQUEST scores. *Journal of Proteome Research*, **2**, 137–46.
103. Cannon, W.R., Jarman, K.H., Webb-Robertson, B.J. *et al.* (2005) Comparison of probability and likelihood models for peptide identification from tandem mass spectrometry data. *Journal of Proteome Research*, **4**, 1687–98.
104. Cristianini, N. and Shawe-Taylor, J. (2000) *An Introduction to Support Vector Machines and Other Kernel-based Learning Methods*, Cambridge University Press, Cambridge.
105. Vapnik, V.N. (1995) *The Nature of Statistical Learning Theory*, Springer, New York.
106. Topchy, A.B., Jain, A.K. and Punch, W. (2004) A mixture model for clustering ensembles. Proceedings of the Fourth SIAM International Conference on Data Mining. Society for Industrial and Applied Mathematics, Philadelphia (ed. M.W. Berry), pp. 379–90.
107. Zeng, Y., Tang, J., Garcia-Frias, J. and Gao, G.R. (2002) An adaptive meta-clustering approach: combining the information from different clustering results. *Proceedings IEEE Computer Society Bioinformatics Conference*, **1**, 276–87.
108. Barutcuoglu, Z., Schapire, R.E. and Troyanskaya, O.G. (2006) Hierarchical multi-label prediction of gene function. *Bioinformatics*, **22**, 830–6.
109. Kano, M., Tsutsumi, S., Kawahara, N. *et al.* (2005) A meta-clustering analysis indicates distinct pattern alteration between two series of gene expression profiles for induced ischemic tolerance in rats. *Physiological Genomics*, **21**, 274–83.
110. Kasturi, J. and Acharya, R. (2005) Clustering of diverse genomic data using information fusion. *Bioinformatics*, **21**, 423–9.
111. Havre, S.L., Webb-Robertson, B.J., Shah, A. *et al.* (2005) Bioinformatic insights from metagenomics through visualization. *Proceedings IEEE Computational Systems Bioinformatics Conference (CSB '05)*, pp. 341–50.
112. Fagan, A., Culhane, A.C. and Higgins, D.G. (2007) A multivariate analysis approach to the integration of proteomic and gene expression data. *Proteomics*, **7**, 2162–71.
113. Hwang, D., Rust, A.G., Ramsey, S. *et al.* (2005) A data integration methodology for systems biology. *Proceedings of the National Academy of Sciences of the United States of America*, **102**, 17296–301.
114. Gilchrist, M.A., Salter, L.A. and Wagner, A. (2004) A statistical framework for combining and interpreting proteomic datasets. *Bioinformatics*, **20**, 689–700.
115. Huttenhower, C. and Troyanskaya, O.G. (2006) Bayesian data integration: a functional perspective. *Computer Systems Bioinformatics Conference*, **4**, 341–51.
116. Troyanskaya, O.G., Dolinski, K., Owen, A.B. *et al.* (2003) A Bayesian framework for combining heterogeneous data sources for gene function prediction (in Saccharomyces cerevisiae). *Proceedings of the National Academy of Sciences of the United States of America*, **100**, 8348–53.
117. Lanckriet, G.R., De Bie, T., Cristianini, N. *et al.* (2004) A statistical framework for genomic data fusion. *Bioinformatics*, **20**, 2626–35.
118. Crick, F.H. (1958) On protein synthesis. *Symposium of the Society of Experimental Biology*, **12**, 138–63.

Section 2
Analysis via Automation

3

High-Throughput DNA Sequencing

Tarjei S. Mikkelsen

Broad Institute of MIT and Harvard, Cambridge, USA

Today, the determination of the complete DNA sequence of an organism has become a fundamental part of biological inquiry. Initially motivated by the Human Genome Project, tremendous effort has been expended into the development of cost-efficient, high-throughput DNA sequencing instruments capable of decoding any genome of interest. In this chapter, we review the current state of traditional dideoxy sequencing workflows, and the development of next-generation technologies that are poised to revolutionize genomics.

3.1 Traditional Dideoxy (Sanger) Sequencing

Efficient methods for determining the nucleotide sequence of a DNA polymer were first demonstrated in 1977, when Allan Maxam, Walter Gilbert and Fred Sanger independently published descriptions of sequencing methods that relied on gel electrophoresis to resolve DNA fragments encoding sequence information at base pair resolution [1, 2]. While Maxam–Gilbert sequencing initially became the most widely used methodology, Sanger's dideoxy sequencing method eventually proved to be more practical, and has been used in the vast majority of sequencing projects over the past three decades.

The classic implementation of dideoxy sequencing requires four separate reactions each containing multiple copies of a single-stranded DNA template, short DNA primers complementary to one site in the template, DNA polymerase, four radiolabeled deoxy nucleotides (dATP, dCTP, dGTP and dTTP), and a relatively low concentration of one of four dideoxy nucleotides (ddATP, ddCTP, ddGTP or ddTTP). The primer initiates the polymerase synthesis of DNA strands complementary to the template. Dideoxy nucleotides

Automation in Proteomics and Genomics: An Engineering Case-Based Approach Edited by Gil Alterovitz, Roseann Benson and Marco Ramoni

lack the 3′-hydroxyl group required to form phosphodiester bonds between adjacent nucleotides during extension, and therefore serve as chain terminators. In each of the four sequencing reactions, the end products will be an ensemble of radiolabeled DNA strands with lengths that correspond to the locations of one of the four nucleotides in the original DNA template. After heat denaturation, the different products can be separated by length and visualized by gel electrophoresis, allowing inference of the template sequence over several hundred base pairs following the primer site.

3.2 Automated Dideoxy Sequencing

Initially, a highly informative – but labor intensive – process that yielded at most a few hundred bases of sequence information per experiment, dideoxy sequencing has been successfully scaled to a level where reading every one of the three billion bases in the human genome several times over in a matter of months has become feasible, at least at specialized sequencing centers. Getting to this point involved the extensive modifications of the original chemistry to make it more amenable to automation. In particular, an avoidance of radiolabeling by introducing fluorescently labeled dideoxy terminators [3] and compatible DNA polymerases [4] proved to be a key innovation. Major investments were also made in robotics and parallelized sample preparation.

Because, in practice, the dideoxy sequencing process is limited to reading less than 1000 base pairs from any one template, indirect strategies are required to infer the contiguous sequence of larger, naturally occurring DNA polymers, such as each chromosome in a genome. At present, the dominant strategy for sequencing a new genome is whole-genome shotgun (WGS) sequencing [5,6], where mechanical shearing is used to fragment a genome and each fragment is then ligated into a common plasmid vector. DNA templates correctly inserted into the vectors are selected and amplified by bacterial cloning, isolated, subjected to the dideoxy reaction using universal primers annealing to the plasmid, and then sequenced using automated gel electrophoresis. Obtaining ~650 bp of information from 20–40 million such fragments is sufficient to infer, computationally, the contiguous sequence of a human-sized genome. Variations of this workflow have also recently been developed for the high-throughput generation and sequencing of targeted PCR (polymerase chain reaction) products (e.g. to sequence only the coding exons in a particular genome). Today, all of the key steps in this workflow can be automated [7,8].

3.2.1 Colony Picking

DNA templates to be sequenced are typically generated in a complex library, for example by mechanical shearing of genomic DNA. The isolation and amplification of individual DNA templates can be achieved by dispersing individual bacteria transformed or infected with template-containing vectors on agar plates with growth medium and allowing clonal colonies to form. An automated system consisting of a CCD camera, a robotic arm and a plate-handling device is then used to pick colonies, using disposable or sterilized pins, and to deposit such colonies into 96- or 384-well sample collection plates containing liquid media. Sophisticated image analysis algorithms are used to automatically identify the correctly sized and useful colonies, such that typical systems can operate unattended and isolate thousands of colonies each hour. Following isolation, each individual DNA template is next amplified by clonal expansion of the picked colonies, typically in overnight cultures.

3.2.2 Template Preparation and Sequencing Reactions

The preparation of pure DNA templates for the sequencing reactions can be performed by robotic workstations that combine automated injection systems to deliver lysis buffers and other reagents, filter systems or magnetic separators to capture purified DNA, plate washers and sealers, and pick-and-place robots in various configurations to minimize manual intervention.

A recent alternative amplification and template preparation approach is the TempliPhi system (GE Healthcare), which uses φ29 DNA polymerase to isothermally amplify circular DNA templates, such as plasmids, by using rolling circle amplification. This system can be applied directly to isolated bacterial colonies, or to saturated cultures, in principle without the need for separate DNA isolation or purification.

Amplified DNA templates can next be prepared for sequencing by robotic plate and liquid-handling systems that automatically aliquot templates into 96- or 384-well sequencing plates, add primers, polymerase, nucleotides and other required reagents for a total reaction volume of a few microliters, and finally heat-seal the plates.

The actual sequencing reactions are carried out in the prepared plates on automated thermocyclers. Multiple rounds of primer annealing, extension and denaturation linearly increase the number of terminated strands, to maximize downstream sensitivity. Specialized thermostable polymerases have been engineered to efficiently incorporate dye-labeled dideoxy nucleotides through these cycles. The products of the reaction cycles are finally transferred to the actual sequencing instrument.

3.2.3 Sequencing

During the two decades following the invention and commercial development of automated fluorescence DNA sequencers in the mid-1980s, there was a steady evolution of increasingly sophisticated and robust dideoxy-based sequencing instruments. In the first decade, the most commonly used automated instruments were based on slab gel electrophoresis, where each instrument was loaded with a large gel plate, with multiple samples being separated in parallel on the gel in either one or four lanes, depending on the fluorescence system used. At their peak, slab gel systems yielded throughput from tens of thousands to a few hundred thousand bases of sequence information per day. However, the manual loading of new gel slabs proved to be a common bottleneck, and eventually the slab gel approach was largely replaced by instruments that process individual samples in capillaries filled with gel polymers. This method not only reduced reagent consumption dramatically but was also amenable to automated gel replenishment and unattended operation over extended time periods.

In all recent implementations of the dideoxy process, each sequencing sample is prepared in a single primer extension reaction with four species of dideoxy nucleotides labeled with one of four different fluorescent dyes that have identical excitation wavelengths, yet unique emission spectra (Figure 3.1). As the sequencing reaction fragments pass through the capillaries in the order determined by their sizes, a single-wavelength laser excites the dyes, and the resulting emission spectra are analyzed to infer the corresponding nucleotides.

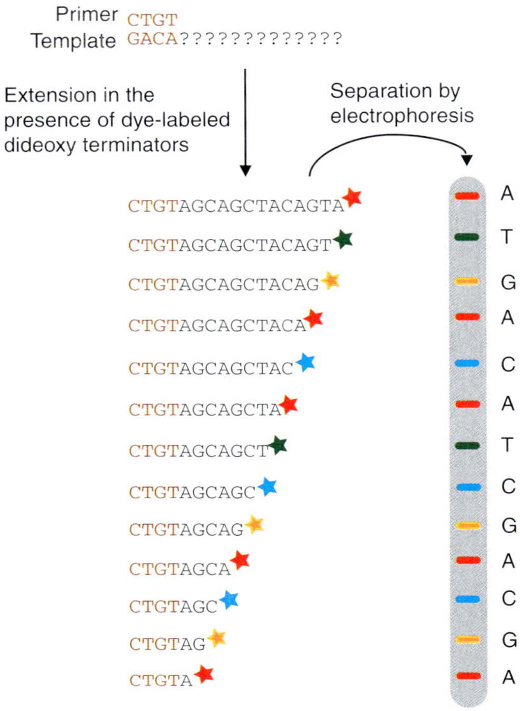

Figure 3.1 *Overview of the dideoxy (Sanger) sequencing process. DNA polymerase is used to synthesize complementary copies of the template from a common primer sequence. Random incorporation of dye-labeled dideoxy terminators generates a pool of differently sized extension products. Gel separation of the extension products allows inference of the template sequence*

Currently, the *de facto* standard dideoxy sequencing instrument in most large academic and commercial sequencing centers is the Applied Biosystems 3730xl capillary electrophoresis sequencer. This system features parallel, four-color gel electrophoresis and laser interrogation in 96 capillaries, and can produce up to 2.1 million bases of sequence information per day, depending on its configuration. Automation features include temperature control, reagent handling and gel replenishment for up to 48 h of unattended operation, integrated robotics and a barcode reader for handling of up to 16 different 384-well sample plates, and integrated base calling.

3.2.4 LIMS and Supply Chain Management

While being far less visible than the robotic platforms and sequencing instruments in a typical automated genome center or core facility, laboratory information management systems (LIMS) and supply chain management systems are essential for the effective management of these complex sequencing processes and the resulting data that are generated.

For example, at the Broad Institute of MIT and Harvard – which is one of the world's largest sequencing centers – approximately 2.2 million individual dideoxy sequencing reactions were carried out per week in 2007, generating data for more than 30 distinct scientific projects. Preparing these reactions and performing the related work required the availability over 900 different types of reagent and other materials, including the weekly preparation of more than 500 l of different solutions and 2000 agar plates, and the labeling of approximately 1.0×10^4 384-well plates. If run-out of any one critical reagent were to occur, the production loss would be on the order of \$100 000–\$200 000 per day. Given that the center operates at full capacity, it would not be possible to recover this lost output; accordingly, the center relies heavily on customized inventory tracking systems and the long-range forecasting and material planning capabilities provided by such systems to maintain consistent costs and throughput.

3.3 Next-Generation Sequencing Technologies

Dideoxy-based sequencing has displayed remarkable staying power and capacity for optimization. Similar to the production of semiconductors, the cost of sequencing has decreased exponentially over the past two decades [9]. However, with the stated post-Human Genome Project goal of sequencing any human genome for \$1000 [10], the consensus in the DNA sequencing field is that, in order to achieve the additional orders of magnitude improvements in throughput and cost required to reach this milestone, an entirely new generation of automated sequencing technologies must be developed. Driven by this goal and the ever-increasing demand for sequencing capacity, commercially available sequencing technologies have recently undergone a not-so-quiet revolution. Few of these technologies were technically 'novel' at the time they were picked up for commercialization, in the sense that they had been developed and explored in varying detail in academic settings, in some cases for several decades. However, significant engineering challenges must be overcome and workflow optimization must be carried out to achieve the full potential of any given technology. These improvements are typically driven by commercial demand. While the state of next-generation sequencing technology is expected to be in flux for a number of years, some trends are emerging today.

3.3.1 Cyclic Array Sequencing

All commercially available non-Sanger sequencing instruments at the time of writing are based on a single unifying principle termed cyclic array sequencing (Figure 3.2) [11]. In this approach, high throughput and decreased costs are achieved by using a single reagent volume to simultaneously infer the sequence of millions (potentially billions) of DNA features immobilized on a two-dimensional array. Depending on the specific implementation, each DNA feature may be a single molecule, or an ensemble of identical molecules in close spatial proximity generated by an *in vitro* amplification step. The DNA features may be deposited on the array either in an ordered grid, or be randomly dispersed. Sequencing takes place in progressive cycles where, in each cycle, an enzymatic process is used to interrogate one nucleotide position in each of the DNA features in parallel. The outcome of each interrogation cycle is reported by the production of light or incorporation

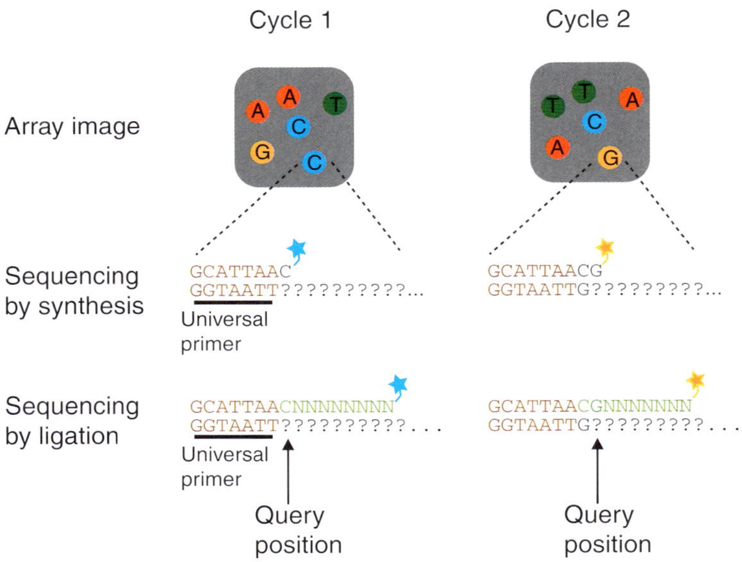

__Figure 3.2__ Overview of cyclic array sequencing. Template DNA to be sequenced is first immobilized to a two-dimensional (random or ordered) array. In sequencing by synthesis (top row), DNA polymerase extends a common primer sequence by cyclically incorporating labeled nucleotides. In sequencing by ligation (bottom row), DNA ligase is used to incorporate oligonucleotides with a dye-label that corresponds to the nucleotide at the specific position to be interrogated. The template sequences are inferred from images taken after each extension or ligation step

of a fluorescent group, and captured by CCD-based imaging of the array. After multiple sequencing cycles, the location and composition of each DNA feature can be inferred from analysis of the full series of imaging data.

The specific details of DNA feature generation, deposition and interrogation differ significantly between current instrument designs.

3.3.2 Pyrosequencing of Emulsion PCR Features

Developed by 454 Life Sciences and later Roche, this was the first next-generation sequencing approach to become commercially available [12].

The DNA features are generated by emulsion PCR and immobilized on the surface of micrometer-scale beads. Emulsion PCR [13] works on the same principle as traditional PCR, but is performed in a water-in-oil emulsion that serves to generate millions of isolated reaction chambers. A library of DNA templates are fitted with common flanking adapters and titrated in the emulsion such that each reaction chamber can be expected to contain a single DNA template. Two universal primers, one of which is attached on paramagnetic beads, initiate the PCR reactions. The end result after thermal cycling is that each reaction chamber contains one bead to which a large number of identical DNA templates are attached.

In order to facilitate sequencing, the bead-attached amplification products are deposited across millions of picoliter-scale wells etched into the surface of a fiber optic bundle. The concentration of beads is titrated to maximize throughput while minimizing the number of wells with multiple beads. Sequence interrogation is completed using the pyrosequencing method [14]. In each cycle, a single nucleotide is introduced into the common reaction volume and polymerase-mediated incorporation events are detected by monitoring luciferase-based light generation upon pyrophosphate release. Parallel incorporation across identical templates on a single bead amplifies the signal for robust detection. Any unincorporated nucleotides are then removed and the process is repeated.

Instruments using this pyrosequencing approach have been shown routinely to generate several million sequence reads of 100–200 bp each, and have been used in multiple different scientific applications. One known limitation of this approach is a relatively low accuracy on sequence intervals where the same base is repeated several times (homopolymers). This stems from the fact that every base in this interval will be filled in by the polymerase in the same cycle. While the amount of light generated from each well is quantitatively correlated with the number of incorporation events, it has proven difficult to achieve accuracies comparable to dideoxy-sequencing in such intervals. The cost per read is roughly an order of magnitude lower than for capillary-based sequencing, although the shorter read length means that the cost per high-quality base is less dramatic.

3.3.3 Sequencing of Emulsion PCR Features by Ligation

This approach has been used in the design of both commercial (the Agentcourt/Applied Biosystem SOLiD system) and 'open-source' sequencing instruments (Harvard University/Danaher). Like the 454/Roche system, DNA features are initially prepared by bead-based emulsion PCR, after which the amplified beads are randomly distributed on a glass slide and immobilized by a thin layer of polyacrylamide gel, or by direct covalent attachment to the surface.

In contrast to other cyclic array approaches, sequencing is achieved by sequence-specific ligation rather than polymerase-based extension [15]. In each cycle, an anchor primer is first hybridized to a universal adapter sequence on each DNA template. Next, the slide is exposed to a population of fluorescently labeled degenerate nonamers (single-stranded 9 bp DNA sequences). The nonamer population is designed such that the attached fluorophore identifies the base at one particular position within it. The ligase discriminates for sequence complementarily up to some distance from the ligation site, ensuring that nonamers with one of the fluorophores are preferentially ligated to each DNA feature. After ligation, the array is imaged in four colors; the ligation products of the anchor primers and 9-mers are then stripped from the beads and the process is repeated.

This sequencing-by-ligation approach has been shown to yield accurate sequence data for at least 6–7 bp next to a ligation site. By reading each adapter-flanked DNA template from both ends, this yields at least 12–14 bp of sequence information. Via sequencing paired end-tags from circularized DNA templates, at least 24–28 bp of information can be obtained from each DNA fragment in a library of interest. Sophisticated ligation chemistry improvements on the ABI/SOLiD system promise to improve the contiguous read length to at least 35 bp. Due to the small features sizes (1 μm) the system has the potential to interrogate more than one billion features on a single slide, which would represent at least one order of magnitude improvement in throughput beyond that already realized by next-generation instruments.

3.3.4 Sequencing of Bridge PCR Features by Synthesis

Commercialized by Solexa and later Illumina, this was the second next-generation sequencing approach to become available on a commercial basis.

The DNA features are generated by bridge PCR directly on a glass slide that has been separated into multiple lanes for parallel sample handling. Bridge PCR [16] works by immobilizing two universal primers to the glass surface. The primers are complementary to adaptors ligated onto each DNA fragment to be sequenced, and serve to capture the fragments on the surface. Upon thermal cycling with all non-DNA reagents moving freely in the aqueous phase, DNA features corresponding to a cluster of ~1000 identical DNA templates are 'grown' on the surface. After amplification, one of the two primers is released from the slide, resulting in only one of the two amplicon strands remaining in each cluster.

Sequencing is achieved by the cyclic polymerase-based incorporation of fluorescently labeled nucleotides, starting from a universal sequencing primer. Reversible terminators ensure that only one nucleotide is incorporated in each cycle. After the removal of any unincorporated nucleotides, the array is imaged in four colors, allowing identification of the identity of one base in each cluster. The reversible terminator group is subsequently cleaved from the clusters, and the process is repeated by extending the previous synthesis products, allowing interrogation of the next base in each of the DNA features.

The Illumina system has been demonstrated routinely to generate 27 to 50 bp sequence reads from at least 40 million features in a single instrument run. The sequence accuracy tends to decrease rapidly beyond this length due to a loss of template after each cycle, and dephasing with clusters. However, read lengths on the order of 100 bp are thought to be realistic with improved chemistry and optimized image-analysis algorithms. Improved optics and brighter fluorophores are also expected to improve throughput by increasing the feature density and reducing the run times.

3.3.5 Sequencing of Single-Molecule Features by Synthesis

The ultimate realization of high-density cyclic array sequencing will be the interrogation of single-molecule DNA features. Variations of this approach are currently at various stages of commercial development by companies such as Helicos and Pacific Biosciences.

In the approach developed by Helicos [17], individual adapter-flanked DNA templates are first immobilized on a quartz slide. To sequence each template, fluorescently labeled universal primers are first hybridized to the templates and imaged to identify the location of each DNA feature. In the subsequent cyclic steps, fluorescently labeled nucleotides are incorporated by a polymerase, imaged, and then inactivated by photobleaching. Observations of single-molecule fluorescence are made with a conventional microscope equipped with total internal reflection illumination, which reduces background fluorescence. In addition, single-pair fluorescence resonance energy transfer (spFRET) is used to minimize noise. The first incorporated nucleotide is labeled with a donor fluorophore (Cy3), and subsequent nucleotides by an acceptor (Cy5). Excitation of the donor leads to fluorescence from acceptors within a limited spatial range that avoids any unincorporated nucleotides on the slide. Photobleaching of the incorporated acceptors does not affect the donor fluorophore, rendering it active for the next cycle. At the time of writing, proof-of-concept demonstrations of this approach had reached read lengths of 5–6 bp.

In the approach developed by Pacific Biosciences [18], conventional microscopy-based observation is eschewed in favor of zero-mode waveguides consisting of subwavelength holes in a metal film placed on a fused silica slide. The enzymes are absorbed on the bottom of the waveguides in the presence of a high concentration of fluorescently labeled nucleotides. The unique optical properties of the waveguides allow polymerase-based sequencing by synthesis to be carried out and observed with a low background in zeptoliter volumes. If successful, this method – or variations of it – have the potential for massive parallelization.

In addition to cyclic array sequencing, a number of other sequencing technologies are currently being explored and have reached varying stages of maturity and throughput.

3.3.6 Dideoxy Sequencing-Based Microfluidics

While many truly alternative approaches to sequencing are being explored, significant efforts have also been put into adapting microfabrication technology developed by the semiconductor industry to miniaturize the traditional dideoxy sequencing process. The goal of this approach is to increase throughput and achieve significant cost savings over traditional sequencing instruments by increasing analysis speed, minimizing reagent consumption, simplifying sample preparation, and reducing the physical footprint of the instruments.

Traditional capillary sequencers typically perform reactions in microliter volumes. Using polymer-filled channels cut into silicon wafers, gel electrophoresis has been performed at the nanoliter scale, which translates to a significant reduction in sample requirements [19]. More integrative approaches involve the combination of all three dideoxy-sequencing steps: thermal cycling, sample-purification and capillary electrophoresis in one microfabricated bioprocessor capable of sequencing 556 contiguous bases at >99% accuracy from only one femtomole of DNA template [20]. Subsequent improvements in polymer chemistry have facilitated the sequencing of >600 nucleotides on an integrated chip in a few minutes [21], which is significantly faster than the 1–2 h typical of commercially available instruments. It is projected that a further optimization of this approach may achieve a 400-fold decrease in reagent consumption and an 800-fold decrease in DNA template requirements over current dideoxy sequencing processes.

While the commercialization of novel microscale sequencing has lagged behind cyclic array sequencing technologies, significant progress has been made and a large potential market remains open. A major advantage of the dideoxy sequencing-based microfluidics approach over current cyclic array sequencers is the retention of well-understood and proven chemistry, and the demonstrated capability to achieve read lengths and raw sequence quality comparable that achieved with traditional sequencing instruments. Extended read lengths simplify the downstream analysis of virtually all sequencing-based assays, and may be essential for some applications. Simplified 'lab-on-a-chip' sequencing instruments may also allow individual laboratories to acquire high-throughput DNA sequencing capabilities that are at present largely limited to specialized centers and core facilities, thereby improving turnaround times and accelerating the scientific process. Such systems may also be suitable for diagnostic sequencing at the point-of-care for clinical applications, and in the field for biodefense and environmental monitoring applications.

3.3.7 Sequencing by Hybridization

This approach relies on the specific pairing between complementary DNA strands [22]. The identity of an unknown DNA sequence is inferred from its pattern of hybridization against a collection of probes with known sequence. Because molecular specificity can be achieved for DNA polymers as short as 11 bp [23], the known probe set can easily be synthesized on a solid support by using established oligonucleotide array technology. Chemically labeled DNA fragments of unknown sequence can then be hybridized to such arrays, and their binding patterns read using automated, quantitative laser scanning microscopy.

While *de novo* sequencing by hybridization to custom-designed 'universal' arrays has been demonstrated on a limited scale [24], the most promising implementation of the approach has been the 'resequencing' of genomic DNA to discover genetic variation between individuals of the same species [25]. In this implementation, hybridization probes are designed based on a known reference genome. For each position that is to be queried, four probes that differ only at their central position (one for each possible nucleotide) are deposited on the sequencing array. The base(s) at that position in an unknown sample can be inferred from differential hybridization to the four probes.

The current limitations of hybridization-based methods include the poor resolution of insertion and deletion mutations, difficulties arising from heterozygosity and repetitive sequences in diploid animal and plant genomes, and the inability to improve accuracy through redundant sequence coverage. Nonetheless, the rapidly decreasing costs and increasing density achieved for array syntheses may preserve sequencing by hybridization as a viable alternative for some high-throughput DNA resequencing applications.

3.3.8 Sequencing by Mass Spectrometry

Another alternative technology explored for DNA sequencing is that of matrix-assisted laser desorption/ionization time-of-flight mass spectrometry (MALDI-ToF MS) [26]. The use of MS has been extended to a variety of DNA-based assays, including *de novo* sequencing, genotyping and the quantification of allele ratios, the detection of insertion and deletion mutations associated with microsatellite repeat instability and short tandem repeats, and the quantification of DNA methylation.

In the context of DNA sequencing, MALDI-TOF MS would serve as an alternative fragment separation method, analogous to the gel electrophoresis step in traditional dideoxy sequencing. Single-stranded polymers are fragmented to 3–29 bp, deposited on a crystal matrix, ionized by a laser, and then passed to a detector. Separation is achieved based on the mass-dependent time of flight, and the original sequence is inferred from the fragment ensemble detected. The potential advantages of MALDI-TOF MS over electrophoresis are an increased resolution, high speed, and an absence of the 'compression zone' artifacts that are common in the gel electrophoresis analysis of repetitive sequences. Current disadvantages include short read lengths (<100 bp) without any significant increase in throughput, and the relatively low stability of DNA under MALDI conditions (which is usually addressed by *in vitro* reverse transcription of the template into RNA prior to analysis). Unless novel approaches are developed to address read length and throughput, sequencing by mass spectrometry is likely to remain a small-niche application.

3.3.9 Sequencing by Exonuclease Digestion

In this approach, the DNA molecules to be sequenced would be transcribed with a polymerase that allows the incorporation of fluorescently-labeled nucleotides (one color for each base) [27]. The resultant DNA products would then be trapped in a capillary with continuous buffer flow and digested with an exonuclease. As the released nucleotides flow downstream from the trapped polymer, their identity and sequence would be identified in real time by single-molecule fluorescence. The main engineering challenges to be overcome before this method becomes practical is the achievement of complete incorporation of the fluorescent bases and the removal of fluorescent impurities; alternatively, it might be possible to implement the direct detection of unmodified nucleotides.

3.3.10 Sequencing by Nanopore Threading

Nanopore-based sequencing is an elegant concept whereby a single DNA strand is threaded through a protein-based or synthetic membrane pore by electrophoresis [27]. The expected base-specific fluctuations in conductance or other membrane properties are registered and used to infer sequence information.

In theory, nanopore sequencing offers tremendous advantages over the above-described technologies, as it would be essentially reagent-free, very rapid, and also provide extremely long, contiguous reads from single molecules. However, beyond simple demonstration experiments this approach has yet to be used successfully for DNA sequencing. Two fundamental engineering challenges remain: (i) the robust discrimination of different bases as they pass through the pore; and (ii) a reduction of the threading rate to a level where this discrimination can be performed for individual bases in a complex DNA sequence.

One proposed solution to improve base discrimination and resolution in nanopore sequencing is to introduce an intermediate step where the original DNA template is replaced by a 'design polymer' (LingVitae). This would be achieved by converting each base to a longer string of bases, and perhaps further derivatizing the polymer with bulky side groups. However, the introduction of extra enzymatic steps would inevitably reduce both yield and throughput; in addition, the ability to sequence single molecules directly from a biological sample would be required in order to realize the full potential of this nanopore sequencing technology.

3.3.11 Sequencing by Scanning Probes

Scanning probe microscopy can also be used to provide atomic resolution data. It has been proposed that the primary sequence could be read directly from a DNA polymer using this technology, although practical success has been elusive [28]. Since current scanning probe technology is not capable of resolving the internal structure of biomolecules, the majority of informative base pairs cannot be read from a double-stranded DNA helix, and single-stranded DNA is notoriously difficult to maintain in conformations that do not obscure the secondary structure. In one recently proposed hybrid scanning probe/threading implementation, atomic force microscopy was used to pull individual DNA strands through a probe-mounted ring, and sequence information inferred from base-specific fluctuations of the resulting molecular friction. If ever successfully developed, scanning probe technology would allow the reading of single-molecule DNA sequences, in the most literal sense possible.

3.4 The DNA Sequencer as a General-Purpose Laboratory Tool

The launch – and continued improvement – of commercial and academic next-generation sequencing instruments have generated tremendous excitement in the DNA-sequencing field. New applications that were effectively out of reach with traditional dideoxy sequencers are now continually being developed and explored. These include the deep sequencing of viral quasi-species (such as the population of HIV viruses in a single infected patient), the meta-genomic identification and monitoring of difficult-to-culture microbes in the soil, sea or human gut, and comprehensive discovery of rare genetic variants in case-control cohorts for disease gene mapping, and in tumor samples for the identification of mutations underlying carcinogenesis.

Importantly, the promise of next-generation sequencing technologies extends far beyond the deeper and more comprehensive cataloging of genomes and genetic variation. Automated sequencing can, in principle, be used to capture the result of any assay for which the end product is a collection of DNA molecules that encodes its outcome. It does not matter whether the DNA is naturally occurring or is the product of a designed enzymatic or chemical process. As long as the sequence reads are long enough to capture the information encoded in each DNA molecule, and the throughput is sufficient to sample the collection to the required depth, then a sequencing instrument could a high-resolution digital measurement of the assay result. Because, in the past, throughput has been the limiting factor for most such assays, the next-generation sequencing instruments will be poised to transform DNA sequencing from specialized instruments to a general-purpose laboratory tool in much the same way that a microscope or a gel electrophoresis apparatus is used to visualize the results of a wide variety of different assays. Current applications typically fall into one of a few major categories (Figure 3.3).

3.4.1 Counting

In this application category, DNA fragments in the sequenced library are identified and grouped by comparing their partial – or, in some cases, complete – primary sequence to a reference database. Depending on the particular assay, the database may contain known genome sequences, transcripts or even pools of synthesized DNA sequences. The end result is a digital count of the number of DNA fragments in the library representing each subinterval or entity in the reference database. The interpretation of these counts depends on the particular assay.

The simplest counting applications involve the enumeration of DNA sequences in a naturally occurring sample – for example, reverse-transcribed messenger RNA or microRNA isolated from one or more cells or tissues of interest. In this case, the number of times that each transcript is represented in the sequenced library provides a measure of its expression level in the assayed cells. The obtained sequence information may be used exclusively for counting known transcripts, or also for the discovery of novel transcripts or splice variants.

Other counting applications involve enumerating DNA sequences in a sample that has been enriched using affinity-based selection from a natural or synthetic fragment pool,

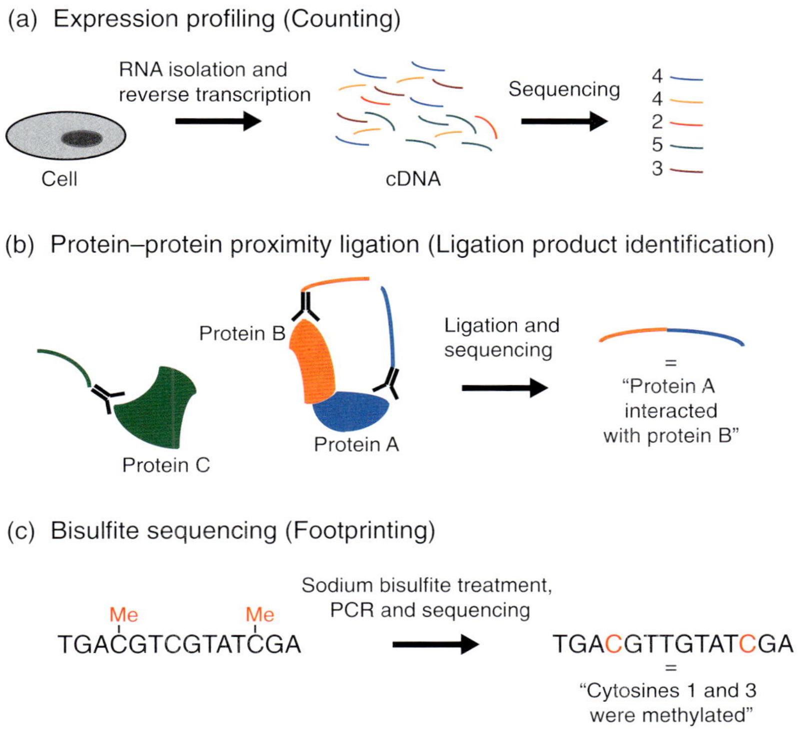

Figure 3.3 *Examples of applications for high-throughput DNA sequencers. Sequencing a complex pool of DNA molecules, such as a cCNA library (a), can be used to identify the relative abundance of each species. Sequencing a pool of ligation products generated in proximity assays (b) can be used to infer the presence or absence of specific molecular interactions. Sequencing chemically treated DNA can be used to infer the location epigenetic modifications, such as cytosine methylation (c)*

or other types of manipulation. The quantitative enrichment of some DNA sequences relative to the preselected pool provides information about genomic location or molecular specificity. Examples include sequencing the free ends of genomic DNA exposed after nuclease treatment to identify accessible regions in chromatin (DNAse-Seq [29]); sequencing DNA enriched after immunoprecipitating fragmented chromatin with antisera specific to a particular histone modification or DNA-binding protein (ChIP-Seq; see Section 3.6); and sequencing DNA fragments or aptamers which are selectively bound by a particular transcription factor or which recognize a specific target molecule *in vitro* (SELEX-SAGE; [27]).

3.4.2 Ligation Product Identification

This is a variation of the counting application category that involves generating and enumerating DNA sequences that did not necessarily exist as contiguous fragments prior to the assay. Information is encoded in the particular ligation products created by the assay.

Two demonstrated examples are chromosome conformation capture and protein–protein proximity ligation assays.

3.4.2.1 Chromosome Conformation Capture

Chromosome conformation capture represents a promising method for interrogating the three-dimensional structure of genomic DNA within cells [30]. Such structural information can yield valuable insight into the functional organization of chromosomes and long-range interactions between genes and regulatory elements. In this assay, genomic DNA is chemically crosslinked while still inside intact nuclei; the DNA is then isolated, digested with a restriction enzyme, and diluted in a large volume of buffer. Ligase is next introduced, which has the effect of preferentially ligating any DNA ends that are in close proximity within the large buffer volume (i.e. those that were crosslinked to each other because they were in close proximity in the original nuclei). In the traditional implementation, quantitative PCR with site-specific primers is then used to query for the presence or absence of a limited number of possible interactions from a predetermined set. By using high-throughput DNA sequencing, all ligation products can – in principle – be enumerated, thus making it possible to infer the complete set of interacting DNA segments.

3.4.2.2 Protein–Protein Proximity Ligation

Protein–protein proximity ligation is an analogous method designed to detect protein interactions [31]. In this assay, synthetic DNA segments are covalently attached to antibodies that are specific to proteins of interest. The DNA–antibody constructs are used to probe a fixed tissue sample or cell extract. When the ligase is subsequently introduced, it preferentially connects DNA segments that have been brought into close proximity because the antibodies to which they are attached have bound the same molecular complex. Identification of the ligation products by sequencing provides information about the presence and relative abundance of different protein–protein interactions.

3.4.3 Footprinting

This application category involves changing the sequence of DNA fragments by chemical modification prior to sequencing. Information is derived by comparing the modified DNA sequence to the original known reference sequence. Also known as 'chemical sequencing', this method is closely related to the Maxam–Gilbert method of DNA sequencing that was developed in parallel with Sanger dideoxy sequencing during the 1970s.

The most widely used footprinting assay is DNA methylation analysis by bisulfite treatment [32]. The covalent modification of nucleotides, in particular by addition of a methyl group to the cytosine moieties, is a common epigenetic regulatory mechanism across a wide range of species. An understanding of the distribution and functional impact of DNA methylation in the human genome is an important challenge in developmental biology and cancer research. Treating genomic DNA with sodium bisulfite has the effect of converting all cytosines to uracils by deamination, unless they are protects by a methyl group. DNA methylation therefore leaves a 'footprint' in bisulfite-treated DNA, which can be located and analyzed by sequencing (uracils are usually converted to thymines by PCR to facilitate the use of unmodified DNA sequencing processes).

Footprinting can also be used to detect protein–DNA interactions *in vivo* [33]. In one common implementation, the cells are first exposed to strong ultraviolet (UV) light. The exposure of chromatin to UV light leads to DNA damage and strand breakage in predictable patterns, but proteins bound to the DNA can provide site-specific protection. Thus, the location of protein–DNA interactions can be inferred by the location of the DNA breakpoints. Traditionally, this has been accomplished with the ligation-mediated PCR amplification of a single site, followed by gel electrophoresis, but high-throughput sequencing provides the option of identifying breakpoints by sequencing the end of each isolated fragment. Although UV footprinting does not reveal the identity of the bound proteins, it provides a far higher resolution of the binding events than can be obtained by immunoprecipitation assays.

3.4.4 Combination Assays

The applications described above are not mutually exclusive and can be mixed and matched as necessary. For example, DNA fragments enriched by a protein-specific antibody could subsequently be subjected to bisulfite treatment, in order to analyze specifically any DNA methylation patterns at molecules bound by the protein. Alternatively, sequence-specific circularization by the ligation of DNA constructs with known tags, followed by the degradation of linear molecules and quantification by sequencing of the remaining tags, could be used for the extremely sensitive detection of mutations or pathogens [34]. Clearly, many additional variations and novel applications are likely to emerge as the next-generation sequencing market matures.

3.5 Case Study: ChIP-Seq

One of the first counting applications to be developed for the next-generation sequencing technologies was that of ChIP-Seq (Figure 3.4) [35]. In this assay, chromatin (the genomic DNA and proteins bound to it in the nucleus) is first extracted from the cells or tissues, and fragmented. An antibody or other affinity reagent is then used to enrich for fragments containing an epitope of interest (such as a histone variant or a transcription factor) by immunoprecipitation. The constituent DNA is then isolated from the enriched chromatin fraction, sequenced, and aligned to a reference genome sequence. A genomic region is inferred to be associated with the targeted epitope if the number of aligned reads across it is significantly higher than would be expected if the total number of reads obtained were randomly distributed across the genome.

ChIP-Seq has been used successfully to map histone modifications and transcription factors across large mammalian genomes in various cell types and states. These maps can be used to annotate active functional regions, such as cell type-specific *cis*-regulatory elements, to elucidate the molecular basis of developmental potency and commitment, and to identify epigenetic lesions in cancer and other disease states. Two key parameters for the assay are the number of sequence reads required to comprehensively map a given epitope across a genome, and the fraction of the genome that can support unique alignments of reads of a given size.

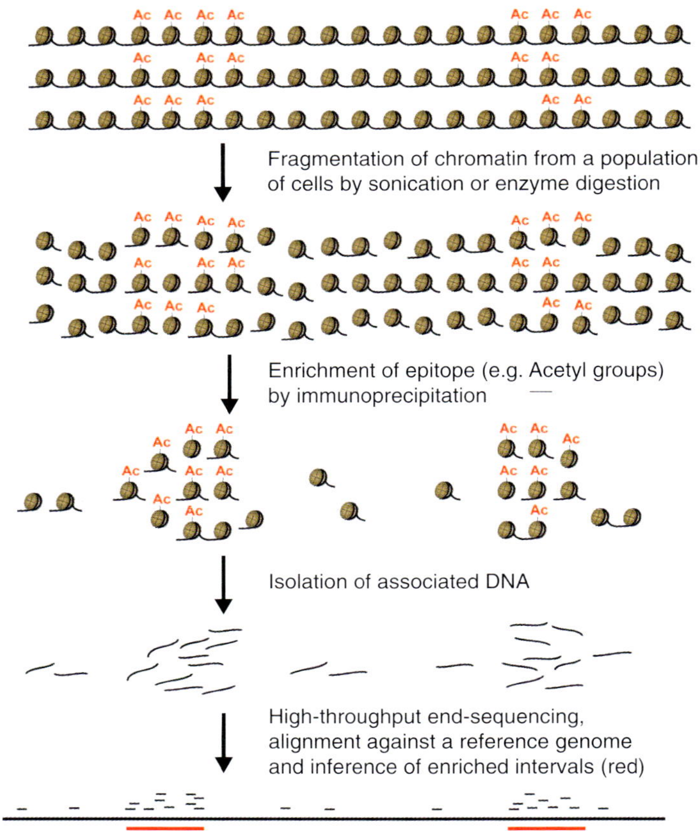

Figure 3.4 *Overview of the ChIP-Seq process. Chemically crosslinked chromatin (consisting of genomic DNA and closely bound histones) is first mechanically or enzymatically sheared. Genomic DNA fragments bound to a protein epitope of interest (such as a covalent histone modification) are enriched by immunoprecipitation, isolated, and identified by sequencing*

The number of sequence reads required to map epitope enrichment can be estimated from a simple model. Suppose that the reference genome is divided into N nonoverlapping bins of fixed size, that a fraction f of these bins contains the epitope, and that one performs ChIP-Seq with an antibody that enriches the sequence in these bins by a factor of e relative to a nonspecific background. If one collects a total of R sequence reads, then the number of reads in a bin should approximately follow a Poisson distribution, with mean eM for bins containing the epitope and M for the other bins, where M is given by Equation 3.1:

$$M = \frac{R}{N(ef + (1 - f))} \tag{3.1}$$

The specificity and sensitivity of ChIP-Seq, conditional on the total number of reads obtained, can be estimated from the overlap of the two Poisson distributions. For example, suppose that an epitope is present across 1% of the human genome (total length ca. 3×10^9 bases), and can be enriched 20-fold relative to the nonspecific background by an antibody. Mapping this epitope with 95% specificity and 95% sensitivity into bins of 500 bp would require approximately 2×10^6 reads. Increasing the resolution to 200 bp would require approximately 5×10^6 reads. Epitopes that enrich less efficiently require more reads (e.g. a 10-fold enrichment and 200 bp resolution would require ~10 million reads). Fortunately, the high throughput of the next-generation sequencing technologies will make obtaining such coverage a feasible prospect.

Most animal and plant genomes, including human, contain large amounts of transposable elements, microsatellites and other repetitive sequences. A short sequence read from a fragment of such a genome cannot always be unambiguously aligned back to a unique location. The effective genome coverage of ChIP-Seq – that is, the fraction that can be interrogated for enrichment by uniquely aligning reads – depends on the length k of the sequence reads, the amount of repetitive sequences, and the alignment algorithm used. In one ChIP-Seq alignment algorithm the first- and second-best alignment of each read are identified (as measured by the number of mismatches between the read and the reference sequence). Reads are considered uniquely aligned and kept for analysis if they have no alternative alignment with $\leq d$ additional mismatches. The exact 'coverable' regions can be determined empirically by aligning every fragment of length k in the reference sequence back to the entire genome, and marking those that are uniquely aligned. For example, if $k = 27$ and $d = 2$ and any 500 bp interval in which at least half of the constituent 27-mers are unique is considered coverable, then ~70% of the human genome can be interrogated. Slightly longer read lengths (36 bp) or paired reads can provide over 80% coverage.

Although ChIP-Seq and related assays are already relatively mature applications for the next-generation sequencing technologies, a number of technical challenges and opportunities for improvement remain. First, the number of sequence reads required – and hence the cost-effectiveness of the ChIP-Seq assay – is heavily dependent on the enrichment level obtained by the affinity reagent used. The lack of specific, strong or renewable affinity reagents has long been a barrier to study novel proteins or chromatin modifications, and automated screening or novel technologies and reagent classes may be required to overcome this limitation. Second, the efficient preparation of enriched DNA is challenging due to the low yield of current immunoprecipitation techniques (often, only 1 ng or less can be obtained from millions of cells), while many biologically interesting cell populations – such as specific regions of developing embryos, adult stem cells and early-stage tumors – are difficult to obtain in large numbers. Thus, automated and miniaturized technologies may help to improve yields. Finally, improved engineering solutions are required to reduce the equipment, reagent and labor costs.

It follows that, as these challenges are met, the ability to map any given protein–DNA interaction, both cheaply and quickly, in any cell type, in any laboratory, using off-the-shelf sequencers as a general-purpose tool, promises to revolutionize developmental biology, cancer research and regenerative medicine.

References

1. Maxam, A.M. and Gilbert, W. (1977) A new method for sequencing DNA. *Proceedings of the National Academy of Sciences of the United States of America*, **74**(2), 560–4.

2. Sanger, F., Nicklen, S. and Coulson, A.R. (1977) DNA sequencing with chain-terminating inhibitors. *Proceedings of the National Academy Sciences of the United States of America*, **74**(12), 5463–7.
3. Prober, J.M., Trainor, G.L., Dam, R.J. *et al.* (1987) A system for rapid DNA sequencing with fluorescent chain-terminating dideoxynucleotides. *Science*, **238**(4825), 336–41.
4. Tabor, S. and Richardson, C.C. (1995) A single residue in DNA polymerases of the *Escherichia coli* DNA polymerase I family is critical for distinguishing between deoxy- and dideoxyribonucleotides. *Proceedings of the National Academy of Sciences of the United States of America*, **92**(14), 6339–43.
5. Sanger, F., Coulson, A.R., Hong, G.F. *et al.* (1982) Nucleotide sequence of bacteriophage lambda DNA. *Journal of Molecular Biology*, **162**(4), 729–73.
6. Myers, G. (1999) Whole-genome DNA sequencing. *Computing in Science and Engineering*, **1**, 33–43.
7. Meldrum, D. (2000) Automation for genomics, part one: preparation for sequencing. *Genome Research*, **10**(8), 1081–92.
8. Meldrum, D. (2000) Automation for genomics, part two: sequencers, microarrays, and future trends. *Genome Research*, **10**(9), 1288–303.
9. Collins, F.S. (2003) Genome research: the next generation. *Cold Spring Harbor Symposium on Quantitative Biology*, **68**, 49–54.
10. Mardis, E.R. (2006) Anticipating the 1,000 dollar genome. *Genome Biology*, **7**(7), 112.
11. Shendure, J.A., Porreca, G.J. and Church, G.M. (2008) Overview of DNA sequencing strategies. *Current Protocols in Molecular Biology*, **Chapter 7**, Unit 7 1.
12. Margulies, M., Egholm, M., Altman, W.E. *et al.* (2005) Genome sequencing in microfabricated high-density picolitre reactors. *Nature*, **437**(7057), 376–80.
13. Tawfik, D.S. and Griffiths, A.D. (1998) Man-made cell-like compartments for molecular evolution. *Nature Biotechnology*, **16**(7), 652–6.
14. Ronaghi, M., Karamohamed, S., Pettersson, B. *et al.* (1996) Real-time DNA sequencing using detection of pyrophosphate release. *Analytical Biochemistry*, **242**(1), 84–9.
15. Shendure, J., Porreca, G.J., Reppas, N.B. *et al.* (2005) Accurate multiplex polony sequencing of an evolved bacterial genome. *Science*, **309**(5741), 1728–32.
16. Fedurco, M., Romieu, A., Williams, S. *et al.* (2006) BTA, a novel reagent for DNA attachment on glass and efficient generation of solid-phase amplified DNA colonies. *Nucleic Acids Research*, **34**(3), e22.
17. Braslavsky, I., Hebert, B., Kartalov, E. *et al.* (2003) Sequence information can be obtained from single DNA molecules. *Proceedings of the National Academy of Sciences of the United States of America*, **100**(7), 3960–4.
18. Levene, M.J., Korlach, J., Turner, S.W. *et al.* (2003) Zero-mode waveguides for single-molecule analysis at high concentrations. *Science*, **299**(5607), 682–6.
19. Emrich, C.A., Tian, H., Medintz, I.L. *et al.* (2002) Microfabricated 384-lane capillary array electrophoresis bioanalyzer for ultrahigh-throughput genetic analysis. *Analytical Chemistry*, **74**(19), 5076–83.
20. Blazej, R.G., Kumaresan, P. and Mathies, R.A. (2006) Microfabricated bioprocessor for integrated nanoliter-scale Sanger DNA sequencing. *Proceedings of the National Academy of Sciences of the United States of America*, **103**(19), 7240–5.
21. Fredlake, C.P., Hert, D.G., Kan, C.W. *et al.* (2008) Ultrafast DNA sequencing on a microchip by a hybrid separation mechanism that gives 600 bases in 6.5 minutes. *Proceedings of the National Academy of Sciences of the United States of America*, **105**(2), 476–81.
22. Drmanac, R. and Drmanac, S. (2001) Sequencing by hybridization arrays. *Methods in Molecular Biology*, **170**, 39–51.

23. Wallace, R.B., Shaffer, J., Murphy, R.F. *et al.* (1979) Hybridization of synthetic oligodeoxyribonucleotides to phi chi 174 DNA: the effect of single base pair mismatch. *Nucleic Acids Research*, **6**(11), 3543–57.
24. Drmanac, R., Drmanac, S., Strezoska, Z. *et al.* (1993) DNA sequence determination by hybridization: a strategy for efficient large-scale sequencing. *Science*, **260**(5114), 1649–52.
25. Patil, N., Berno, A.J., Hinds, D.A. *et al.* (2001) Blocks of limited haplotype diversity revealed by high-resolution scanning of human chromosome 21. *Science* **294**(5547), 1719–23.
26. Ragoussis, J., Elvidge, G.P., Kaur, K. *et al.* (2006) Matrix-assisted laser desorption/ionisation, time-of-flight mass spectrometry in genomics research. *Public Library of Science Genetics*, **2**(7), e100.
27. Bayley, H. (2006) Sequencing single molecules of DNA. *Current Opinion in Chemical Biology*, **10**(6), 628–37.
28. Pope, L.H., Davies, M.C., Roberts, C.J. *et al.* (1998) DNA analysis with scanning probe microscopy. *Analytical Communications*, **35**, 5H–7H.
29. Crawford, G.E., Holt, I.E., Whittle, J. *et al.* (2006) Genome-wide mapping of DNase hypersensitive sites using massively parallel signature sequencing (MPSS). *Genome Research*, **16**(1), 123–31.
30. Dostie, J., Richmond, T.A., Arnaout, R.A. *et al.* (2006) Chromosome Conformation Capture Carbon Copy (5C): a massively parallel solution for mapping interactions between genomic elements. *Genome Research*, **16**(10), 1299–309.
31. Landegren, U., Schallmeiner, E., Nilsson, M. *et al.* (2004) Molecular tools for a molecular medicine: analyzing genes, transcripts and proteins using padlock and proximity probes. *Journal of Molecular Recognition*, **17**(3), 194–7.
32. Frommer, M., McDonald, L.E., Millar, D.S. *et al.* (1992) A genomic sequencing protocol that yields a positive display of 5-methylcytosine residues in individual DNA strands. *Proceedings of the National Academy of Sciences of the United States of America*, **89**(5), 1827–31.
33. Pfeifer, G.P. and Tommasi, S. (2000) In vivo footprinting using UV light and ligation-mediated PCR. *Methods in Molecular Biology*, **130**, 13–27.
34. Akhras, M.S., Thiyagarajan, S., Villablanca, A.C. *et al.* (2007) PathogenMip assay: a multiplex pathogen detection assay. *PLoS ONE*, **2**(2), e223.
35. Mikkelsen, T.S., Ku, M., Jaffe, D.B. *et al.* (2007) Genome-wide maps of chromatin state in pluripotent and lineage-committed cells. *Nature*, **448**(7153), 553–60.

4

Modeling a Regulatory Network Using Temporal Gene Expression Data: Why and How?

Sophie Lèbre[1] and Gaëlle Lelandais[2]

[1] *Centre for Bioinformatics, Division of Molecular Biosciences, Imperial College London*
[2] *Equipe de Bioinformatique Génomique et Moléculaire, INSERM U726, Université Paris Diderot, Paris, France*

4.1 Introduction

In recent years, the rapid development of sequencing methods and computer technology has led to the complete DNA sequencing and annotation of many important model organisms [1]. In order to understand the functioning of an organism, the next major step is to identify which genes are expressed, under which conditions, and to what extent. Gene expression is a complex process that is regulated at several stages in the synthesis of proteins; hence, the identification of genes, the products of which function together in the cell, is a major task of postgenomic approaches. Fundamental questions regarding the topology of networks such as the protein interactome [2], metabolome [3] or transcriptional regulation networks [4, 5] have been addressed. As traditional gene-by-gene approaches have proved to be insufficient, new methods and technologies have been developed such that all of these components can be analyzed simultaneously. This chapter focuses on the analysis of transcriptional regulatory networks, as they play an essential role in the cell. In order to control the expression of specific genes according to specific environmental conditions, multiple regulatory systems that comprise many components and are connected through interlocking positive and negative feedback loops, are required. In this chapter, we describe the experimental approaches used to measure whole-genome expression,

Automation in Proteomics and Genomics: An Engineering Case-Based Approach Edited by Gil Alterovitz, Roseann Benson and Marco Ramoni

and outline the methods used to analyze gene expression profiles, focusing mainly on the clustering algorithms used to identify groups of coexpressed genes. More refined mathematical approaches to model regulatory networks based on graphical modeling are also presented, and a distinction is made between static and dynamic modeling applications. Modeling assumptions are also discussed, suggesting how results may differ according to the method used. The chapter concludes with an introduction to the various dimension reduction approaches and their corresponding software for regulatory network inference, when the number of genes far exceeds the number of measurements.

4.2 Experimental Approaches to Measure Whole-Genome Expression

4.2.1 RNA: Gene Expression

'Gene expression' can be defined as the process by which the DNA sequence of a gene is converted into a functional gene product, such as mRNA and, ultimately, into a protein. A central goal of molecular biology is to understand the regulation of mRNA and protein synthesis, and their reactions to external and internal cellular signals. This regulation incorporates several mechanisms at each step of the gene expression process, including mechanisms for controlling transcription initiation, RNA splicing, mRNA transport, translation initiation, post-translational modifications, or the degradation of mRNA/protein. In this context, the regulation of mRNA transcription represents an important preliminary step towards the precise coordination of all these regulatory processes. Specific proteins – known as transcription factors – are able to bind to regulatory regions along the DNA and hence play a key role by modulating the transcription of the genes that they control. An understanding of the nature of these complex biological processes thus requires the precise observation of spatiotemporal gene expression patterns. The experimental approaches that have been developed to measure mRNA levels on a quantitative basis can be divided into two categories, namely low-throughput and high-throughput. Low-throughput technologies allow the expression of one gene to be measured at a time, using Northern blot analysis or real-time polymerase chain reaction (RT-PCR). Although the study of gene expression on a one-by-one basis provides a wealth of biological insights, the desire to fully understand genomic sequences in an organism led in time to the development of high-throughput technologies, whereby the expression of all genes could be studied at once. The term 'genomic' differs from 'genetic', in that it does not relate to a gene in isolation, but rather at how many genes work together to produce phenotypic effects. In that respect, DNA microarrays serve as high-capacity systems for monitoring the expression of many genes in parallel [6]. Although other parallelization technologies do exist (examples include SAGE (Serial Analysis of Gene Expression) [7] or SuperSAGE technologies [8]), for the sake of clarity only DNA microarrays will be presented in the following section. Finally, whichever technology is used to obtain quantitative measurements of gene expression, all of the bioinformatical methodologies presented in this chapter can be applied to analyze the data and identify regulatory associations between genes.

4.2.2 Whole-Genome Expression Profiling Through Microarray Technology

More than ten years after its initial development, DNA microarray technology is still undergoing a rapid evolution [6, 9]. Indeed, today it is one of the essential approaches for

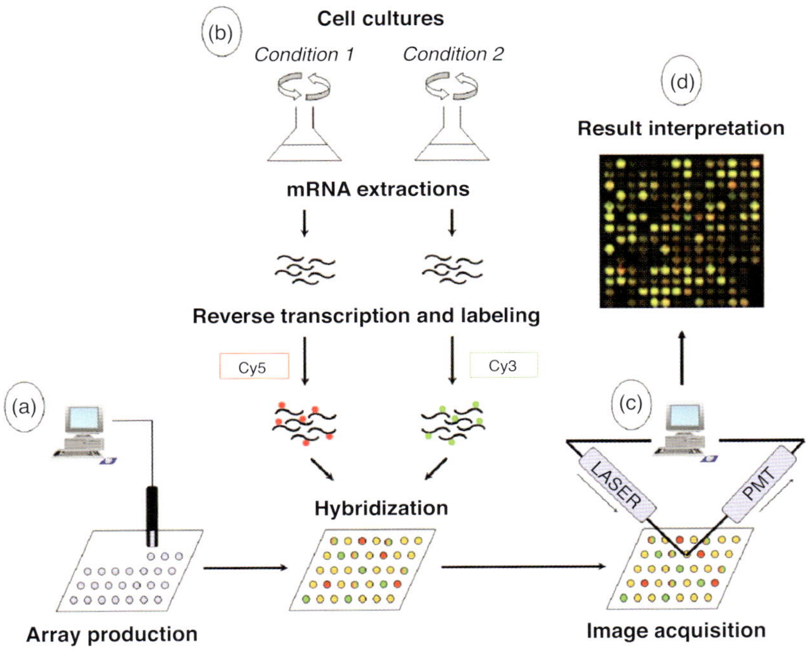

Figure 4.1 An overview of the major steps in using a microarray. (a) Array production consists of spotting the DNA probes (either PCR products or oligonucleotides) onto the glass surface of the array with a spotting robot; (b) mRNAs are first extracted from the cell cultures (or tissues) of interest. Two samples are obtained and labeled with two fluorescent molecules, usually Cy3 and Cy5. The samples are finally hybridized to the array simultaneously; (c) An image of the surface of the hybridized array is produced using a scanner. DNA probes that are bound with labeled nucleic acid molecules fluoresce, when excited by light of an appropriate wavelength. The measured fluorescence should be proportional to the quantity of mRNA initially situated in the studied samples; (d) The final step consists of analyzing and interpreting the fluorescence measures obtained for each DNA probes located on the array. A normalization procedure is generally required to resolve the systematic errors and bias introduced during the experiment

high-throughput analysis for the provision of a rough 'snapshot' of the transcriptome state – that is, the expression level of all genes expressed in a cell at any one given time. An overview of the major steps in using a microarray is presented in Figure 4.1 (a complete description of DNA microarray technology can be found in Ref. [10]).

A DNA microarray consists of a solid surface, onto which DNA molecules (either oligonucleotides or cDNAs) are immobilized in a predefined organization (Figure 4.1a). Each DNA molecule is specific to one gene, the expression of which must be monitored. Thus, the purpose of a microarray is to detect the presence and abundance of labeled nucleic acids in a biological sample (Figure 4.1b and c). Today, thousands – or even tens of thousands – of DNA molecules can be spotted onto a microscope slide, and the relative expression levels of each gene can be determined by measuring the fluorescence intensity of labeled mRNA hybridized to the arrays, allowing the measurement of RNA levels for the complete set of transcripts of an organism. A specific feature of microarray technology is the level of statistics and bioinformatics required at all stages of the procedure and, in particular, the interpretation of quantitative expression measurements obtained for each gene (Figure 4.1d) when using specific methodologies and automation approaches to answer biological questions. These particular issues are highlighted in the following sections, namely to identify expression relationships between genes, and to model regulatory networks using temporal gene expression data.

4.3 Temporal Gene Expression Data and Analysis of Relationships Between Genes

4.3.1 Principle

DNA microarrays provide the opportunity of measuring, simultaneously, the expression of several thousand genes within a cell, and thus to observe transcriptome states under various cellular conditions. If a single microarray experiment is performed, then a 'static view' of transcriptome states under particular cellular conditions is obtained. However, if several microarray experiments are performed over time, it is possible to follow the modifications of the gene expression measurements, such that a 'dynamic view' of transcriptome states is acquired. Each gene is characterized by a 'gene expression profile' – the successive expression measurements observed with a series of microarrays (Figure 4.2a). When describing the chronology of transcriptional events, time series microarray experiments represent a valuable sources of information for the study of a wide range of biological processes, including cell cycle analyses [11, 12], responses to environmental stresses [13, 14] and developmental studies [15]. A variety of temporal gene expression data mining methods exist, in addition to approaches aimed at understanding their biological meaning. One of the most successful of these methods is based on the assumption that genes with similar expression profiles are more likely to be involved in the same biological process. For example, if a single regulatory system controls two genes, then the genes would be expected to be coexpressed – that is, to exhibit similar expression profiles. In fact, there is evidence that functionally related genes are often coexpressed [16], and the identification of genes which behave in a similar or coordinate manner is a challenging task. In this section, we present details of several analytical techniques designed to identify patterns of expression and to create biologically relevant clusters of genes. The section is organized into three parts:

1. 'Similarity of gene expression profiles', examines different methods for quantifying resemblances between two sets of gene expression measurements (Figure 4.2b).
2. 'Clustering methods' introduce the most commonly used approaches for identifying groups of closely related genes (Figure 4.2c). Hierarchical approaches link genes with similar expression profiles to form a 'tree structure' (much like a phylogenetic tree), whereas partitioning approaches are methods of nonhierarchical clustering that require the number of clusters in advance.
3. 'Specificities for temporal gene expression data' discusses the limitations of classical clustering approaches when analyzing temporal gene expression data, and hence, the need to use more sophisticated approaches that take into account the temporal dependency between gene expression measurements.

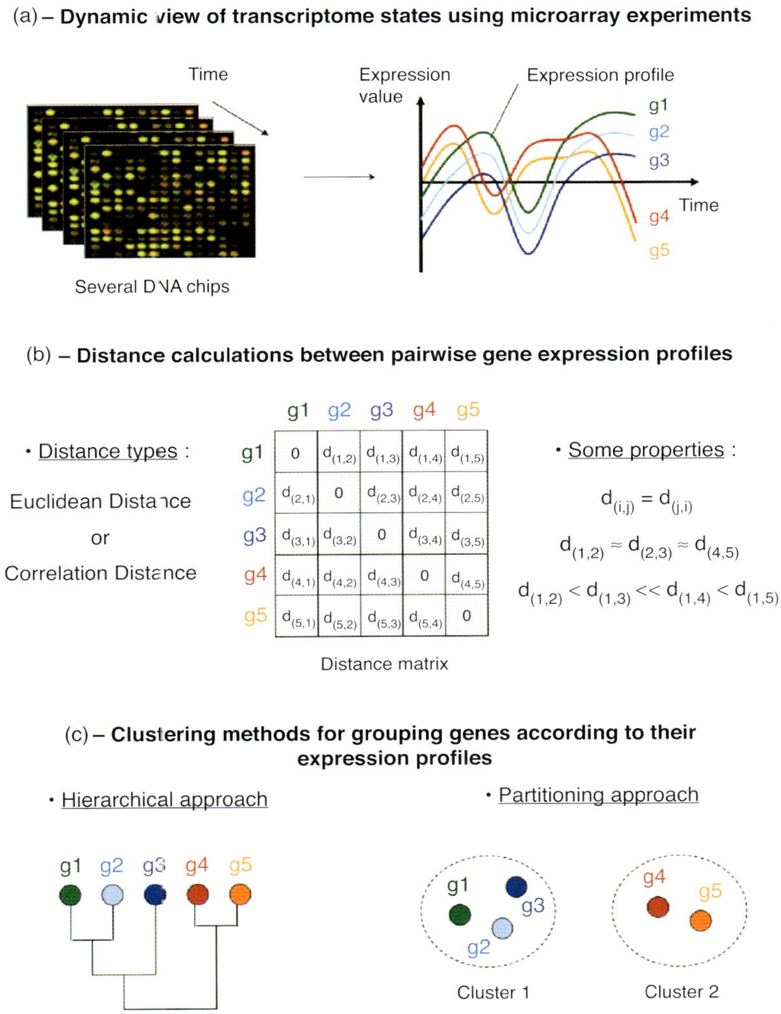

Figure 4.2 Analysis of relationships between gene expression profiles. (a) Using data obtained with successive microarray experiments, expression profiles can be drawn up for each gene located on the microarray. These expression profiles provide a dynamic view of the transcriptional changes that occur in a particular cellular process. As an illustration, five gene expression profiles are represented here. Genes 1, 2 and 3 (g1, g2, g3) exhibit similar variations of expression measurement among time, whereas genes 4 and 5 (g4, g5) exhibit similar patterns of expression together but are different from these of genes 1, 2 and 3; (b) Distance matrix between all gene pairs. The distance measure (Euclidean or correlation distances) quantifies the 'resemblance' between pair wise gene expression profiles; (c) Clustering methods allow the classification of genes according to their expression profiles, using the distance measures calculated in (b). Two types of approaches are illustrated here: hierarchical and partitioning methods (see the main text for details)

4.3.2 Notations

By using a series of n microarray experiments, a 'gene expression matrix' can be drawn up where the rows correspond to individual genes, the columns are individual experiments, and the cells contain a measure of the gene activity. It should be noted that the series of experiments are either time-course gene expression data or gene expression values measured under different experimental conditions. In this matrix, denoted by $X = (X_t^i)_{\substack{1 \le i \le p \\ 1 \le t \le n}}$, let X_t^i be the gene expression level of the i-th gene in the t-th experiment, for $i = 1, 2, \ldots, p$ and $t = 1, 2, \ldots, n$; where p denotes the total number of genes (e.g. the entire genome of an organism) and n the number of experiments (or time points). Therefore, each gene can be given a coordinate (its expression profile), which is, for the i-th gene, the vector $X^i = (X_1^i, \ldots, X_t^i, \ldots, X_n^i)$ and for the j-th gene the vector $X^j = (X_1^j, \ldots, X_t^j, \ldots, X_n^j)$.

4.3.3 Similarity of Gene Expression Profiles

One of the main advantages of microarrays is their ability to be used to identify relationships between genes – that is, to classify genes that behave in a similar or coordinate manner. To perform computational analysis, the prerequisite is to transform the intuitive notion of 'similarity' into quantitative measures, and to do this, classically a distance measure between expression profiles is applied. Two types of distances are presented here ('correlation distance' and 'Euclidean distance'), although many others have been reported in the literature [17].

4.3.3.1 Correlation Distance

The correlation distance derives from the correlation coefficient – that is, a statistical concept that quantifies the strength and direction of a linear relationship between two sets of measurements. More precisely, if we denote two sets of gene expression measurements by $X^i = (X_1^i, \ldots, X_t^i, \ldots, X_n^i)$ and $X^j = (X_1^j, \ldots, X_t^j, \ldots, X_n^j)$, the correlation coefficient r is given by the following formula:

$$r(X^i, X^j) = \frac{n \sum_{t=1}^{n} X_t^i X_t^j - \sum_{t=1}^{n} X_t^i \sum_{t=1}^{n} X_t^j}{\sqrt{\left(n \sum_{t=1}^{n} (X_t^i)^2 - \left(\sum_{t=1}^{n} X_t^i\right)^2\right)\left(n \sum_{t=1}^{n} (X_t^j)^2 - \left(\sum_{t=1}^{n} X_t^j\right)^2\right)}} \tag{4.1}$$

To satisfy some theoretical properties required to define a distance measure [10], the correlation distance (d) between gene expression profiles X^i and X^j is finally set by:

$$d(X^i, X^j) = 1 - r(X^i, X^j) \tag{4.2}$$

It should be noted that the correlation coefficient r only takes a value from between -1 and $+1$, implying that correlation distance (d) takes a value from between 0 and 2. A value of 0 represents a perfect positive correlation between the gene expression profiles (Figure 4.3a), a value of 1 indicates no correlation, and a value of 2 indicates anticorrelation – that

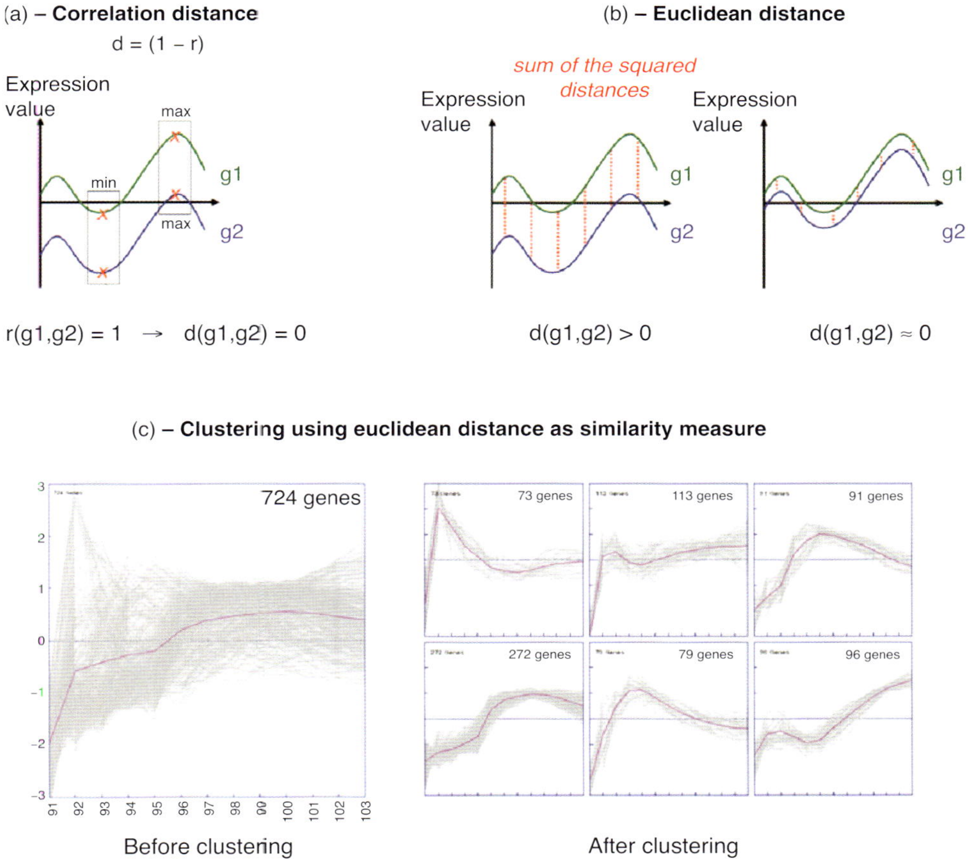

Figure 4.3 *Distance measures between gene expression profiles. Distance calculation between two gene expression profiles (colored in green and blue), using correlation distance (a) or Euclidean distance (b) (see main text for more details). This example highlights the specific properties for each distance. The green and blue profiles are two different profiles with the same shape, but with different magnitudes. They appear to be identical using correlation distance ($d = 0$), whereas they appear to be distant with the Euclidean distance ($d > 0$); (c) Clustering of temporal gene expression profiles, using k-means algorithm and Euclidean distance as similarity measure. The microarray measurements were obtained from the study of the sporulation process in yeast* Saccharomyces cerevisiae *[18]*

is, opposite expression profiles. When expression is high for one gene (respectively low), the other gene also exhibits high expression (respectively low).

4.3.3.2 Euclidean Distance

The Euclidian distance is another measure of the relationship between two gene expression profiles, and is based on the sum of the squared distances of two vector values

(Figure 4.3b). Using the same notations as before, the Euclidean distance between two profiles X^i and X^j is given by the following equation:

$$d(X^i, X^j) = \sqrt{\sum_{t=1}^{n} (X_t^i - X_t^j)^2} \tag{4.3}$$

The Euclidean distance takes a value from between 0 and $+\infty$. A value of 0 means that the two profiles are identical (Figure 4.3b).

4.3.3.3 Differences Between Correlation Distance and Euclidean Distance

Unlike the correlation distance, the Euclidean distance is not scale invariant. Two gene expression profiles with the same shape, but different magnitudes, will appear to be different with the Euclidean distance ($d > 0$), whereas they will appear to be identical with the correlation distance ($d = 0$) (Figure 4.3a and b). One the other hand, when using the correlation distance, flat patterns – that is gene expression profiles for which it is difficult to distinguish expression signal from microarray background noise – can be strongly correlated with highly variable (and hence biologically meaningful) gene expression profiles. Such an erroneous association is not possible using Euclidean distance. The choice of an appropriate distance is therefore a critical step in the computational analyses of gene expression profiles, and depends on the intuitive biological notion of 'coexpression' between two genes.

4.3.4 Clustering Methods

4.3.4.1 Principle

Clustering can be defined as the process of separating a set of objects into several subsets, on the basis of their similarity (see Figure 4.3c for an illustration). When analyzing gene expression data, the 'objects' are the genes, while 'similarity' is based on the distance calculation between their corresponding gene expression measurements (see Section 4.3.3). The aim of clustering methods is therefore to identify clusters of genes that exhibit both internal cohesion (the intracluster variability is low) and external isolation (the intercluster distances are high). In the literature, two major types of clustering methods have been proposed, namely 'hierarchical' and 'partitioning'. Hierarchical methods start with each gene considered as a separate cluster after which, at each successive step in the clustering procedure, two of the clusters are merged together until only one cluster, which then comprises the entire dataset, remains (Figure 4.4a). In contrast, partitioning methods produce distinct nonoverlapping clusters where each gene is allocated to the cluster with which it is most similar, using a distance criterion between gene expression measurements (Figure 4.4b).

4.3.4.2 Hierarchical Methods

Hierarchical clustering is a methodology that arranges the gene expression profiles into a tree structure (much like a phylogenetic tree), so that similar profiles appear close together in the tree and dissimilar profiles are farther apart. The hierarchical clustering procedure is summarized by five steps (Figure 4.4a):

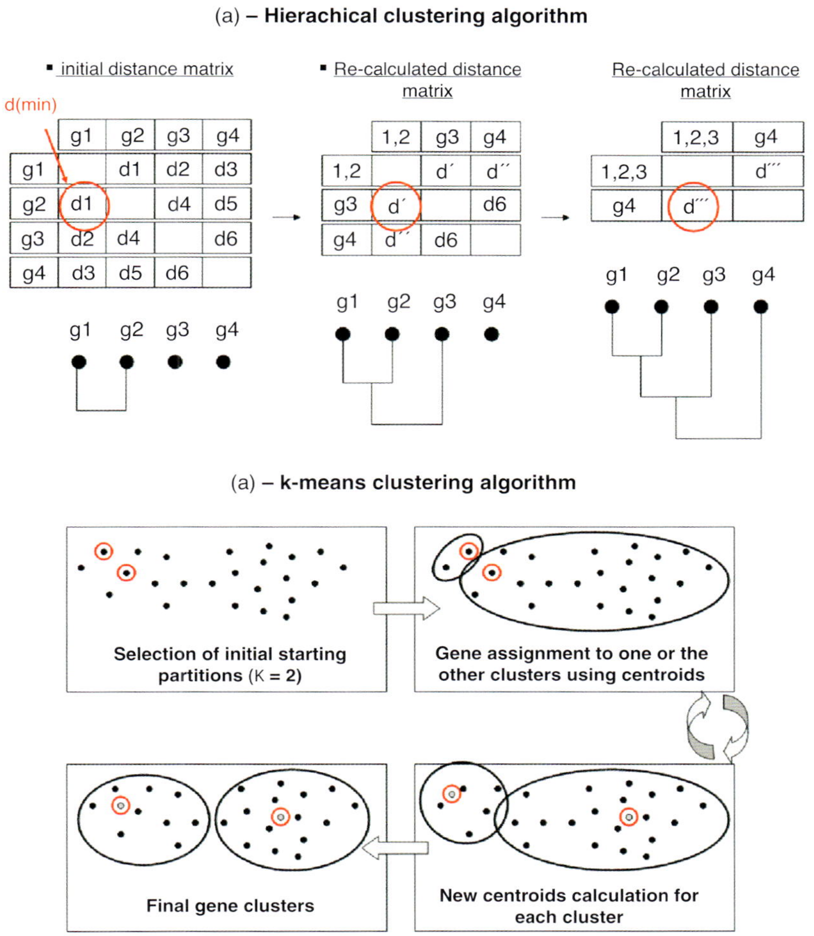

Figure 4.4 *Illustration of two clustering algorithms. (a) Hierarchical clustering algorithm and (b) k-means algorithm. Complete descriptions of the hierarchical and k-means algorithms can be found in the main text*

1. Distances between all genes pairs are calculated using distance measures such as 'correlation distance' or 'Euclidean distance'.
2. The resulting distance matrix is inspected in order to find the smallest distance value between expression profiles.
3. The corresponding genes are joined together in the tree and form a new cluster.
4. Distances between the newly formed cluster and the other genes are recalculated.
5. Steps 2, 3 and 4 are repeated until all genes and clusters are linked in a final tree.

The analysis of microarray data using the hierarchical clustering algorithm quickly became one of the preferred clustering approaches, due to an original representation of the results, initially proposed in Ref. [19]. The idea is to represent the reordered gene

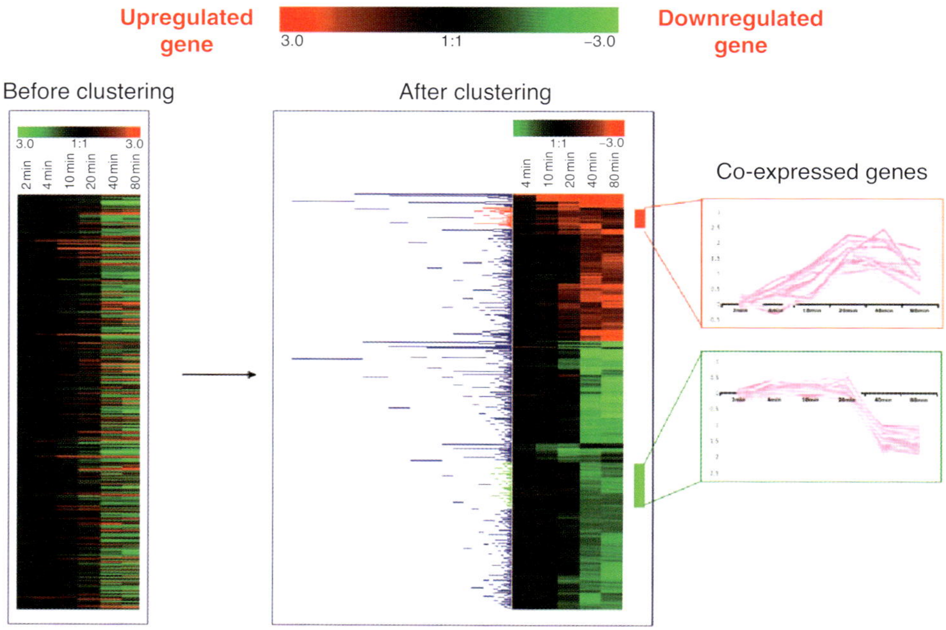

Figure 4.5 *Graphical representation of hierarchical clustering results. Gene expression profiles are represented using the color code described in Ref. [19]. The color scales range from saturated green (maximum negative value) to saturated red (maximum positive value). Cells with an expression measurement of 0 (genes unchanged) are colored black, increasingly positive values with red intensity, and increasing negative values with green intensity. Each gene is represented by a single row of colored boxes; each time is represented by a single column. Two separate clusters are indicated by colored bars and by identical coloring of the corresponding region of the dendogram. These comprised genes for which expression is very similar during the time course*

expression profiles using a color code that quantitatively and qualitatively reflects the original microarray measurements: green for negative values (down-regulated genes) and red for positive values (up-regulated genes) (Figure 4.5). The tree structure represents complex gene expression data that, through statistical organization and graphical display, allows biologists to assimilate and explore the data in a natural and intuitive manner.

4.3.4.3 Partitioning Methods

The objective of partitioning methods is to produce distinct nonoverlapping clusters. For a given number of genes, p, and a given number of clusters, q, the number of possible partitions is finite but extremely large. As it is unworkable to investigate each possible partition to find the most advantageous, a solution consists of choosing a clustering criterion or 'cost function' that will guide the search for a better partition. In that spirit, the k-means algorithm has been proposed [20]. This is summarized by the following six steps (Figure 4.4b):

1. The number of clusters q is chosen.
2. Each gene expression profile is randomly assigned to one of the q clusters.
3. Centroids are calculated for each cluster; these correspond to the average expression values, taking into account the gene repartitions proposed in step 2.
4. For each gene to be classified, the distances between its gene expression profile and the centroids of each cluster are calculated.
5. Genes that exhibit a smaller distance with a centroid of a different cluster from the cluster to which it belongs are assigned to a new cluster.
6. Centroids are recalculated, taking into account the new gene partition, and steps 4 and 5 are finally repeated until no gene changes cluster membership. In the k-means algorithm, the cost function is related to the intracluster variability and decreases at each repetition of the partitioning process [20].

4.3.4.4 *Differences Between Partitioning Methods and Hierarchical Clustering*

Partitioning methods can be distinguished from hierarchical clustering on the basis of several characteristics. First, the number of clusters must be specified in advance, which implies that the user must perform several attempts to correctly estimate the number of clusters. Second, partitioning methods require the selection of an initial starting partition. Some algorithms (e.g. k-means) use randomly selected data elements as starting partitions [20], while others allow the user to specify starting seeds [21]. As a consequence, different runs of the k-means procedure can produce slightly different results, whereas hierarchical clustering approach is completely deterministic. Finally, there is no hierarchy or relationship between clusters; the clusters are simply groups of similar gene expression profiles.

4.3.4.5 *Other Methods*

In the past, numerous approaches to cluster gene expression profiles have been proposed [22]. Yet, despite there being differences between algorithms, the clustering process can be reviewed, taking into account the following five major steps:

Step 1: Selection of the genes to be clustered according to their gene expression measurements. Among all genes for which expression data are available (generally more than several thousand with microarray technology), only those for which significant changes in mRNA levels are observed for several microarray experiments are analyzed. Other genes, the expression of which varies only slightly (or not at all) across the set of conditions, may reflect microarray background noise rather than any relevant biological expression variations. Their elimination simplifies the downstream analyses (there are less genes to classify) and precluded inconsistent clustering results, as these flat patterns can exhibit significant correlation with almost anything.

Step 2: Choice of a distance measure to quantify the similarity between gene expression profiles. In the previous section, several distance measures that reflect the degree of closeness or separation between gene expression profiles were introduced. As two gene expression profiles can be more or less similar when using one or the other distance measure, the choice of an appropriate distance criterion is a critical step that can influence the final clustering results. Which distance will provide the best similarity measure in uncertain; different measures have different strengths and weaknesses, and can be combined to produce different results (some frequent distances are described in Section 4.3.3).

Step 3: Choice of a clustering algorithm. Several clustering approaches were presented in the previous subsections, all of which have been designed to identify different types of cluster structures. Hence, the concept of 'cluster' is fundamental at this stage of the analysis [23]. Several definitions exist, and each can be valid under particular conditions [24]. As each method attempts to find the best partition, using a specific definition of cluster structure, there is no guarantee that a given algorithm will find the optimal partition in the data. Evaluating the biological relevance of the results obtained using a particular clustering algorithm is therefore a challenging task and generally needs further information than expression measurements (see Step 5).

Step 4: Determination of the number of clusters required for optimal partition of genes. This problem has received increased attention in the clustering literature in recent years. For example, procedures exist to test whether a significant cluster structure has been found in the gene expression dataset in comparison with a randomly generated cluster partition [25].

Step 5: Interpret, test and replicate the resulting cluster analysis. Interpretation of the cluster within the applied context requires biological knowledge and expertise. In that respect, the use of functional gene annotation such as gene ontology (GO) can be particularly useful. GO is a structural network consisting of defined terms and relationships which, between them, describe three attributes of gene products: molecular function, biological process and cellular components [26]. The identification of a cluster of genes in which a particular GO term is over-represented has raised interesting questions for further experimental analyses [27].

Although variations on this five-step procedure may be necessary to fit a particular application, this sequence represents the critical steps in a cluster analysis.

4.3.4.6 Specificities for Temporal Gene Expression Data

Clustering methods based on pairwise distance calculations may yield many biological insights, but are not optimal for analyzing temporal gene expression datasets. Indeed, correlation and Euclidean distances make the implicit assumption that the data at each time point are collected independent of each other, thus ignoring the sequential nature of temporal gene expression data. In order to overcome this major limitation, a number of clustering algorithms designed specifically for time series gene expression data have been suggested [28–30]. These include clustering based on dynamics of the expression patterns [31] or clustering using continuous representation of the profiles [32]. In one case [33], a time translational matrix is used to model the temporal relationships between different modes of the singular value decomposition (SVD). In Ref. [32], Bar-Joseph *et al.* used a statistical spline estimation to represent time-series gene expression profiles as continuous curves, taking into account the actual duration that each time point represents. The major drawback of such algorithms is that relatively long time series datasets are required (generally more than ten time points), these algorithms not being appropriate for shorter time series. As more than 80% of all time series datasets available in the Standford MicroArray Database [34] contain less than eight points [35], specific clustering approaches were recently proposed. For instance, Ernst *et al.* [35] proposed a methodology that focuses on modeling and analyzing the temporal aspects of short time series. The main idea is first to select a set of profiles covering the entire space of possible gene expression profiles that can be generated in the experiment. Next, each gene is assigned to one of the profiles, and an enrichment of genes in each of the profiles is finally computed to determine profile significance. Short time series expression datasets present unique challenges due to the large number of genes sampled and the small number of values for each gene, but have raised fascinating biological questions. The coexpression of genes of known function with poorly characterized or novel genes may provide interesting information concerning the function of many genes for which information is not available currently.

4.4 Modeling a Regulatory Network

4.4.1 Principle

Gene expression programs depend on the recognition of specific promoter sequences by transcriptional regulatory proteins called transcription factors. A ‘regulatory network’ can be defined as the set of genes, the expression of which is modulated by one (or several) transcription factor(s). The deciphering of a regulatory network consists of: (i) identifying genes that compose the network; (ii) defining the interaction between regulators and target genes; and (iii) understanding the functioning of the network under physiological conditions.

Today, the elucidation of the dynamic behavior of transcriptional regulatory networks represents one of the most significant challenges in systems biology. This is clearly a challenging problem, because biological processes are controlled by multiple interactions over time, between hundreds of genes. Clustering approaches based on similarity measures between gene expression profiles are useful for discovering genes that are coregulated, and represent an important preliminary step towards elucidating the transcriptional regulation processes. However, a more ambitious goal consists of modeling and recovering gene regulation phenomena, by seeking genetic relationships such as ‘gene *i* activates (or inhibits) the expression of gene *j*’. It is also desirable to capture more complex scenarios [5] such as auto-regulations, feed-forward or multicomponent regulatory loops (see Figure 4.6a for an illustration of this point).

Methods for inferring and modeling regulatory networks must strike a balance between the model complexity and the limitations of the available data. For instance, microarray data alone provides only a partial picture of the regulatory events, as it does not reflect effects such as post-translational modifications or cellular localizations. The ideal model must be sufficiently complex to accurately describe the system, dealing with a number of genes, which is extremely large compared to the number of available expression measurements. A simple genetic network modeling (static or dynamic) is exposed in the aim of extracting the underlying genetic interactions. References to more complex models are also mentioned.

4.4.2 Static Modeling

In the literature, regulatory gene networks were initially described by using static modeling approaches – that is, approaches that do not take into account the temporal dependency between gene expression measurements. Each expression measurement is considered

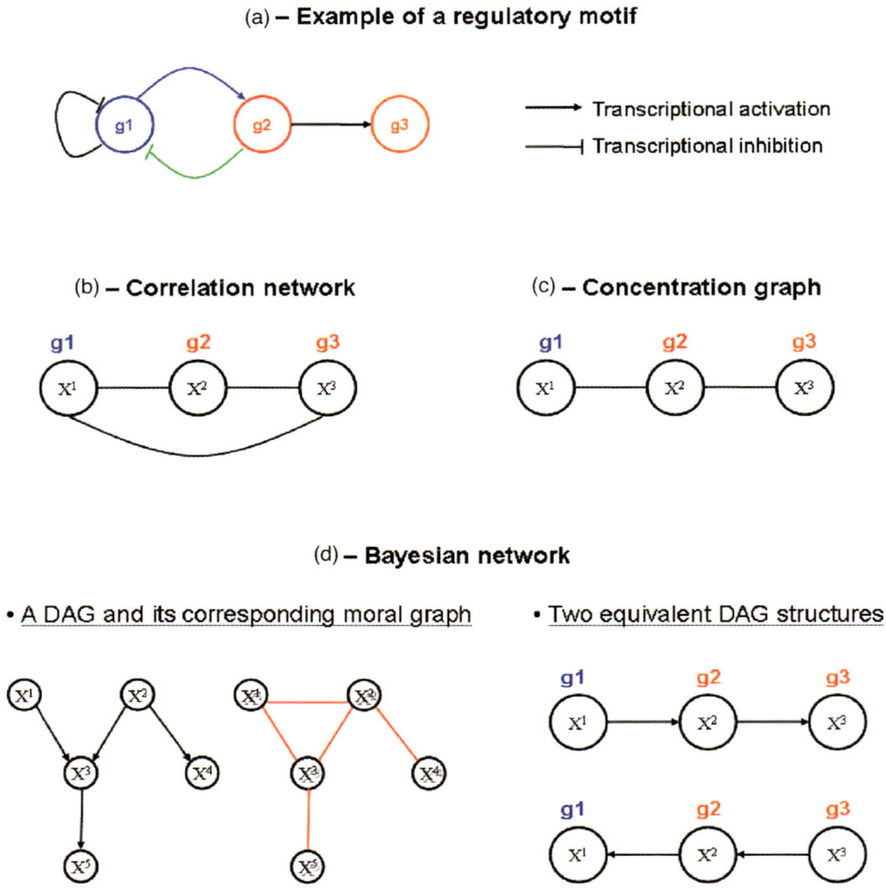

Figure 4.6 *Static modeling of regulatory networks. (a) An example of a regulatory network with both transcriptional activation and inhibition. Genes 1 and 2 (g1, g2) are transcription factors, as they are able to modulate the expression of other genes. This regulatory motif will serve as an illustration for the presentation of several static and dynamic network models; (b) Correlation network expected from the regulatory motif presented in (a). An edge is drawn between X^1 and X^3 as these variables are expected to be correlated and are both correlated with variable X^2 (indirect dependency); (c) Concentration graph expected from the regulatory motif shown in (a). The advantage of the concentration graph is that there is no edge between X^1 and X^3. Concentration graphs only describe conditional dependencies given all the variables represented in the graph. Here, variables X^1 and X^3 are expected to be conditionally independent given the variable X^2; (d) Left: an example of a directed graph containing no cycle. Such directed acyclic graph (DAG) structure is necessary to define a Bayesian network. The associated moral graph allowing the derivation of conditional independencies between variables is also represented. For instance, the conditional independence $X^3 \perp X^4 | X^2$ can be derived from this moral graph, as node X^2 blocks all paths from node X^3 to node X^4. Right: two different DAGs describing the conditional independence between variables X^1 and X^3 given variable X^2 ($P(X^1|X^2, X^3) = P(X^1|X^2)$). Indeed, the joint distribution factorizes both as $P(X^1, X^2, X^3) = P(X^3|X^2)P(X^2|X^1)P(X^1)$ or as $P(X^1, X^2, X^3) = P(X^1|X^2)P(X^2|X^3)P(X^3)$*

as a sample of the same process, and repeated time measurements are assumed to be independent. Even though the temporal information of the data is not exploited as fully as possible, static modeling remains of interest when analyzing expression dependency between genes.

4.4.2.1 Correlation Networks

One of the first tools used to describe interaction between genes is the 'correlation network' [36]. Also referred to as 'relevance network' [37] or 'covariance graph' [38], the correlation network is a nondirected graph which describes the pairwise correlation between variables. In the case of gene regulatory network modeling, each gene i is represented by a node. An undirected edge is then drawn between two genes, i and j, whenever their expression levels X^i and X^j are correlated – that is, when their correlation coefficient $r(X^i, X^j)$ (Equation (4.1)) significantly differs from zero. As an illustration, Figure 4.6b displays the correlation network expected from the regulatory motif presented Figure 4.6a. Assuming that each variable X^i follows a Gaussian distribution, the correlation network describes the set of the dependency relationships between expression levels of the p genes. There is no edge between two variables whenever these two variables are independent.

A major drawback of correlation network models is that the observation of correlation between two variables may come from the linkage with other variables. Two variables can be correlated when considering them separately, but not correlated conditionally on some other variables. Such indirect correlation relationships may generate spurious edges in the correlation network. As an illustration, the correlation network displayed in Figure 4.6b has a spurious edge between the variables X^1 and X^3. Indeed, given that the expression levels of gene 1 (g1) and gene 3 (g3) are both correlated with the expression level of gene 2 (g2), the correlation between X^1 and X^3 is expected to be significant but is not relevant considering the underlying regulatory motif shown Figure 4.6a.

4.4.2.2 Concentration Graphs (Graphical Gaussian Models)

Assuming that the vector variable $(X^i)_{1\leq i\leq p}$ representing the expression levels of p genes follows a multivariate Gaussian distribution of mean μ and covariance matrix Σ,

$$(X^i)_{1\leq i\leq p} \sim \mathcal{N}(\mu, \Sigma) \tag{4.4}$$

graphical Gaussian models (GGMs) describe conditional independency between the variables through the 'concentration graph' [39]. Also known as the 'covariance selection model', the concentration graph describes only the direct dependency relationships between gene expression measurements, given the whole set of observed genes. An undirected edge is drawn between two genes i and j whenever their expression levels X^i and X^j are *conditionally* dependent, taking into account the remaining gene expression levels; this means that there is an edge between genes i and j whenever the partial correlation $p_{ij} = r(X^i, X^j|\{X^k; k \neq i, j\})$ between the variables X^i and X^j given the $(p-2)$ remaining variables $\{X^k; k \neq i, j\}$ differs significantly from zero.

The matrix of partial correlation coefficients $P = (p_{ij})$ is related to the inverse Σ^{-1} of the covariance matrix Σ. The partial correlation coefficient between variables X^i and X^j given the $(p-2)$ remaining variables is null whenever the element $\Sigma^{-1}_{[i,j]}$ in the ith row and jth column of the inverse of the covariance matrix is null,

$$r(X^i, X^j|\{X^k; k \neq i, j\}) = 0 \quad \Leftrightarrow \quad \Sigma^{-1}_{[i,j]} = 0. \tag{4.5}$$

Contrary to the correlation network, the concentration graph enables elimination of the spurious edges due to indirect relationships between variables. As a result, the concentration graph inferred from the regulatory motif of Figure 4.6a is expected not to have an edge between variables X^1 and X^3 as these variables are conditionally independent given X^2 (see Figure 4.6c). However, the concentration graph does not offer a fully accurate description of the interactions; in particular, no direction is given to the interactions and some motifs containing cycles (see Figure 4.6a for instance) cannot be properly represented.

4.4.2.3 Bayesian Networks

Unlike previous graphical models, 'Bayesian networks' model directed relationships between genes [40,41]. Based on a probabilistic measure, a Bayesian network representation of a model is defined by a 'directed acyclic graph (DAG)', that is, a graph G that does not contain cycles (Figure 4.6d, left). Let us call the 'parents' of a node X^i in graph G, denoted by $pa(X^i, G)$, the set of variables having an edge pointing towards the node X^i. A Bayesian network is entirely defined by a DAG G and the set of conditional probability distributions of each variable given its parents in G. To summarize, a stochastic process X admits a Bayesian network representation according to a DAG G, whenever its probability distribution $P(X)$ factorizes as a product of the conditional probability distribution of each variable X^i given its parents in G; that is,

$$P(X) = \prod_{i=1}^{p} P(X^i|pa(X^i, G)). \tag{4.6}$$

However, the acyclicity constraint in static Bayesian networks is a serious restriction given the expected structure of genetic networks. Moreover, interpretation of the edges and their directions must be made carefully. Indeed, some differing DAG structures are equivalent in terms of dependency between the represented variables. For example, a Bayesian network model for three variables X^1, X^2, X^3 such that the variables X^1 and X^3 are 'conditionally independent' given X^2, denoted by $X^1 \perp X^3 \big| X^2$ (as expected from the regulation motif of Figure 4.6a), can be defined by any of the two DAGs shown in Figure 4.6d (right). Moreover, it is important to bear in mind that, for any Bayesian network model defined by a DAG G, the dependency properties must be derived from the 'moral graph' G^m. The moral graph G^m is obtained from G by first 'marrying' the parents; that is, an undirected edge is drawn between each pair of parents of each variable X^i, after which the directions of the original edges of G are deleted (Figure 4.6d, moral graph). We call a 'path' any succession of nodes linked by an edge (for instance X^2, X^3, X^4 in the moral graph of Figure 4.6d). Conditional independencies are then derived from G^m as follows: whenever all paths from node X^3 to node X^4 proceed via node X^2 in the moral graph G^m, then variables X^3 and X^4 are conditionally independent given variable X^2 (for more details, see the directed global Markov property in Ref. [39]). The interpretation of the edges in a DAG G defining a Bayesian network must be made carefully, although this modeling enables

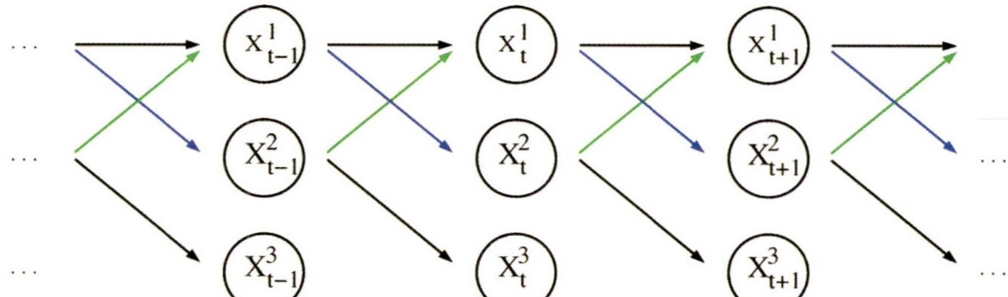

Figure 4.7 *Dynamic network equivalent to the regulation motif in Figure 4.6a. Each vertex X^i_t represents the expression level of gene i at time t. The transcriptional activation of gene 2 by gene 1 in Figure 4.6a is described by edges pointing out from nodes X^1_{t-1} towards nodes X^2_t for all $t > 1$. The direction according to time guarantees the acyclicity of this graph and hence allows the definition of a DBN*

the derivation of conditional independencies in static modeling. Nevertheless, dynamic modeling methodologies, such as dynamic Bayesian networks (DBNs) (as presented in Section 4.4.3) take into account time dependencies between gene expression measurements and allow the modeling of cyclic regulatory motifs. Moreover, the interpretation of the edges of DBNs is straightforward.

4.4.3 Dynamic Bayesian Network (DBN) Modeling

Until now, many dynamic approaches have been proposed to model genetic regulatory networks, such as Boolean networks [42, 43], differential equations [42, 44], DBNs [45] or neural networks [46]. Among these approaches, DBNs – which have attracted great interest in the field of systems biology – were first introduced for the analysis of gene expression time series by Friedman *et al.* [45] and Murphy and Mian [47]. In DBNs, each gene is no longer represented by a single node, but rather by a node for each time point of the experiment. Moreover, the regulatory relationships are assumed to be time-delayed: a dynamic network (Figure 4.7) is obtained by unfolding in time the initial cyclic motif in Figure 4.6a. The direction according to time guarantees the acyclicity of this dynamic network, and consequently this 'directed acyclic graph' (DAG) allows the definition of a Bayesian network (see Section 4.4.2 for more detail). In the DAG defining such a DBN, an edge is drawn between two successive variables, for example X^1_{t-1} and X^2_t, whenever these two variables are conditionally dependent given the remaining variables observed at time $t-1$: $\{X^i_{t-1}; 2 \leq i \leq p\}$ (see Definition – Theorem 1 below). This property is derived from the theory of graphical models for DAGs [39], and allows an extension of the principle of the concentration graph (describing conditional independencies) to a dynamic framework. Although the nature (activation or inhibition) of the regulation in the biological motif does not appear in the DAG, it can nonetheless be derived from the sign (positive or negative) of the model parameter estimates. It should be noted, however, that dynamic modeling is very dependent on the time point measurements sampling. Indeed, a regulatory relationship that actually occurs at a time scale that is shorter than the time sampling may be not detectable from the data or could be misinterpreted.

4.4.3.1 Assumptions

Here, sufficient conditions are introduced such that the probability distribution of process X admits a DBN representation defined by a DAG G (e.g. as in the dynamic network of Figure 4.7). The first assumption is that the observed process X is first-order Markovian; that is, given the past of the gene expression level process $X_{1:t-1}$, the expression level of a gene at time t depends only on the gene expression levels observed at the previous time $(t-1)$.

Assumption 1: The stochastic process X is first-order Markovian, that is for all $t \geq 3$, *the variables* X_t *are conditionally independent from the past variables* $X_{1:t-2}$ *given the variables observed at the previous time point* X_{t-1}, *which is written:* $\forall t \geq 3,\ X_t \perp X_{1:t-1} | X_{t-1}$.

The second assumption is that the variables observed simultaneously are conditionally independent, given the past of the process; in other words, time measurements are considered close enough so that a gene expression level X_t^i measured at time t is better explained by the previous time expression levels X_{t-1} than by some current expression level X_t^j.

Assumption 2: For all $t \geq 1$, *the random variables* $\{X_t^i; \forall 1 \leq i \leq p\}$ *are conditionally independent given the past of the process* $X_{1:t-1}$, *that is,* $\forall t \geq 1, \forall 1 \geq j, X_t^i \perp X_t^j | X_{1:t-1}$.

Shortly, Assumptions 1 and 2 allow the existence of a DBN representation according to a DAG G that only contains edges pointing out from a variable observed at some time $(t-1)$ towards a variable observed at next time t (no edges between simultaneously observed variables). All in all, to restrict the dimension, this DBN model assumes a constant time delay for all regulatory relationships (defined by the time points sampling). It is possible to add simultaneous interactions, or a longer time delay by allowing the existence of edges between variables observed either at the same time (i.e. $X_t^1 \rightarrow X_t^2$) or with a longer time delay (i.e. $X_{t-2}^1 \rightarrow X_t^2$). However, the dimension of the model increases exponentially with the number of authorized time delays, which we can hardly afford given the amount of time points. Finally, DAG G is unique whenever the expression profiles of the p genes are linearly independent (see Assumption 3) – that is, whenever none of the profiles can be written as a linear combination of the others.

Assumption 3: The expression profiles of the p genes form a set of linearly independent vectors.

Whenever these three assumptions are satisfied, the probability distribution of process X admits a DBN representation as exposed in the following theorem.

Definition-Theorem 1: (DBN representation [48]) Whenever Assumptions 1, 2 and 3 are satisfied, the probability distribution of process X admits a DBN representation according to DAG G, *the edges of which describe exactly the full order conditional dependencies between successive variables* X_{t-1}^j *and* X_t^i *given the remaining variables observed at time* $t-1$ *denoted by* $X_{t-1}^{-j} = \{X_{t-1}^k; 1 \leq k \leq p, k \neq j\}$. *For Gaussian variables, the set of edges of DAG* G *defining a DBN is*

$$G = \left\{ X_{t-1}^j \rightarrow X_t^i;\ \forall 1 \leq i, j \leq p,\ \ r\left(X_{t-1}^j, X_t^i \middle| X_{t-1}^{-j}\right) \neq 0 \right\}. \tag{4.7}$$

A large majority of gene expression time series contain no or very few repeated measurement(s) of the expression level of the same gene at a given time. Hence, to carry out an estimation it is often assumed that the process is homogeneous across time (Assumption 4).

Assumption 4: The process is homogeneous across time: any edge is present during the whole process.

This consists of considering that the system is governed by the same rules during the whole time course. Then, $(n-1)$ is observed in repeated measurements of the expression level of each gene at two successive time points. Note that this is a strong assumption, which is not always satisfied but is necessary for estimation in most cases.

4.4.3.2 DBN Modeling for a Multivariate Auto-Regressive Process

A particular case of the DBN modeling discussed in the previous section is introduced here when considering linear relationships – that is, the following first order auto-regressive model (AR(1)).

AR(1) model with diagonal error covariance matrix:

$$\forall t > 1,\ X_t = AX_{t-1} + B + \varepsilon_t, \qquad \varepsilon_t \sim \mathcal{N}(0, \Sigma). \tag{4.8}$$

where $A = (a_{ij})_{1\le i,j\le p}$ is a $p \times p$ matrix, $B = (b_i)_{1\le i\le p}$ is a p-dimensional real vector and $\varepsilon_t = (\varepsilon_t^i)_{1\le i\le p}$ is a p-dimensional Gaussian vector with zero mean and covariance matrix $\Sigma = (\sigma_{ij})_{1\le i,j\le p}$ such that $\sigma_{ij} = 0$ for all $i \neq j$. The errors ε_t^i and ε_t^j of two different variables i and j are not correlated.

For an illustration, this AR(1) model was used by Opgen Rhein and Strimmer [49] to analyze starch metabolism of *Arabidopsis thaliana*. Multivariate AR(1) modeling assumes homogeneity across time (constant matrix A) and linearity of the dependency relationships. Moreover, conditional on the past of the process, the random vector X_t only depends on the random vector X_{t-1} observed at time $(t-1)$, then Assumption 1 is satisfied. Finally, Assumption 2 is satisfied whenever the error covariance matrix Σ is diagonal. Considering noncorrelated measurement errors between distinct genes is a strong assumption, especially since microarray data contain several sources of noise. Nevertheless, assuming Σ to be diagonal is still reasonable after a normalization procedure.

For an illustration, any AR(1) process whose error covariance matrix Σ is diagonal and where matrix A has the following form (where a_{11}, a_{12}, a_{21} and a_{32} refer to nonzero coefficients),

$$A = \begin{pmatrix} a_{11} & a_{12} & 0 \\ a_{21} & 0 & 0 \\ 0 & a_{32} & 0 \end{pmatrix}$$

admits a DBN representation according to the dynamic network of Figure 4.7a ($p = 3$). Thus, the nonzero coefficient a_{21} corresponds to the edges pointing out from X_{t-1}^1 toward X_t^2 for all $t \ge 2$. Indeed, according to the AR(1) model defined by matrix A, we have $X_t^2 = a_{21}X_{t-1}^1 + \varepsilon_t^2$.

4.4.3.3 More DBN Approaches

Various other DBN representations based on different probabilistic models have been proposed in the literature. Both, Ong *et al.* [50] and Zou and Conzen [51] used discrete DBN models: the level expression of each gene is assigned to a binary variable (each gene is considered to be either 'On' (1) or 'Off' (0); that is, transcribed or not). Perrin *et al.* [52] and Wu *et al.* [53] added hidden states to describe gene regulation (Hidden Markov Models (HMMs), also called State Space Models (SSMs). Rangel *et al.* [54] and Beal *et al.* [55] also used SSMs but added feedback from the previous time step gene expression. Other researchers such as Imoto *et al.* [56] and Kim *et al.* [57] used nonparametric additive regression models, while Sugimoto and Iba [58] applied nonparametric additive regression to the difference expression levels between successive time points in an approach called 'Dynamic Differential Bayesian networks' (see also Kim *et al.* [57] for a review of several of these methods).

4.5 Automation Methods for Inferring Regulatory Networks: the Curse of Dimension

4.5.1 Problem

In most microarray gene expression data, there is a small number of measurements n and a large number of variables p. The inference of either the correlation network or the concentration graph requires computation of the covariance matrix, Σ. However, the standard theory to estimate Σ can be exploited only when $n \gg p$ (which ensures that the sample covariance matrix is positive definite [39]). In the same way, for DBN inference assuming an AR(1) model, standard parameter estimation methods can only be used when $n \gg p$. Then, the use of regularized estimators is absolutely essential. The dimension reduction approaches discussed here improve estimation efficiency, which allows the 'curse of dimension' inherent to gene expression data ($n \ll p$) to be handled. First, a positive definite estimate of Σ can be computed with shrinkage estimation. 'Shrinkage' estimates of partial correlation coefficients and regression coefficients can also be derived. A standard procedure ('Lasso' regression) and an heuristic approach based on 'partial order dependencies' can be exposed, both of which allow selection to be carried out within the putative edges of either a concentration graph or a DBN model.

4.5.2 Shrinkage Estimation

4.5.2.1 Definition

Assuming centered data $X = (X^1, \ldots, X^p)$ for p variables (columns), the unbiased empirical estimator of the covariance matrix is $S = \frac{1}{n-1}X^T X$, where X^T refers to the transpose of matrix X. $S = (s_{ij})$ is known to be inefficient for a small number of observations ($n \ll p$). An efficient estimator of Σ can be furnished by shrinking the empirical correlations $r_{ij} = \frac{s_{ij}}{\sqrt{s_{ii}s_{jj}}}$ between gene expression levels towards 0 and the empirical variances s_{ii} against their median $s_{\text{median}} = \underset{1 \leq i \leq p}{\text{Median}}(s_{ii})$. This guarantees the positive definiteness of the estimated covariance matrix [59]. The components of a shrinkage estimates S^* are obtained as follows,

$$s_{ij}^* = r_{ij}^* \sqrt{s_i^* s_j^*} \quad \text{with} \quad \begin{cases} r_{ij}^* = (1 - \lambda_1^*) r_{ij} \\ s_i^* = \lambda_2^* s_{\text{median}} + (1 - \lambda_2^*) s_{ii} \end{cases} \tag{4.9}$$

where the particular choice of the shrinkage intensities λ_1^* and λ_2^* is aimed at minimizing the overall mean squared error.

4.5.2.2 *Usage*

4.5.2.2.1 *Correlation Network*

Shrinkage estimates of both covariance and correlation matrix can be computed with the function 'cov.shrink' of the R package 'corpcor' implemented by Schäfer and Strimmer [59]. Further information about the R programming language can be found in Ref. [60].

4.5.2.2.2 *Concentration Graph*

By basing on the shrinkage estimates S^*, Schäfer and Strimmer [59] propose a model selection procedure for concentration graph. The whole procedure (shrinkage estimation of the partial correlation matrix and edge selection) is implemented in the R package 'GeneNet' [61].

4.5.2.2.3 *DBN*

Small sample shrinkage estimates of the coefficients of a multivariate AR(1) process can be obtained by appropriately substituting the empirical covariance by the shrinkage covariance ([49], the R code is available at http://strimmerlab.org/software.html).

4.5.2.3 *Conclusions*

The shrinkage approach improves the global estimation precision of the correlation, partial correlation or partial regression coefficients in comparison with standard methods. Then, by ordering the edges by decreasing coefficients, edge selection can be carried out. Multiple testing correction can be performed with the local false discovery rate (FDR) approach introduced by Schäffer and Strimmer [59].

4.5.3 Lasso Regression

4.5.3.1 *Definition*

The 'LASSO' for Least Absolute Shrinkage and Selection Operator [62] is a constrained estimation procedure, which tends to produce some coefficients that are exactly zero. Variable selection is then straightforward: only nonzero coefficients define significant dependency relationships. When considering the DBN modeling for AR(1) (Equation (4.8)), the gene expression level of each gene i is described by the next linear regression model,

$$\forall t \geq 2, \; X_t^i = b_i + \sum_{i=1}^{p} a_{ij} X_{t-1}^j + e_t^i, \quad \text{where} \quad e_t^i \sim \mathcal{N}(0, \sigma_{ii}). \tag{4.10}$$

The LASSO estimates $\left(\hat{b}_i, (\hat{a}_{ij})_{1\leq j\leq p}\right)$ are obtained by minimizing the residual sum of squares subject to the sum of the absolute values of the coefficients being less than a constant c, as follows,

$$\left(\hat{b}_i, (\hat{a}_{ij})_{1\leq j\leq p}\right) = \arg\min\left(\sum_{t=2}^{n} x_t^i - b_i - \sum_{j=1}^{p} a_{ij}x_{t-1}^j\right)^2. \tag{4.11}$$

The value of constant c is usually chosen by cross-validation.

4.5.3.2 *Usage*

LASSO estimation can be carried out with the LARS software, as developed by Efron *et al.* [63] for R and Splus programming languages. The LASSO estimates are used straightforwardly to infer dynamic Bayesian networks for an AR(1) process: nonzero coefficients correspond to an edge of the DAG defining the DBN.

LASSO regression also allows concentration graphs inference, as proposed by Meinhausen and Bühlmann *et al.* [64], given that the regression coefficient satisfies,

$$|r(X^i, X^j|\{X^k; k \neq i, j\})| = \sqrt{a_{ij}a_{ji}} \tag{4.12}$$

where a_{ij} (resp. a_{ji}) refers to the regression coefficient defined by the following 'static' linear regression model explaining gene i (resp. gene j) expression level at time t, by considering all gene $j \neq i$ expressions observed at the same time t,

$$\forall t \geq 1, X_t^i = b_i + \sum_{\substack{j=1\\ j\neq i}}^{p} a_{ij}X_t^j + e_t^i, \quad \text{where} \quad e_t^i \sim \mathcal{N}(0, \sigma_{ii}). \tag{4.13}$$

Note also that the 'Inferelator' is a procedure proposed by Bonneau *et al.* [65] for deriving regulatory networks which combines LASSO regression with the integration of genome annotation. The R code for the Inferelator is freely available upon request from the authors.

4.5.3.3 *Conclusions*

The LASSO is very easy to use and allows straightforward edge selection (nonsignificant coefficients are automatically set to zero). It should be noted, however, that the LASSO regression is performed successively for each gene (using Equation (4.10)), and tends to maintain the number of parents of each node uniformly small (instead of keeping small the comprehensive number of edges only). Indeed, the regulatory network to be inferred is globally sparse, but not uniformly sparse.

4.5.4 Partial-Order Conditional Independencies

4.5.4.1 *Definition*

Another powerful approach that recently attracted much attention for inferring both concentration graphs and DBNs is based on the consideration of zero- and first-order conditional independencies. The idea is to approximate the concentration graph (describing full-order conditional independencies) by the graph $G^{(0-1)}$ describing zero- and first-order conditional independence [66]. In the graph $G^{(0-1)}$, an edge between the variables X^i and X^j is drawn if – and only if – together correlation $r(X^i, X^j)$ and all first-order correlations $r(X^i, X^j|X^k)$ between these two variables differ from zero – that is, the graph $G^{(0-1)}$ contains the following set of undirected edges,

$$G^{(0-1)} = \{(X^i, X^j);\ \forall 1 \leq i, j \leq p,\ r(X^i, X^j) \neq 0 \quad \text{and} \quad \forall k \neq i, j,\ r(X^i, X^j|X^k) \neq 0\} \tag{4.14}$$

where $r(X^i, X^j|X^k) = \frac{r(X^i,X^j)-r(X^i,X^k)r(X^j,X^k)}{\sqrt{[1-(r(X^i,X^k))^2][1-(r(X^j,X^k))^2]}}$ is the partial correlation between X^i and X^j given X^k.

Hence, whenever the possible correlation between two variables X^i and X^j can be entirely explained by the effect of some variables X^k, no edge is drawn between the nodes i and j. This procedure allows a drastic dimension reduction: by using first-order conditional correlations, estimation can be carried out accurately even with a small number of observations. Even though the graph of zero- and first-order conditional independence differs from the concentration graph in general, it still reflects some measure of conditional independence. Consequently, several automated procedures for regulatory network inference use low-order independence. Wille and Bühlmann [66] have shown, through simulations, that the graph $G^{(0-1)}$ offers a good approximation of sparse concentration graphs. Castelo and Roverato [67] extended the approach to q-th order partial independence graphs for ($q > 1$). In such q-th order partial independence graphs $G^{(q)}$, there is no edge between two genes i and j if, for all subsets of l variables ($0 \leq l \leq q$), the l-th order partial correlation $r(X^i, X^j|\{X^{k_1}, \ldots, X^{k_l}\})$ is null. More recently, this approach using low-order independence has been adapted for DBN inference (see Chapter 3 in Ref. [48]).

4.5.4.2 *Usage*

4.5.4.2.1 *Concentration Graph*

Castelo and Roverato [67] exposed both a sharp analysis of their properties and an efficient estimation based on q-th order partial correlation inference which is available in the R package 'qp'.

4.5.4.2.2 *DBN*

Using a first-order partial dependency approximation, the *R* package 'G1DBN' developed by Lebre and Chiquet [68] allows the inference of DBNs through a two-step procedure: (i) infer the first-order partial dependency DAG $G^{(1)}$; and (ii) infer the full-order dependency DAG G from $G^{(1)}$.

4.5.4.3 *Conclusions*

The consideration of low-order partial independence is an heuristic approach based on an approximation. However, basing on low-order conditional independence represents a drastic dimension reduction in comparison with full-order independence testing, and makes testing much more accurate.

4.5.5 Further Graph Inference Approaches

The growing interest in genetic regulatory networks modeling has instigated the development of numerous modeling alternatives. Among others, Toh and Horimoto [69] have introduced a method which combines cluster analysis with GGM modeling, while Wu *et al.* [69] have proposed an interactive analysis of gene interactions based on GGM where the user may interactively analyze, modify and explore the inferred concentration graph. Wang *et al.* [70] proposed first, to predict genetic regulatory networks with a GGM, and second to quantify the effects of different experimental treatment conditions on gene expression by using a graphical log-linear model (GLM). This approach is implemented in the Matlab toolbox Mgraph.

Liang *et al.* [71] introduced the REVEAL algorithm for Boolean networks inference based on mutual information, while Murphy [72] has proposed several Bayesian structure learning procedures for Bayesian network (static or dynamic) in the open-source Matlab package BNT (Bayes Net Toolbox).

Eventually, for DBN inference, Ong *et al.* [50] reduced the dimension of the problem by considering prior knowledge on operons; Zou and Conzen [51] limited the potential regulators to the genes with either earlier or simultaneous expression changes and estimate the transcription time lag; Nachman *et al.* [73] have modeled regulatory relationships with sigmoids (instead of linear regression); Stuart *et al.* [74] proposed to infer module networks, where variables in each module share the same parents; and Beal *et al.* [55] and Luna *et al.* [75] respectively have proposed a variational Bayesian method for DBN inference.

4.6 Conclusions

While deciphering the structure and organization of gene regulatory networks is still in its infancy, one of the main obstacles is the difficulty of choosing the appropriate experimental data, together with the appropriate computational approaches. Functional genomics has yielded experimental techniques that allow interactions between genes to be elucidated in a large-scale manner. An example is the use of DNA microarrays to monitor gene expression over time. In this chapter, methodologies dedicated to the analysis of temporal gene expression data were presented, together with their application, in order to answer several biological questions. Whilst each of these approaches has its merits, none of them is sufficient in and of itself. In the future, regulatory network models will benefit from improvements in experimental data diversity (transcriptome, proteome, metabolome, etc.). The trend is clearly towards the aggregation of multiple sources of biological information in an effort to foster an understanding of the biology of studied systems. Indeed, the integration of these sources represents one of the greatest challenges faced by computational biologists today.

List of Abbreviations

RNA	ribonucleic acid
mRNA	messenger RNA
DNA	deoxyribonucleic acid
cDNA	complementary DNA

RT-PCR	real-time polymerase chain reaction
SAGE	serial analysis of gene expression
GO	gene ontology
GGMs	graphical Gaussian models
BNs	Bayesian networks
DAG	directed acyclic graph
DBNs	dynamic Bayesian networks
AR	auto-regressive model
HMM	hidden Markov models
SSM	state space models
FDR	false discovery rate
LASSO	least absolute shrinkage and selection operator.

References

1. Liolios, K., Mavromatis, K., Tavernarakis, N. and Kyrpides, N.C. (2008) The Genomes On Line Database (GOLD) in 2007: status of genomic and metagenomic projects and their associated metadata. *Nucleic Acids Research*, **36**(Database issue), D475–9.
2. Bu, D., Zhao, Y., Cai, L. *et al.* (2003) Topological structure analysis of the protein-protein interaction network in budding yeast. *Nucleic Acids Research*, **31**(9), 2443–50.
3. Jeong, H., Tombor, B., Albert, R. *et al.* (2000) The large-scale organization of metabolic networks. *Nature*, **407**(6804), 651–4.
4. Guelzim, N., Bottani, S., Bourgine, P. and Kepes, F. (2002) Topological and causal structure of the yeast transcriptional regulatory network. *Nature Genetics*, **31**(1), 60–3.
5. Lee, T.I., Rinaldi, N.J., Robert, F. *et al.* (2002) Transcriptional regulatory networks in *Saccharomyces cerevisiae*. *Science*, **298**(5594), 799–804.
6. Schena, M., Shalon, D., Davis, R.W. and Brown, P.O. (1995) Quantitative monitoring of gene expression patterns with a complementary DNA microarray. *Science*, **270**(5235), 467–70.
7. Wang, S.M. (2007) Understanding SAGE data. *Trends in Genetics*, **23**(1), 42–50.
8. Matsumura, H., Bin Nasir, K.H., Yoshida, K. *et al.* (2006) SuperSAGE array: the direct use of 26-base-pair transcript tags in oligonucleotide arrays. *Nature Methods*, **3**(6), 469–74.
9. Lee, N.H. and Saeed, A.I. (2007) Microarrays: an overview. *Methods in Molecular Biology*, **353**, 265–300.
10. Stekel, D. (2003) *Microarray Bioinformatics*, Cambridge University Press.
11. Spellman, P.T., Sherlock, G., Zhang, M.Q. *et al.* (1998) Comprehensive identification of cell cycle-regulated genes of the yeast *Saccharomyces cerevisiae* by microarray hybridization. *Molecular Biology of the Cell*, **9**(12), 3273–97.
12. Rustici, G., Mata, J., Kivinen, K. *et al.* (2004) Periodic gene expression program of the fission yeast cell cycle. *Nature Genetics*, **36**(8), 809–17.
13. Gasch, A.P. and Werner-Washburne, M. (2002) The genomics of yeast responses to environmental stress and starvation. *Functional and Integrative Genomics*, **2**(4–5), 181–92.
14. Lucau-Danila, A., Lelandais, G. *et al.* (2005) Early expression of yeast genes affected by chemical stress. *Molecular and Cellular Biology*, **25**(5), 1860–8.
15. Arbeitman, M.N., Furlong, E.E., Imam, F. *et al.* (2002) Gene expression during the life cycle of *Drosophila melanogaster*. *Science*, **297**(5590), 2270–5.
16. Hughes, T.R., Marton, M.J., Jones, A.R. *et al.* (2000) Functional discovery via a compendium of expression profiles. *Cell*, **102**(1), 109–26.

17. Sturn, A. (2001) Cluster analysis for large scale gene expression studies. PhD Thesis, Institute for Biomedical Engineering, Graz University of Technology, Graz, Austria.
18. Chu, S. *et al.* (1998) The transcriptional program of sporulation in budding yeast. *Science*, **282**, 669–705. [Erratum (1998) Science, 282, 1421].
19. Eisen, M.B., Spellman, P.T., Brown, P.O. and Botstein, D. (1998) Cluster analysis and display of genome-wide expression patterns. *Proceedings of the National Academy of Sciences of the United States of America*, **95**(25), 14863–8.
20. Forgy, E. (1965) Cluster analysis of multivariate data: efficiency vs. interpretanility of classifications. *Biometrics*, **21**, 768–9.
21. Lelandais, G., Marc, P., Vincens, P. *et al.* (2004) MiCoViTo: a tool for gene-centric comparison and visualization of yeast transcriptome states. *BMC Bioinformatics*, **5**, 20.
22. Belacel, N., Wang, Q. and Cuperlovic-Culf, M. (2006) Clustering methods for microarray gene expression data. *Omics*, **10**(4), 507–31.
23. Datta, S. and Datta, S. (2003) Comparisons and validation of statistical clustering techniques for microarray gene expression data. *Bioinformatics*, **19**(4), 459–66.
24. Yin, L., Huang, C.H. and Ni, J. (2006) Clustering of gene expression data: performance and similarity analysis. *BMC Bioinformatics*, **7**(Suppl. 4), S19.
25. Dudoit, S. and Fridlyand, J. (2002) A prediction-based resampling method for estimating the number of clusters in a dataset. *Genome Biology*, **3**(7), RESEARCH0036.
26. Ashburner, M., Ball, C.A., Blake, J.A. *et al.* (2000) Gene ontology: tool for the unification of biology. The Gene Ontology Consortium. *Nature Genetics*, **25**(1), 25–9.
27. Juan, H.F. and Huang, H.C. (2007) Bioinformatics: microarray data clustering and functional classification. *Methods in Molecular Biology*, **382**, 405–16.
28. Kim, J. and Kim, J.H. (2007) Difference-based clustering of short time-course microarray data with replicates. *BMC Bioinformatics*, **8**, 253.
29. Magni, P., Ferrazzi, F., Sacchi, L. and Bellazzi, R. (2008) TimeClust: a clustering tool for gene expression time series. *Bioinformatics*, **24**(3), 430–2.
30. Ernst, J. and Bar-Joseph, Z. (2006) STEM: a tool for the analysis of short time series gene expression data. *BMC Bioinformatics*, **7**, 191.
31. Ramoni, M.F., Sebastiani, P. and Kohane, I.S. (2002) Cluster analysis of gene expression dynamics. *Proceedings of the National Academy of Sciences of the United States of America*, **99** (14), 9121–6.
32. Bar-Joseph, Z., Gerber, G.K., Gifford, D.K. *et al.* (2003) Continuous representations of time-series gene expression data. *Journal of Computational Biology*, **10**(3–4), 341–56.
33. Holter, N.S., Maritan, A., Cieplak, M. *et al.* (2001) Dynamic modeling of gene expression data. *Proceedings of the National Academy of Sciences of the United States of America*, **98**(4), 1693–8.
34. Gollub, J., Ball, C.A., Binkley, G. *et al.* (2003) The Stanford microarray database: data access and quality assessment tools. *Nucleic Acids Research*, **31**(1), 94–6.
35. Ernst, J., Nau, G.J. and Bar-Joseph, Z. (2005) Clustering short time series gene expression data. *Bioinformatics*, **21**(Suppl. 1), i159–68.
36. Steuer, R., Kurths, J., Fiehn, O. and Weckwerth, W. (2003) Observing and interpreting correlations in metabolomic networks. *Bioinformatics*, **19**(8), 1019–26.
37. Butte, A.J., Tamayo, P., Slonim, D. *et al.* (2000) Discovering functional relationships between RNA expression and chemotherapeutic susceptibility using relevance networks. *Proceedings of the National Academy of Sciences of the United States of America*, **97**(22), 12182–6.
38. Cox, D.R. and Wermuth, N. (1996) *Multivariate Dependencies: Models, Analysis and Interpretation*, Chapman & Hall, London.
39. Lauritzen, S.L. (1996) *Graphical Models*, Oxford Statistical Science Series.

40. Friedman, N., Linial, M., Nachman, I. and Pe'er, D. (2000) Using Bayesian networks to analyze expression data. *Journal of Computational Biology*, **7**(3–4), 601–20.
41. de Jong, H. (2002) Modeling and simulation of genetic regulatory systems: a literature review. *Journal of Computational Biology*, **9**(1), 67–103.
42. Kauffman, S.A. (1969) Metabolic stability and epigenesis in randomly constructed genetic nets. *Journal of Theoretical Biology*, **22**, 437–67.
43. Akutsu, T., Miyano, S. and Kuhara, S. (1999) Identification of genetic networks from a small number of gene expression patterns under the Boolean network model. *Pacific Symposium on Biocomputing*, **3**, 17–28.
44. Chen, T., He, H.L. and Church, G.M. (1999) Modeling gene expression with differential equations. *Pacific Symposium on Biocomputing*, **4**, 29–40.
45. Friedman, N., Murphy, K. and Russell, S. (1998) Learning the structure of dynamic probabilistic networks, in Proceedings of the 14th conference on the Uncertainty in Artificial Intelligence. SM, CA, USA, Morgan Kaufmann.
46. Weaver, D.C., Workman, C.T. and Stormo, G.D. (1999) Modeling regulatory networks with weight matrices. *Pacific Symposium on Biocomputing*, **4**, 112–23.
47. Murphy, K. and Mian, M.S. (1999) Modelling gene expression data using dynamic Bayesian networks. Technical report, Computer Science Division, University of California, Berkeley, CA.
48. Lebre, S. (2007) *Stochastic Process Analysis for Genomics and Dynamic Bayesian Networks Inference*, 'PhD Thesis, University of Evry-Val-d'Essonne, France.
49. Opgen-Rhein, R. and Strimmer, K. (2007) Learning causal networks from systems biology time course data: an effective model selection procedure for the vector autoregressive process. *BMC Bioinformatics*, **8**(Suppl. 2), S3.
50. Ong, I.M., Glasner, J.D. and Page, D. (2002) Modelling regulatory pathways in E. coli from time series expression profiles. *Bioinformatics*, **18**(Suppl. 1), S241–8.
51. Zou, M. and Conzen, S.D. (2005) A new dynamic Bayesian network (DBN) approach for identifying gene regulatory networks from time course microarray data. *Bioinformatics*, **21**(1), 71–9.
52. Perrin, B.E., Ralaivola, L., Mazurie, A. *et al.* (2003) Gene networks inference using dynamic Bayesian networks. *Bioinformatics*, **19**(Suppl. 2), ii138–48.
53. Wu, F.X., Zhang, W.J. and Kusalik, A.J. (2004) Modeling gene expression from microarray expression data with state-space equations. *Pacific Symposium on Biocomputing*, **00**, 581–92.
54. Rangel, C., Angus, J., Ghahramani, Z. *et al.* (2004) Modeling T-cell activation using gene expression profiling and state-space models. *Bioinformatics*, **20**(9), 1361–72.
55. Beal, M.J., Falciani, F., Ghahramani, Z. *et al.* (2005) A Bayesian approach to reconstructing genetic regulatory networks with hidden factors. *Bioinformatics*, **21**(3), 349–56.
56. Imoto, S., Kim, S., Goto, T. *et al.* (2003) Bayesian network and nonparametric heteroscedastic regression for nonlinear modeling of genetic network. *Journal of Bioinformatics and Computational Biology*, **1**(2), 231–52.
57. Kim, S., Imoto, S. and Miyano, S. (2003) Inferring gene networks from time series microarray data using dynamic Bayesian networks. *Briefings in Bioinformatics*, **4**(3), 228.
58. Sugimoto, N. and Iba, H. (2004) Inference of gene regulatory networks by means of dynamic differential Bayesian networks and nonparametric regression. *Genome Informatics; International Conference on Genome Informatics*, **15**(2), 121–30.
59. Schafer, J. and Strimmer, K. (2005) A shrinkage approach to large-scale covariance matrix estimation and implications for functional genomics. *Statistical Applications in Genetics and Molecular Biology*, **4**, Article 32.
60. R. [cited; Available from: http://cran.r-project.org; last accessed August 2008].
61. GeneNet. [cited; Available from: http://www.strimmerlab.org/software/genenet/; last accessed August 2008].

62. Tibshirani, R. (1996) Regression shrinkage and selection via the lasso. *Journal of the Royal Statistical Society B*, **58**, 267–88.
63. Efron, B. *et al.* (2004) Least angle regression. *Annals of Statistics*, **32**(2), 407–99.
64. Meinshausen, N. and Buhlmann, P. (2006) High-dimensional graphs and variable selection with the Lasso. *Annals of Statistics*, **34**, 1436–62.
65. Bonneau, R., Reiss, D.J., Shannon, P. *et al.* (2006) The Inferelator: an algorithm for learning parsimonious regulatory networks from systems-biology data sets de novo. *Genome Biology*, **7**(5), R36.
66. Wille, A. and Buhlmann, P. (2006) Low-order conditional independence graphs for inferring genetic networks. *Statistical Applications in Genetics and Molecular Biology*, **5**, Article 1.
67. Castelo, R. and Roverato, A. (2006) Graphical model search procedure in the large p and small n paradigm with applications to microarray data. *Journal of Machine Learning Research*, **7**, 2621–50.
68. G1DBN. [cited; Available from: http://lib.stat.cmu.edu/R/CRAN/src/contrib/Descriptions/G1DBN.html; last accessed August 2008].
69. Toh, H. and Horimoto, K. (2002) Inference of a genetic network by a combined approach of cluster analysis and graphical Gaussian modeling. *Bioinformatics*, **18**(2), 287–97.
70. Wang, J., Myklebost, O. and Hovig, E. (2003) MGraph: graphical models for microarray data analysis. *Bioinformatics*, **19**(17), 2210–11.
71. Liang, S., Fuhrman, S. and Somogyi, R. (1998) Reveal, a general reverse engineering algorithm for inference of genetic network architectures. *Pacific Symposium on Biocomputing*, **3**, 18–29.
72. Murphy, K. (2001) The Bayes Net Toolbox for Matlab. *Computing Science and Statistics*, **33**.
73. Nachman, I., Regev, A. and Friedman, N. (2004) Inferring quantitative models of regulatory networks from expression data. *Bioinformatics*, **20**(Suppl. 1), i248–56.
74. Stuart, J.M., Segal, E., Koller, D. and Kim, S.K. (2003) A gene-coexpression network for global discovery of conserved genetic modules. *Science*, **302**(5643), 249–55.
75. Luna, I.T., Huang, Y., Yin, Y. *et al.* (2007) Uncovering gene regulatory networks from time-series microarray data with variational Bayesian structural expectation maximization. *EURASIP Journal on Bioinformatics and Systems Biology*, 71312.

5

Automated Prediction of Protein Attributes and Its Impact on Biomedicine and Drug Discovery

Kuo-Chen Chou

Gordon Life Science Institute, San Diego, California, USA

5.1 Introduction

Recent advances in large-scale genome sequencing have generated a huge number of protein sequences. For example, whilst the Swiss-Prot database contained only 3939 protein sequence entries in 1986, the number has now increased to 405 506, according to version 56.6 of the UniProtKB/Swiss-Prot released on 16 December 2008. In other words, the number of protein sequences now known is more than 102 times that known about 20 years ago! With such an 'avalanche' of gene products in this post-genomic age, the critical challenge is how to characterize these new-found proteins, both timely and accurately, according to their functional, locational and structural features. This is because these types of features or attributes [1, 2] are very useful for both basic research and drug discovery. For example, when given an uncharacterized protein sequence, how can it be identified as an enzyme or a nonenzyme? And, if it is an enzyme, to which main functional class and subfunctional class does it belong? Is it a membrane protein or a nonmembrane protein? If the former, to which membrane protein type does it belong? Which subcellular location site does the protein reside? Does the protein remain in a single subcellular location, or can it exist simultaneously in or move between two and more subcellular locations? Which part of the protein serves as its signal sequence, and where

Automation in Proteomics and Genomics: An Engineering Case-Based Approach Edited by Gil Alterovitz, Roseann Benson and Marco Ramoni

might it be cleaved by proteases such as HIV protease and SARS enzyme? The list of questions is vast.

Although the answers to such questions can be determined by conducting a variety of biochemical experiments, the straightforward approach of performing experiments is not only time-consuming but also very costly. Consequently, the gap between the number of newly discovered protein sequences and knowledge of their attributes continues to expand. In order to use these new-found proteins for basic research and drug discovery in a timely manner [3–5], it would be highly desirable if such a gap could be bridged by developing effective automated methods for predicting the various attributes of uncharacterized proteins, based on their sequences.

In this chapter, we systematically introduce the recent progress of automated methods with various computational approaches and models. In particular, for those automated methods where web-servers are available, step-by-step instructions are provided describing how these can be used to obtain the requisite data.

5.2 Locational and Functional Characterization

5.2.1 Subcellular Localization

Referred to by many as the 'building block of life', the cell is deemed the most basic structural and functional unit of all living organisms. According to cellular anatomy, a cell is constituted by many different components, compartments or organelles (Figure 5.1), all of which carry out different, specialized tasks. For example, the cell nucleus contains the genetic material (DNA) that governs all functions of the cell, while the cytoplasm, a jelly-like material, takes up most of the cell volume, filling the cell and serving as a 'molecular soup' in which the cell organelles are suspended. The cytoskeleton functions as the cell's scaffold, organizing and maintaining its shape, as well as anchoring the organelles in place. The cell membrane functions as a boundary layer to contain the cytoplasm, the cell wall provides protection from physical injury, and the mitochondrion serves as the cell's 'power generator' playing a critical role in generating energy in the eukaryotic cell.

The most critical survival functions of the cell, however, are effected by its proteins [6,7]. Within it divided by many different compartments or organelles – which usually are referred to as 'subcellular locations' (Figure 5.1) – a cell typically contains approximately one billion (10^9) protein molecules. Each protein molecule has its own location (for a single-location protein) or locations (for a multiple-location or multiplex protein). Thus, one of the fundamental goals in cell biology and proteomics is to identify the subcellular localization and function of those proteins which serve as the cell's 'primary engines'. Information regarding the subcellular locations of the proteins can provide useful insights about their functions. Likewise, in order to understand the intricate pathways that regulate biological processes at the cellular level, knowledge of protein subcellular localization is indispensable.

During the past 16 years, a variety of predictors have been developed to deal with the challenge [8–52]. This section focuses on those predictors that can be used by the vast majority of experimental scientists to easily generate practically more useful data. In order to meet such a requirement, the predictors should have the following features:

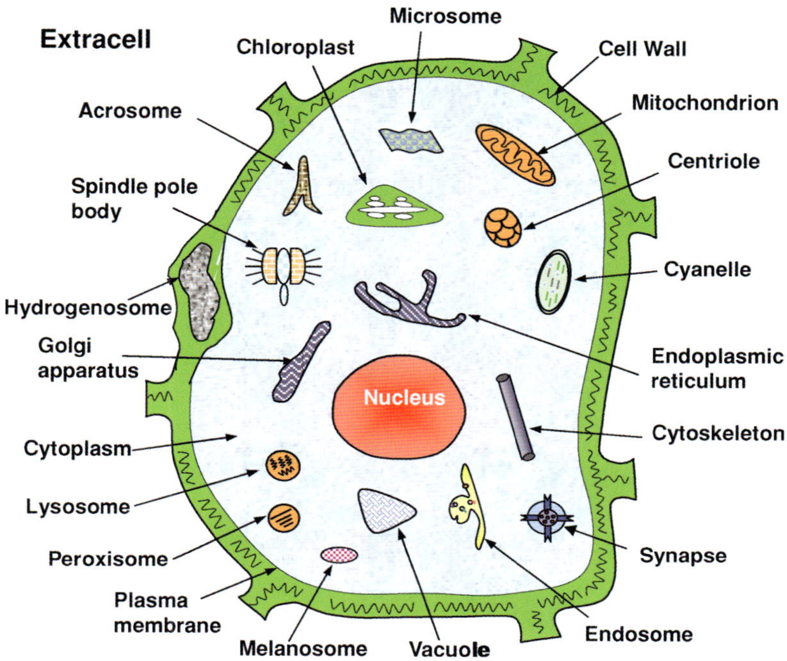

Figure 5.1 *Schematic illustration to show many different components or organelles in a eukaryotic cell. (Reproduced, with permission, from Ref. [59]. Copyright 2007, American Chemical Society)*

(i) a user-friendly web-server that is freely accessible to the public; (ii) a powerful prediction engine to achieve very high accuracy; (iii) a rigorous dataset with wide coverage scope to train and test the prediction engine; and (iv) a considerable flexibility to deal with proteins with multiple location sites or for various specific organisms.

A web-server package called 'Cell-PLoc' [53] was established recently at the web site of http://chou.med.harvard.edu/bioinf/Cell-PLoc/. This is an automated and computational tool that identifies the subcellular locations of uncharacterized proteins based on their sequences, without the need to understand the detailed mathematics. The web-server package contains six predictors, formulated as follows:

$$\textbf{Cell-PLoc} = \begin{cases} \textbf{Plant-PLoc}, \text{for plant proteins covering 11 sites} \\ \textbf{Gpos-PLoc}, \text{for Gram positive proteins covering 5 sites} \\ \textbf{Gneg-PLoc}, \text{for Gram negative proteins covering 8 sites} \\ \textbf{Virus-PLoc}, \text{for virus proteins covering 7 sites} \\ \textbf{Hum-mPLoc}, \text{for human proteins covering 14 sites} \\ \textbf{Euk-mPLoc}, \text{for eukaryotic proteins covering 22 sites} \end{cases} \tag{5.1}$$

Plant-PLoc in Equation 5.1 is specialized for predicting the subcellular localization of plant proteins [45]. To access the Plant-PLoc predictor, open the web page http://chou.med.harvard.edu/bioinf/Cell-PLoc/, and click on the button Plant-PLoc. Figure 5.2 shows the top page of the web server for Plant-PLoc. By clicking the relevant

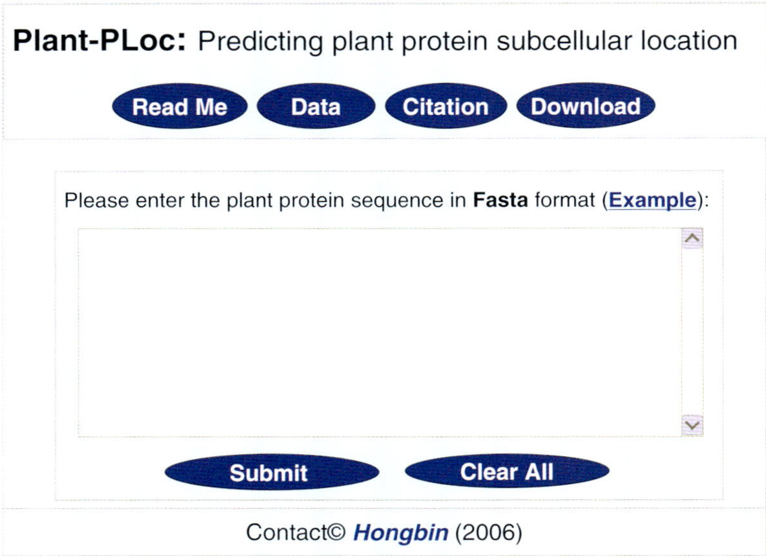

Figure 5.2 *Illustration showing the top page of the web server Plant-PLoc at http://chou.med.harvard.edu/bioinf/plant/*

button, it is possible to browse the desired information; for example, clicking the Read Me button will pop a screen to show the 'Caveat' in using the predictor and its coverage scope. The current version of Plant-PLoc can cover the following 11 subcellular locations: (1) cell wall; (2) chloroplast; (3) cytoplasm; (4) endoplasmic reticulum; (5) extracellular; (6) mitochondrion; (7) nucleus; (8) peroxisome; (9) plasma membrane; (10) plastid; and (11) vacuole. This is illustrated by the figure inserted in the 'Caveat' panel in the Read Me slot at the web site.

By clicking the Citation button, it is possible to find the relevant reports that document the detailed development of Plant-PLoc.

By clicking the Data button, one can find the benchmark dataset used to develop Plant-PLoc predictor.

By clicking the Download button, one can download the results predicted by Plant-PLoc for all the plant protein entries (except those annotated with 'fragment' or those with less than 50 amino acids) in Swiss-Prot database that do not have subcellular location annotations, or are annotated with uncertain terms such as 'probable', 'potential', 'likely', or 'by similarity'. The large-scale predicted results have been deposited in a downloadable file prepared in 'Microsoft Excel' format and 'PDF' format, respectively. To download the former, click Tab Plant-PLoc.xls; to download the latter, click Tab Plant-PLoc.pdf. In order to support the plant genome sequencing projects [54, 55], the large-scale predicted results have been categorized according to their species into these 16 groups: (1) *Arabidopsis*; (2) barley; (3) *Chlamydomonas*; (4) liverwort; (5) maize; (6) mesostigma; (7) pea; (8) potato; (9) rape; (10) rice; (11) soybean; (12) spinach; (13) tobacco; (14) tomato; (15) wheat; and (16) others. To download the categorized results, simply click Tab Plant-PLoc category.xls. Note that the above large-scale results predicted by Plant-PLoc will be updated periodically to include new entries of plant proteins and reflect the continuous development of Plant-PLoc. The large open window at the center of Figure 5.2 is for users to type or paste in the sequence of a query protein for prediction. The sequence should be in FASTA format, as can be seen by clicking the Example button. After entering the input into the window, click the Submit button to obtain the predicted result. For a step-by-step guide on how to use the web server, see Ref. [53].

Gpos-PLoc [56] is specialized for predicting the subcellular localization of Gram-positive bacterial proteins, and its current version covers the following five sites: (1) cell wall; (2) cytoplasm; (3) extracellular; (4) periplasm; and (5) plasma membrane.

Gneg-PLoc [57] is specialized for predicting the subcellular localization of Gram-negative bacterial proteins, and its current version covers the following eight sites: (1) cytoplasm; (2) extracellular; (3) fimbrium; (4) flagellum; (5) inner membrane; (6) nucleoid; (7) outer membrane; and (8) periplasm.

Virus-PLoc [58] is specialized for predicting the subcellular localization of virus proteins within host and virus-infected cells. The current version encompasses the following seven sites: (1) cytoplasm; (2) endoplasmic reticulum; (3) extracellular; (4) inner capsid; (5) nucleus; (6) outer capsid; and (7) plasma membrane.

By following the same procedures as described for Plant-PLoc above, users can access and draw on Gpos-PLoc, Gneg-PLoc and Virus-PLoc, respectively, according to their needs.

Hum-mPLoc and Euk-mPLoc are specialized for predicting the subcellular localization of human proteins [46] and eukaryotic proteins [59], respectively. The 'm' right before 'PLoc' stands for the first character of 'multiple', meaning that the corresponding predictor can be used to deal with proteins with both single and multiple subcellular locations. Although most proteins reside in one subcellular location, some may simultaneously exist at, or move between, two or more different subcellular locations [60]. For instance, according to the Swiss-Prot database (version 50.7, released 19 September 2006), among the 6408 human protein entries that have experimentally observed subcellular location annotations, 973 ($\approx$15%) have multiple location sites; among the 33 925 eukaryotic protein entries that have experimentally observed subcellular location annotations, 2715 have multiple location sites, meaning about 8% bearing the multiplex feature. Proteins with multiple locations or dynamic features of this type are particularly interesting, as they may have some very special biological functions intriguing to both basic research and drug discovery investigators.

Only two of the previously discussed predictors been developed for dealing with the multiplex proteins because, to date, the observed percentages of multiple-location proteins in the other organisms are still less than 5%. It is anticipated that as more experimental annotation data become available in the future, predictors also capable of dealing with multiplex proteins for other organisms, such as Plant-mPLoc, Gpos-mPLoc, Gneg-PLoc and Virus-mPLoc, will be developed in response.

Hum-mPLoc [46] was developed from Hum-PLoc [61] by enabling it also to deal with multiplex proteins. The current version of Hum-mPLoc covers the following 14 subcellular locations: (1) centriole; (2) cytoplasm; (3) cytoskeleton; (4) endoplasmic reticulum; (5) endosome; (6) extracellular; (7) Golgi apparatus; (8) lysosome; (9) microsome; (10) mitochondrion; (11) nucleus; (12) peroxisome; (13) plasma membrane; and (14) synapse.

Euk-mPLoc [59] was developed from Euk-OET-PLoc [62] and Euk-PLoc [47] by enabling it also to deal with multiplex proteins. The current version of Euk-mPLoc spans the following 22 subcellular location sites: (1) acrosome; (2) cell wall; (3) centriole; (4) chloroplast; (5) cyanelle; (6) cytoplasm; (7) cytoskeleton; (8) endoplasmic reticulum; (9) endosome; (10) extracellular; (11) Golgi apparatus; (12) hydrogenosome; (13) lysosome; (14) melanosome; (15) microsome; (16) mitochondrion; (17) nucleus; (18) peroxisome; (19) plasma membrane; (20) plastid; (21) spindle pole body; and (22) vacuole, as illustrated in Figure 5.1.

For the detailed principles and mathematics of Hum-mPLoc and Euk-mPLoc, the reader is referred to the original reports [46, 59] and to a recent review [52]. By following the same procedures as described for Plant-Ploc, both Hum-mPLoc and Euk-mPLoc can be accessed. Now, the difference is that the predicted result for a query protein by Hum-mPLoc or Euk-mPLoc may contain one or more than one location, but a predicted result by any other predictor in Equation 5.1 will always contain one – and only one – location.

Although the establishment of the aforementioned predictors has involved much complicated mathematics and other knowledge, the corresponding web servers are extremely simple and easy to use. By simply typing or pasting in the query protein entry, the user can secure the desired result in less than 5 seconds [53]. This is particularly useful for the vast majority of experimental scientists who wish to obtain their desired results, without the need to understand the detailed mathematics.

In addition to the user-friendly web-server, each of the six predictors in the Cell-PLoc package has the following features that distinguish themselves from the other existing predictors:

- Wide-coverage and stringent benchmark datasets which cover up to 22 subcellular location sites [59] and in which none of protein samples included has $\geq 25\%$ sequence identity to any other in a same subcellular location to avoid homology bias [46, 56, 57, 61–63];
- Very sophisticated and powerful techniques, such as optimized evidence-theoretic K-nearest neighbor (OET-KNN) classifier [2, 64] and fusion approach [52, 65], were introduced to enhance the prediction accuracy.
- The large-scale predicted results are available through a downloadable file in the web site, as illustrated below.

In order to maximize the convenience for the people working in germane disciplines, each of the six predictors in the Cell-PLoc package has been used to identify all of the Swiss-Prot database protein entries in the corresponding organism (except those annotated with ‘fragment’, or those with fewer than 50 amino acids) that do not have subcellular location annotations or are annotated with uncertain terms such as ‘probable’, ‘potential’, ‘likely’, or ‘by similarity’. These large-scale predicted results can be directly downloaded by clicking the Download button on the top page of each of the six web-servers. These results can serve two purposes: (i) that they are available directly to the users for immediate use; and (ii) to set a preceding mark for future experimental results to examine the accuracy of these web-server predictors.

In total, 334 eukaryotic proteins are listed in Appendix A (Section 5.6) as examples. Their experimental annotated subcellular locations were not available before Swiss-Prot 53.2 was released on 26 June 2007; however, according to the large-scale predicted results by Euk-mPLoc that were submitted for publication on 12 November 2006 as Supporting

Information B in Ref. [59], they were also simultaneously placed in the downloadable file called Tab_Euk-mPLoc at http://chou.med.harvard.edu/bioinf/euk-multi/Download.htm. The subcellular locations of the 334 eukaryotic proteins are presented in column 4 of Appendix A where, in order to facilitate comparison, the corresponding experimental results which became available about seven months later are also listed, in column 5. The data in the table illustrate the following. Of the 334 eukaryotic proteins, 309 are with single location site and 25 with multiple location sites. Of the 309 single location proteins, only 22 were incorrectly predicted; of the 25 multiple location proteins, two (i.e. No. 104 and No. 322) were incorrectly predicted. It is interesting to see that the predicted result for No. 104 was 'Centriole; Nucleus', while the experimental observation was 'Cytoplasm; Nucleus'. The significance is that only one of its two location sites was incorrectly predicted. The predicted result for No. 322 was 'Centriole; Cytoplasm; Nucleus' while the experimental observation was 'Nucleus; Cytoplasm', meaning that both of its observed location sites were correctly predicted, although the site of 'Centriole' was overpredicted. As proved later experimentally, the overall success rate for the 334 proteins is over 93%.

5.2.2 Membrane Protein Type

Given an uncharacterized protein sequence, how can one identify whether it is a membrane protein, or not? And, if it *is* a membrane protein, to which membrane protein type does it belong? It is important to address these problems quickly, because they are closely relevant to the query protein's biological function and to its molecular interaction process within a biological system. Most of the functional units, or organelles, are 'enveloped' by one or more membranes, which form the structural basis for many important biological functions. Although the lipid bilayer is the basic structure of membranes, most of the specific functions of the cell membrane are performed by the membrane proteins (see, e.g. Refs [6, 7]). For example, it is through membrane proteins that molecules can be transported into and out of cells by such methods as proton pumps (see Refs [66a, 66b, 67]) and ion pumps (see Refs [68,69]), channel and carrier proteins (see Refs [70]); that various chemical messages such as nerve impulses and hormone activity can be passed between cells (see Ref. [71]); that cells can be attached to an extracellular matrix in grouping cells together to form tissues; that parts of the cytoskeleton can be attached to the cell membrane in order to provide shape; and that the metabolism process and the body's defense mechanisms can be completed.

The function of a membrane protein is correlated with the type to which it belongs, and membrane proteins possess different types. For instance, transmembrane proteins can either transport molecules across the membrane or function on both of its sides, whereas proteins that function on only one side of the lipid bilayer are often associated exclusively with the lipid monolayer or the protein domain on that side. Therefore, information about membrane protein type can provide useful hints for determining the function of an uncharacterized membrane protein. Furthermore, because of the fluid nature of their infrastructure, membrane proteins can move around the cell membrane to reach where their function is required. Identifying the membrane protein type can shed light upon its brand of motion, which is indispensable for studying the biological process at the cellular level from the dynamic point of view. Consequently, if acquiring the knowledge of the membrane protein type is timely, the pace in determining the function of uncharacterized membrane proteins will be expedited, and it will also help in understanding their action process. With the deluge of protein sequences entering into databanks, and the fact that membrane proteins are encoded by 20–35% of genes [72], it remains a challenge to develop a sequence-based automated method to quickly and effectively identify a new-found protein according to the following two questions: (1) Is it a membrane protein? and (2) if it is, to which type does it belong?

Stimulated by the encouraging results in predicting the structural classification of proteins based on their amino acid composition (AAC) [73–81], the covariant discriminant algorithm was introduced in 1999 [82] to predict the types of membrane protein according to their AAC. However, the AAC does not contain any sequence order information. In order to avoid completely losing the sequence order information, the pseudo amino acid composition (PseAAC) was introduced [20], since which time various prediction methods have been proposed in this area [63, 64, 83–95].

As the concept of PseAAC has been widely used by many investigators to improve the prediction quality of various protein attributes [1, 27, 32, 34, 40, 44, 47–50, 63–65, 87, 89, 90, 92, 93, 95–113], a web-server called PseAAC [114] was recently established at http://chou.med.harvard.edu/bioinf/PseAAC/. With this web-server, users are able to generate different types of PseAACs for a given protein sequence.

In this section, we focus on a recently developed powerful predictor called 'MemType-2L' [115], which covers eight membrane types (most other predictors cover only five to six membrane types) (Figure 5.3). The high success rates yielded by MemType-2L are due to: (i) taking into account the evolution information by representing the protein samples with the pseudo position-specific score matrix (Pse-PSSM) vectors derived from the results generated by PSI-BLAST [116]; and (ii) operating by fusing many powerful individual OET-KNN classifiers [62], so as to minimize both the information-missing problem and the overfitting problem (Figure 5.4a). MemType-2L is a two-layer predictor (Figure 5.4b): the first layer prediction engine identifies a query protein as membrane or nonmembrane; if it is membrane, the process will be automatically continued with the second-layer prediction engine to further identify its type among the following eight categories (Figure 5.3): (1) type I; (2) type II; (3) type III; (4) type IV; (5) multipass; (6) lipid-chain-anchored; (7) glycophosphatidylinositol (GPI)-anchored; and (8) peripheral.

In order to support the people working in the relevant area, the user-friendly, freely accessible web-server for MemType-2L is provided at http://chou.med.harvard.edu/bioinf/MemType.

5.2.3 Enzyme Functional Class

Given a protein sequence, how is it identified as an enzyme or nonenzyme? Also, if it is an enzyme, to which main functional class does it belong? And does it have a subfunctional class? It is important to address these problems because they are closely correlated with the biological function of an uncharacterized protein and its acting object and process [117]. Although the answers to these questions can be found by conducting various biochemical experiments, such an approach is both time-consuming and costly. Hence, during the past five years a number of predictors have been developed to deal with these problems [88, 107, 118–123].

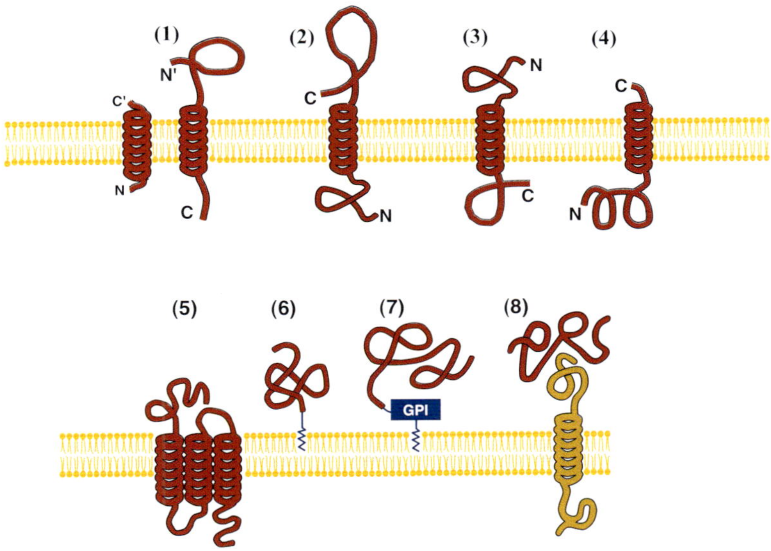

Figure 5.3 *Schematic illustration showing the eight types of membrane protein. (1) Type I transmembrane; (2) type II; (3) type III; (4) type IV; (5) multipass transmembrane; (6) lipid-chain-anchored membrane; (7) GPI-anchored membrane; and (8) peripheral membrane. As shown in the figure, types I, II, III and IV are all of single-pass transmembrane proteins; see Ref. [252] for a detailed description concerning their difference. (Reproduced, with permission, from Ref. [115])*

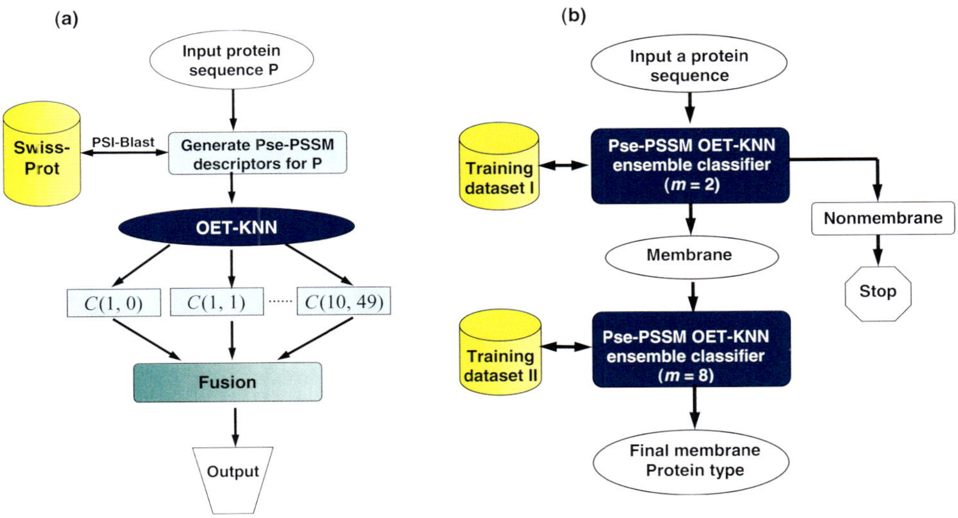

Figure 5.4 *A flowchart to show: (a) the Pse-PSSM OET-KNN ensemble classifier; and (b) the MemType-2L. (Reproduced, with permission, from Ref. [115])*

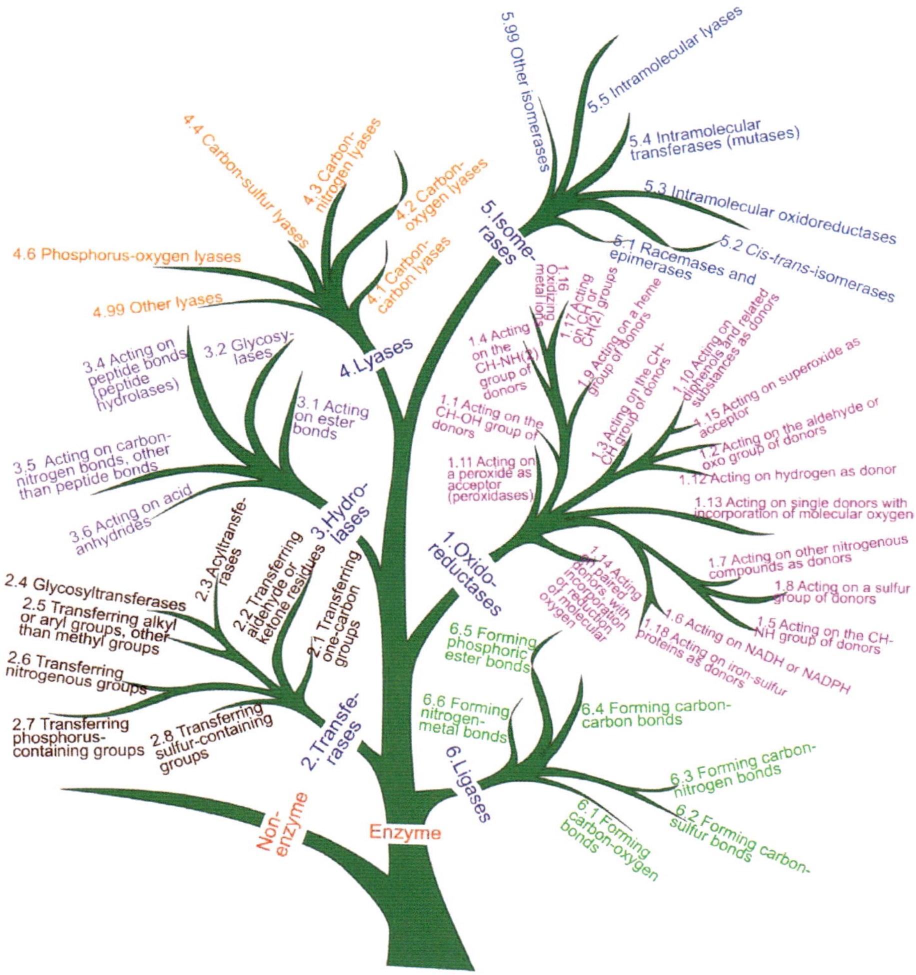

Figure 5.5 *A schematic drawing using tree branches to classify enzyme and nonenzyme, as well as the six main functional classes of enzymes and their subclasses. (Reproduced, with permission, from Ref. [123])*

Among the aforementioned predictors, the recently developed 'EzyPred' [123] distinguished itself with the following features:

- Wider and deeper coverage. EzyPred covers not only all six enzyme main-functional classes [124], but also many of their subfunctional classes (see Figure 5.5).
- Higher expected accuracy. EzyPred is formed by fusing many powerful individual OET-KNN classifiers [62] based on the FunD (functional domain) approach and the Pse-PSSM approach, respectively (Figure 5.6a). The former is closely related to the functions of proteins [125], while the latter can incorporate their evolution information [116]. This type of hybridization approach yields very high success rates [123].

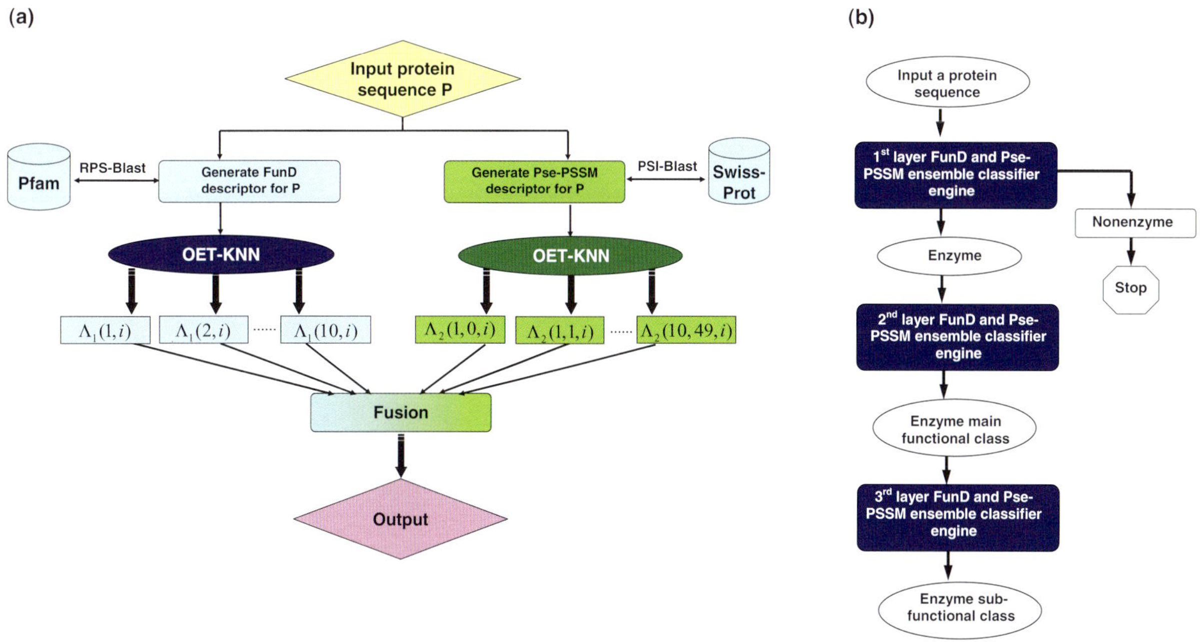

Figure 5.6 *A flowchart to show: (a) how to fuse the FunD approach and Pse-PSSM approach into a prediction engine; and (b) how the top-down approach of the three-layer predictor works. (Reproduced, with permission, from Ref. [123])*

- A user-friendly web-server, which is freely accessible at http://chou.med.harvard.edu/bioinf/EzyPred/ by the public. EzyPred is very easy to use, being designed with a top-down approach scheme. It is a three-layer predictor: the first layer prediction engine serves to identify a query protein as enzyme or nonenzyme; the second layer identifies the main functional class, and the third layer the subfunctional class (Figure 5.6b). Within 90 s of submitting the sequence of a query protein into its input box, EzyPred will determine whether the query protein is an enzyme or nonenzyme and, if it is an enzyme, to which main-functional class and subfunctional class it belongs. For a detailed description and algorithm of EzyPred, the reader is referred to Ref. [123].

5.2.4 Protease Type

Proteases, which are also referred to as proteinases or peptidases [126], are proteolytic enzymes that are essential for the synthesis of all proteins. They control protein size, composition, shape, turnover and ultimate destruction, and account for approximately 2% of the human genome and 1–5% of the genomes of infectious organisms [127]. According to the recent inference by Rawlings [128], the number of proteases might actually be at least double what has previously been believed. In regulating most physiological processes by controlling the activation, synthesis and turnover of proteins, proteases play pivotal regulatory roles in the conception, birth, digestion, growth, maturation, aging and death of all organisms (see Refs [3, 129–136]). Proteases are also essential in viruses, bacteria and parasites for their replication and the spread of infectious diseases; in all insects, organisms and animals for the effective transmission of disease; and in human and animal hosts for the mediation and sustenance of diseases. The actions of proteases are exquisitely selective (see Refs [137, 138]), with each protease being responsible for splitting very specific sequences of amino acids under a preferred set of environmental conditions. According to their catalytic mechanisms, proteases are classified into the following six types: (1) aspartic; (2) cysteine; (3) glutamic; (4) metallo-; (5) serine; and (6) threonine. As the different types of proteases have different functions and biological processes, it is important for both basic research and drug discovery to consider the following two problems. First, given the sequence of a protein, how is it identified as a protease or nonprotease? Second, if it is a protease, then to which protease type does it belong?

Although the two problems can be solved by various experimental means, again, this will be both time-consuming and costly. Recently, two approaches were developed to rectify these problems. The first method is called 'FunD-PseAA' [139], and is based on a strategy that involves hybridizing the functional domain composition [25] and PseAA composition [20]; the other method – known as 'GO-PseAA' [140] – is based on a strategy of hybridizing the gene ontology database [141] [119, 142] and PseAA composition [20]. Both methods have shown much promise. For GO-PseAA [140], the overall expected success rate in identifying a protein as protease or nonprotease was about 91%, and that of a protease type about 85%. For FunD-PseAA [139], the corresponding rates were 92 and 94%, respectively.

5.2.5 GPCR Type

One of the largest gene families in the human genome is that encoding the G-protein-coupled receptors (GPCRs), with approximately 450 genes identified to date. GPCRs are plasma membrane receptors with a trademark of seven-transmembrane helices

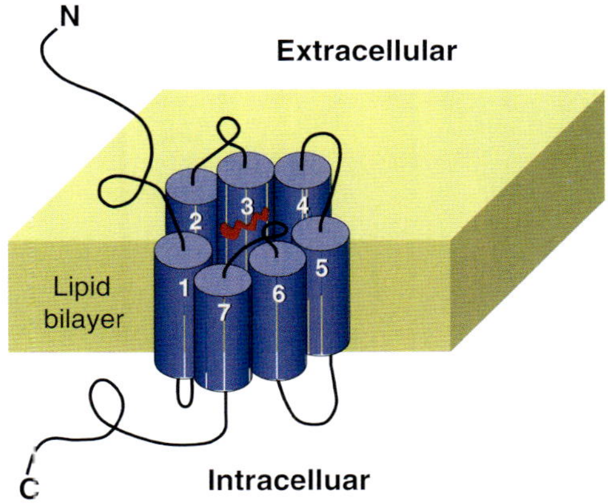

Figure 5.7 *Schematic representation of a GPCR with a trademark of seven-transmembrane helices, depicted as cylinders and connected by alternating cytoplasmic and extracellular hydrophilic loops. The seven-helix bundle thus formed has a central pore on its extracellular surface. The red entity located in the central pore represents a ligand messenger. (Reproduced, with permission, from Ref. [147])*

(Figure 5.7), and play a key role in cellular signaling networks that regulate various physiological processes, such as vision, smell, taste, neurotransmission, secretion, inflammatory, immune responses, cellular metabolism and cellular growth. These proteins are important for understanding human physiology and disease and, indeed, much effort in pharmaceutical research have been targeted at understanding their structure and function. The pathways involving GPCRs are the targets of hundreds of drugs, including antihistamines, neuroleptics, antidepressants and antihypertensives. GPCRs also mediate the actions of certain medications used to treat disorders as diverse as cardiovascular disease, drug dependency and mental illness [143].

As membrane proteins GPCRs are difficult to crystallize, and most will not dissolve in normal solvents. Hence, at this juncture very few GPCR structures have been determined. In contrast, more than thousand GPCR sequences are known, and many more are expected to be known in the near future. Likewise, the functions of many GPCRs are not known, and determining their ligands and signaling pathways is both time-consuming and costly. These difficulties have both motivated and challenged the development of an 'evolutionary pharmacology', where we need a computational method which can predict the classification of the families and subfamilies of GPCRs based on their primary sequences to enable drug classification.

During the past five years or so, several methods have been proposed in this regard [144–149]. Some have been developed to identify the main functional classes of GPCRs (see Ref. [147]), and others for the subfunctional classes (see Ref. [145]) although, owing to a lack of sufficient statistical data, the prediction coverage is quite limited. For instance, according to the G protein-coupled receptor database (GPCRDB) [150, 151], GPCRs are classified into the following six main functional classes: 'rhodopsin like'; 'secretin like'; 'metabotrophic/glutamate/pheromone' class; 'fungal pheromone'; 'cAMP receptors'; and 'Frizzled/Smoothened family'. The prediction method in Ref. [147] can

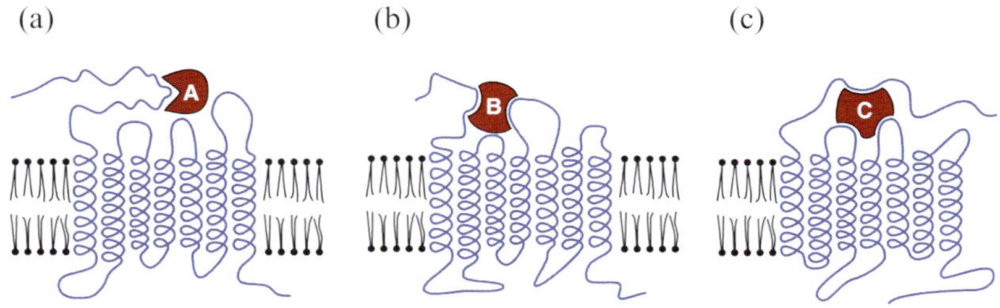

Figure 5.8 Schematic drawing to show three different main families of GPCRs. (a) Class A-'rhodopsin like'; (b) class B-'secretin like'; (c) class C-'metabotrophic/glutamate/pheromone'. (Reproduced, with permission, from Ref. [147])

cover only the first three main functional classes (see Figure 5.8), while in Ref. [145] only seven subfunctional classes of the rhodopsin-like GPCR family are covered (as shown in Figure 5.9). Very recently, based on the cellular automaton image [238a, b] and the gray-level-co-occurrence matrix approach, a web-server known as 'GPCR-CA' was established at the web site http://218.65.61.89:8080/bioinfo/GPCR-CA. This is able to identify whether a query protein is GPCR or nonGPCR and, if it is a GPCR, which of the six main-functional classes to which it belongs will also be identified as documented in [239].

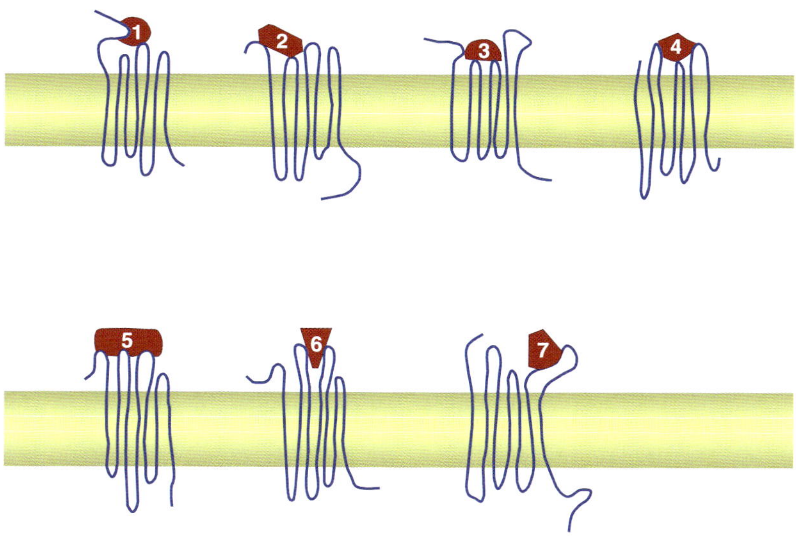

Figure 5.9 Schematic drawing showing the different subtypes of rhodopsin-like GPCR, where the receptors binding with ligands 1, 2, 3, 4, 5, 6 and 7 represent the adrenoceptor-type, chemokine-type, dopamine-type, neuropeptide-type, olfactory-type, rhodopsin-type and serotonin type, respectively. (Reproduced, with permission, from Ref. [145])

5.3 Signal Peptide and Protease Cleavage Site

The identification of cleavage sites in proteins is an important topic, because it is closely relevant to both basic research and drug discovery problems, as illustrated below.

5.3.1 Signal Peptide

A large number of proteins with various essential functions are constantly being constructed within cells and, as nascent proteins, must be transported either out of the cell or to the different compartments (the organelles) within the cell. The main question is, how are these newly made proteins transported across the membrane surrounding the organelles, and how are they directed to their correct location? It has become clear now that, whether a protein will pass through a membrane into a particular organelle, become integrated into the membrane, or be exported out of the cell, is determined by a signal peptide. This is a short sequence of amino acids in a particular order that forms an integral part of the protein. Signal peptides are usually N-terminal extensions that are between three and 60 amino acids long, although they can also be located within a protein or at its C-terminal end [152, 153]. All secreted proteins, as well as many transmembrane proteins, are synthesized with N-terminal signal peptides. Functioning as an 'address tag' or 'zip code' for directing proteins to their correct cellular and extracellular locations (Figure 5.10), signal peptides control the entry of virtually all secretory proteins to the pathway, both

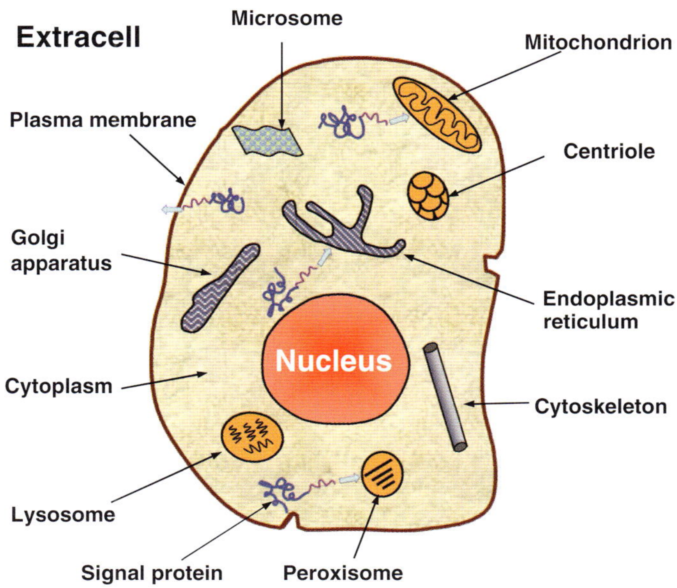

Figure 5.10 Schematic diagram showing how the signal peptides of secretory proteins function as an 'address tag' in directing the proteins to their proper cellular and extracellular locations. The signal peptide sequence is colored in purple, and the mature protein sequence in blue. (Reproduced, with permission, from Ref. [167])

in eukaryotes and prokaryotes [154, 155]. If the signal peptide for a nascent protein were to be changed, the protein in the incorrect cellular location would cause a variety of unusual diseases. For example, a very high level of cholesterol in the blood in some forms of familial hypercholesterolemia is due to deficient transport signals, whereas hereditary diseases such as cystic fibrosis are caused by proteins that do not reach their correct destination. Knowledge of signal peptides can also be used to reprogram cells in a desired way for future cell and gene therapy. To realize this, it is important to identify the signal peptide for a nascent protein. With the 'avalanche' of nascent protein sequences entering into databanks in the post-genomic age, it is desirable to develop an automated method for the rapid and reliable prediction of signal peptides for timely use in basic research and drug discovery [3], and many efforts have been made in this regard [156–166]. Recently, two new signal peptide predictors were developed, and details of these will be briefly introduced below. However, as very few studies have been completed in predicting the signal peptide within a protein or at the C-terminal end, the focus here will be only on the N-terminal signal peptide prediction.

5.3.1.1 Signal-CF

Signal-CF [167] is a two-layered predictor. The first-layer prediction engine is to identify a query protein as secretory or nonsecretory; if it is secretory, then the process will be automatically continued with the second-layer prediction engine to further identify the cleavage site of its signal peptide (Figure 5.11). As mentioned above, the signal peptide of a secretory protein is usually located at its N-terminal end, and will be cleaved off by a signal peptidase once the protein is translocated through a membrane. The cleavage site is commonly symbolized by $(-1, +1)$, namely the position between the last residue of the signal peptide and the first residue of the mature protein, as illustrated in Figure 5.12. It

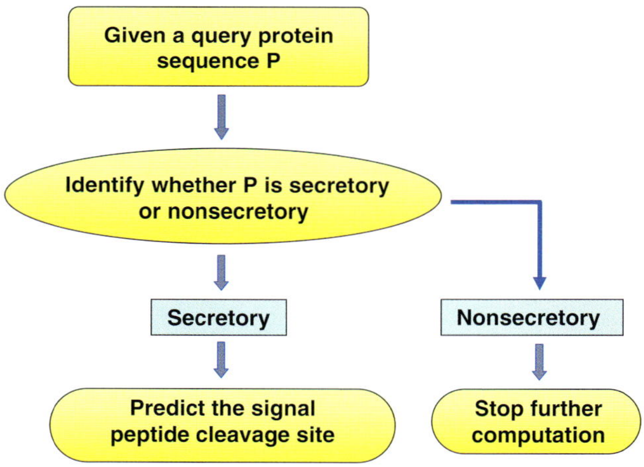

Figure 5.11 *Flowchart showing how the Signal-CF predictor functions in identifying a query protein as secretory or nonsecretory, and in predicting its signal peptide cleavage site if the protein is secretory*

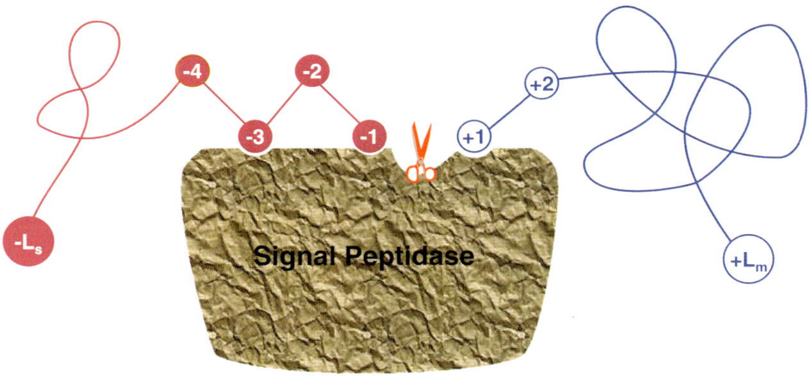

Figure 5.12 *Schematic diagram showing the signal sequence of a protein and how it is cleaved by the signal peptidase. An amino acid in the signal part is depicted as a red circle with a white number to indicate its sequential position, while that in the mature protein depicted as an open circle with a blue number. The signal sequence contains L_s residues and the mature protein L_m residues. The cleavage site is at the position $(-1, +1)$ – that is, between the last residue of the signal sequence and the first residue of the mature protein. (Reproduced, with permission, from Ref. [167])*

can also be seen from the figure that, once the cleavage site is identified, the corresponding signal peptide is automatically known, and vice versa. However, for different secretory proteins the signal peptides are also quite different – not only in sequence components and sequence orders but also in sequence lengths. This was an unavoidable difficulty for all previous methods, and in order to deal with this type of situation the flexible scaled window approach was introduced in Signal-CF by fusing the results derived from many width-different scaled windows through a voting system. Signal-CF has also distinguished itself from many of the previous predictors by explicitly incorporating the subsite coupling effects along a protein sequence. These two remarkable features have defined the name of Signal-CF, where C stands for ‘coupling’ and F for ‘fusion’.

Designed for predicting signal peptides in eukaryotic proteins as well as in Gram-positive and Gram-negative proteins, Signal-CF is freely available as a web-server at http://www.csbio.sjtu.edu.cn/bioinf/Signal-CF or http://chou.med.harvard.edu/bioinf/Signal-CF/. As it can yield highly accurate predicted results in a very short computational time, Signal-CF is particularly useful for large-scale prediction tasks.

5.3.1.2 Signal-3L

Signal-3L [168], which was developed in order to further increase the prediction power and the coverage scope, is a three-layer predictor designed for identifying the signal peptides of human, plant, animal, eukaryotic, Gram-positive and Gram-negative proteins. The target of the first-layer is to identify a query protein as secretory or nonsecretory with the OET-KNN classifier [62] in a Pse-AA composition space [20]. If the protein is identified as secretory, the process will be automatically continued by entering into the second-layer, where a set of candidates for its signal peptide cleavage site are to be selected

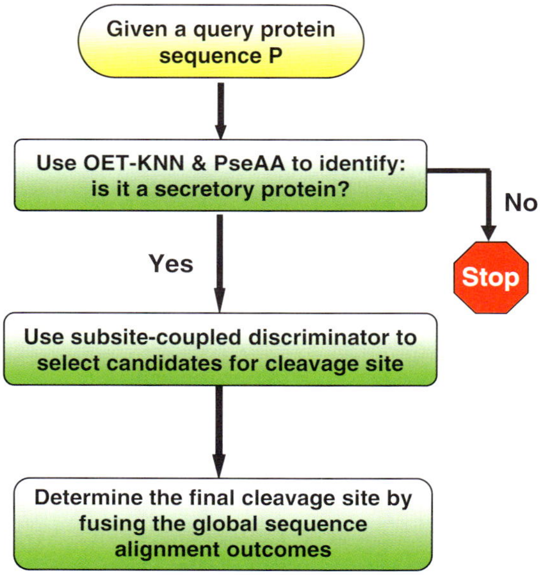

Figure 5.13 *Flowchart showing how the three-layer predictor functions to identify a query protein as secretory or nonsecretory, selecting the candidates of its signal peptide cleavage site if the protein is secretory, and determining the final cleavage site*

with a subsite-coupled discriminator by sliding a scaled window along the protein sequence. The role of the third-layer is finally to determine the unique cleavage site by fusing the global sequence alignment outcome for each of the selected candidates through a voting system. The flowchart in Figure 5.13 shows the process of how the three-layer predictor functions in identifying the signal peptide of a query protein. Signal-3L is accessible to the public as a web-server at http://chou.med.harvard.edu/bioinf/Signal-3L/ or http://www.csbio.sjtu.edu.cn/bioinf/Signal-3L/. Compared to Signal-CF, the computational time of Signal 3-L may be longer, but it will yield slightly more accurate results. In order to maximize convenience for the people working in relevant areas, Signal-3L has been used to predict the signal peptide cleavage sites for all those protein entries in the Swiss-Prot database that are classified as secretory proteins by Signal-3L, but that do not have signal peptide annotations or are annotated with uncertain terms. The results obtained at present have filled the blank area of signal peptide for 4080 human proteins, 3124 plant proteins, 13 527 animal proteins, 6165 other eukaryotic proteins, 5418 Gram-positive proteins and 13 790 Gram-negative proteins. The large-scale results have been deposited in a downloadable file prepared with Microsoft Excel and named 'Tab_Signal-3L.xls'. To download these results, open the web-server Signal-3L, and then click on the Data button. The large-scale results can also be obtained from the Online Supporting Information B of [168].

Both, Signal-CF and Signal-3L can be used to refine the results by other predictors in this area. For instance, listed in Table 5.1 are the signal peptides that were mis-predicted by SignalP-NN and/or SignalP-HMM in the SignalP package [165], yet were corrected by Signal-3L.

Table 5.1 *List of examples showing that signal peptides miss-predicted by SignalP-NN and/or SignalP-HMM are corrected by Signal-3L*

Protein[a]	Experimentally verified signal peptide[a]	SignalP 3.0-NN	SignalP 3.0-HMM	Signal-3L
AAF91396.1	**1-40**	1-37	1-37	**1-40**
DKK1_HUMAN	**1-31**	1-22	1-28	**1-31**
MIME_HUMAN	**1-20**	1-19	1-19	**1-20**
NP_0,57466.1	**1-21**	1-19	1-19	**1-21**
NP_0,57663.1	**1-35**	1-30	1-46	**1-35**
NP_4,43122.2	**1-21**	1-22	1-22	**1-21**
NP_4,43164.1	**1-26**	1-33	1-33	**1-26**
Q6UXL0	**1-28**	1-29	1-29	**1-28**
STC1_HUMAN	**1-17**	1-21	1-18	**1-17**
TRLT_HUMAN	**1-25**	1-24	1-27	**1-25**
CD5L_HUMAN	**1-19**	1-18	**1-19**	**1-19**
EDAR_HUMAN	**1-26**	1-28	**1-26**	**1-26**
FZD3_HUMAN	**1-22**	1-17	**1-22**	**1-22**
IBP7_HUMAN	**1-26**	**1-26**	1-29	**1-26**
KLK3_HUMAN	**1-17**	**1-17**	1-23	**1-17**
NMA_HUMAN	**1-20**	**1-20**	1-26	**1-20**
NP_0,64510.1	**1-22**	**1-22**	1-23	**1-22**
NP_0,68742.1	**1-24**	**1-24**	1-25	**1-24**
NTRI_HUMAN	**1-33**	1-30	1-33	**1-33**
SY01_HUMAN	**1-23**	**1-23**	1-18	**1-23**
TIE1_HUMAN	**1-21**	**1-21**	1-22	**1-21**
TL19_HUMAN	**1-26**	1-23	**1-26**	**1-26**
TR14_HUMAN	**1-38**	1-36	**1-38**	**1-38**
TR19_HUMAN	**1-29**	**1-29**	1-25	**1-29**
XP_1,66856	**1-17**	**1-17**	1-20	**1-17**
XP_2,09141	**1-22**	1-23	**1-22**	**1-22**

[a] Data taken from Ref. [251]. The signal peptides experimentally verified and correctly predicted are in bold-face type colored in blue; those incorrectly predicted are in red.

5.3.2 HIV Protease Cleavage Sites

During the past 15 years, two strategies have often been utilized to identify drugs to treat acquired immunodeficiency syndrome (AIDS). The first strategy targets the HIV (human immunodeficiency virus) reverse transcriptase (see Refs [169–175]), while the second strategy is aimed at the design of HIV protease inhibitors [138, 176–181].

Functioning as a dimer, HIV protease consists of two identical subunits, each having 99 residues, but with only one active site [138, 179]. The essential function of HIV protease is to cleave the precursor polyproteins; a loss of cleavage-ability will halt the life cycle of infectious HIV, which is the culprit [182, 183] of AIDS.

In order to identify effective inhibitors against HIV protease, it is very helpful to understand the polyproteins cleavage mechanism and the 'distorted key' theory [138] approach to the problem, as described below. HIV protease is a member of the highly substrate-selective and cleavage-specific aspartyl proteases. The HIV protease-susceptible sites in a given protein extend to an octapeptide region [184], with its amino acid residues sequentially symbolized by eight subsites, $R_4, R_3, R_2, R_1, R_{1'}, R_{2'}, R_{3'}, R_{4'}$ [185], as shown

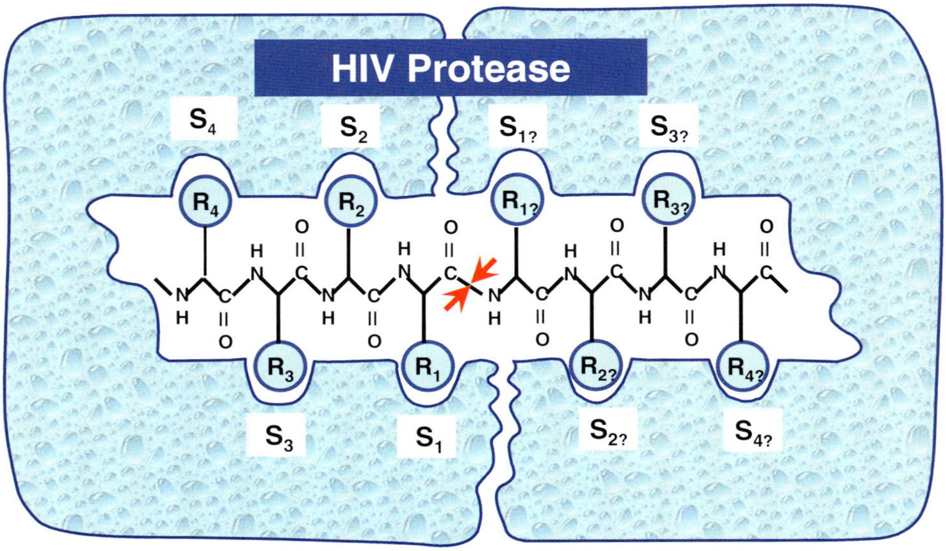

Figure 5.14 Schematic representation of substrate bound to HIV protease based on the analysis of protease–inhibitor crystal structures. The active site of enzyme is composed of eight extended 'subsites', S_4, S_3, S_2, S_1, $S_{1'}$, $S_{1'}$, $S_{2'}$, $S_{3'}$, $S_{4'}$, and their counterparts in a substrate extended to an octapeptide region, sequentially symbolized by R_4, R_3, R_2, R_1, $R_{1'}$, $R_{1'}$, $R_{2'}$, $R_{3'}$, $R_{4'}$, respectively. The scissile bond is located between the subsites R_1 and $R_{1'}$. (Reproduced, with permission, from Ref. [138])

in Figure 5.14. The scissile bond is located between the subsites R_1 and $R_{1'}$. According to the 'lock-and-key' mechanism in enzymology, an HIV protease-cleavable peptide must satisfy the substrate specificity – that is, a good fit for binding to the active site. However, such a peptide, after a modification of its scissile bond with a chemical procedure, will completely lose its cleavability but still be capable of binding to the active site of an enzyme. The molecule, thus modified, can be compared to a 'distorted key', which can be inserted into a lock but can neither open the lock nor be pulled out from it. It is in this way that a molecule modified from a cleavable peptide can spontaneously become a competitive inhibitor against the enzyme. A concept illustration is shown in Figure 5.15, where panel (a) shows the effective binding of a cleavable peptide to the active site of HIV protease, while panel (b) shows that the peptide has become noncleavable after its scissile bond has been modified, even though it can still tightly bind to the active site. Such a modified peptide, or 'distorted key', will automatically become an inhibitor candidate of HIV protease. Even for nonpeptide inhibitors, this can also provide useful insights about the key binding groups, hydrophobic or hydrophilic environment, fitting conformation, and so on. Accordingly, in the search for the potential inhibitors, it is important to discern the type of peptides that can and cannot be cleaved by HIV protease. Although limited within the range of an octapeptide, it is not easy to answer such a question, due to the vast number of possible octapeptides that can be formed from 20 amino acids (approaching $20^8 = 10^{8\log 20} \simeq 2.56 \times 10^{10}$). Whilst the experimental testing of such an astronomic number of octapeptides would be prohibitive, if an effective computational method were to be

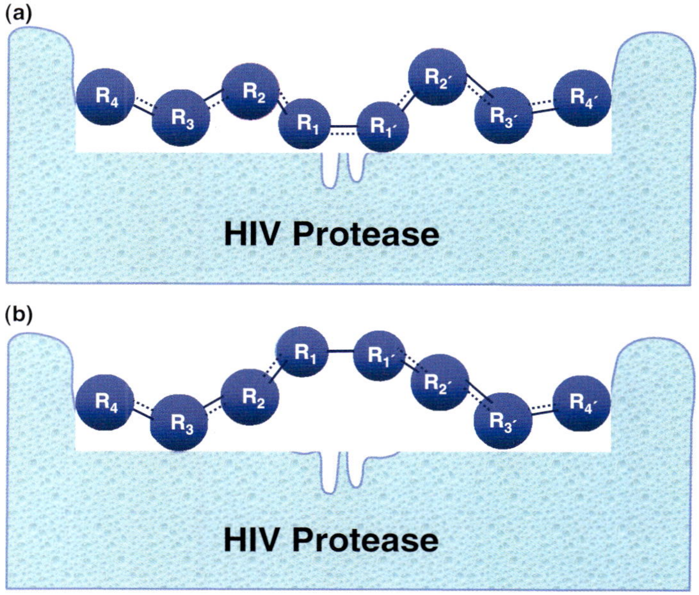

Figure 5.15 *Schematic illustration to show: (a) a cleavable octapeptide is chemically effectively bound to the active site of HIV protease; and (b) although still bound to the active site, the peptide has lost its cleavability after its scissile bond is modified from a hybrid peptide bond [253] to a single bond by some simple routine procedure. The eight residues of the peptide is sequentially symbolized* R_4, R_3, R_2, R_1, $R_{1'}$, $R_{2'}$, $R_{3'}$ *and* $R_{4'}$. *The scissile bond is located between* R_1 *and* $R_{1'}$. *(Reproduced, with permission, from Ref. [13])*

found for predicting the cleavage sites in proteins by HIV protease, then the pace of quest for HIV protease inhibitors would be significantly expedited. During the past decade, a variety of predictive methods have been developed in this regard [137, 178, 186–191], and recently, based on the discriminant function algorithm [138], a web server called HIVcleave [192] was established at the web site http://chou.med.harvard.edu/bioinf/HIV/ or the web site http://www.csbio.sjtu.edu.cn/bioinf/HIV/. For a given protein sequence, HIVcleave can be used to predict its cleavage sites by HIV-1 and HIV-2 proteases, respectively.

5.3.3 SARS Coronavirus Protease Cleavage Sites

SARS (severe acute respiratory syndrome), which was first reported in Asia in February 2003, is a viral respiratory illness caused by a previously unrecognized coronavirus. Patients suffering from SARS present initially with a high fever, sometimes associated with chills or other symptoms, such as headache, body aches and diarrhea, followed by the development of a dry, nonproductive cough that might be accompanied by, or progress to, hypoxia – a condition where insufficient oxygen is being transported to the blood. Most patients subsequently develop pneumonia. It is well known that the life cycle of the replicating SARS coronavirus – the culprit of SARS – is required to pass through a stage in which the viral polyproteins are cleaved by an enzyme known as SARS coronavirus protease [3, 5, 135, 193]. The functional importance of this enzyme in the viral life cycle makes it an attractive target for drugs developed to combat the condition. As the role of the coronavirus protease in SARS is comparable to that of HIV protease in AIDS, the 'distorted key' theory [138] is equally applicable here, and knowledge of the SARS coronavirus protease-mediated cleavage sites in proteins may be valuable when developing anti-SARS drugs. As a consequence, methods to predict such cleavage sites have been developed (see Ref. [194]) and, based on the 'distorted key' theory [138] and predictions of the cleavage sites, several drug candidates have recently been proposed against AIDS and SARS (see Refs [135, 136, 195–205].

5.4 Systems Biology

The field of systems biology, a new and expansive topic, is focused on the systematic study of complex interactions in biological systems. In this brief introduction to the subject, the preliminary discussion is limited to protein–protein interactions and networking couples.

5.4.1 Protein–Protein Interactions

Just as life is full of interactions, proteins – one of the most important elements in living organisms – rarely function in isolation. In order to understand the 'molecular underpinnings' of life, it is first essential to examine the protein–protein interactions through which many functions that are essential to life are manifested. Examples include interactions between different protein subunits as the basis of allosteric changes in oligomers; structural connections between cells being formed through protein–protein interactions; and proteins being directed to the 'correct' compartments of cells by binding to other proteins. In addition, some inhibitors of enzymes are proteins, proteins are modified and degraded by enzymes, protein messengers bind to protein receptors on the outer surfaces of cell membranes to send signals between cells, and protein–protein interactions underlie very large-scale movements in organisms, such as muscle contraction. Protein–protein interactions do indeed affect all processes in a cell!

It has been proposed that all proteins in a given cell are connected through an extensive network, where noncovalent interactions are continuously forming and dissociating [206]. Virtually all cellular processes depend on the precisely orchestrated interactions between proteins.

Imagine a cell in which the specific interactions between proteins suddenly disappeared – the deprived cell would become 'blind' and 'deaf', completely paralytic, and would finally perish. Also, imagine a cell in which many abnormal interactions between proteins suddenly occurred – the unfortunate cell would completely lose control, leading to network confusion and breakdown, because specific and normal protein–protein interactions are involved in almost all physiological processes (see Refs [147, 207]). Thus, the characterization of protein–protein interactions, and an understanding the interaction network, are important with regards to problems ranging from rational drug design (see Refs [3, 208]) to the analysis of metabolic and signal transduction networks (see Refs [209, 210]).

In recent years, much effort has been made in this regard [211–216]. Indeed, the aim of one study [216] was to develop an automated method to identify protein–protein interactions from sequences on a genomic scale. In this study [216], the protein–protein interactions were classified as high, medium and low confidence (note here that 'interactions' should not be interpreted as 'physical binding', but rather as a 'functional association'). The type of work in classifying protein–protein interactions is especially useful when attention moves from traditional methods of investigating individual proteins towards the new frontiers of 'systems biology' and/or 'cellular networking'. In that study [216], the GO-PseAA approach [142, 216, 217] was introduced to formulate the descriptor for the sample of a protein pair, and some encouraging results were obtained.

5.4.2 Networking Couples for Metabolic Pathways

A living organism must be an open and steady-state system rather than a closed and equilibrium system. To maintain order – and hence life – in a universe bent on maximizing disorder, a continuous influx of free energy is indispensable. Metabolism (which is Greek for 'change' or 'overthrow') is the biochemical modification of chemical compounds in living organisms and cells by a series of chemical reactions in order to maintain cell life, growth and division. It is through such metabolic processes that living systems acquire and utilize the free energy they need to perform various functions. Clearly, without metabolism, we would not survive!

Metabolic processes are generally classified as anabolism and catabolism [218]. Anabolism includes the biosynthesis of complex organic molecules and the production of new cell components, usually through processes that demand energy and reducing power obtained from nutrient catabolism, whereas catabolism includes the obtaining of energy and reducing power from nutrients.

Metabolism comprises a set of sophisticated metabolic pathways, which are a series of consecutive enzymatic reactions that produce specific products, and through which the steady state in a living system is maintained. *Cell metabolism* covers all chemical processes within a cell, while *total metabolism* comprises all of the biochemical processes of an organism. Because a living system utilizes many metabolites (i.e. reactants, intermediates and products), the number of metabolic pathways is very large, reflecting the fact that 'life is extremely complicated'. The most important metabolic pathways for humans are [218]:

- Glycolysis, which involves the oxidation of glucose to produce ATP.
- Citric acid cycle (Krebs' cycle) [219], in which acetyl-CoA is oxidized to produce GTP and other valuable energy-intermediates.
- Oxidative phosphorylation, involving the disposal of electrons released by glycolysis and the citric acid cycle (much of the energy released in this process can be stored as ATP).
- Pentose phosphate pathways, in which pentoses are synthesized to release the reducing power needed for anabolic reactions.
- Urea cycle, which involves the disposal of NH_4^+ in less toxic forms.
- Fatty acid β-oxidation, where fatty acids are broken down into acetyl-CoA for use in the Krebs' cycle
- Gluconeogenesis, which involves the synthesis of glucose from smaller precursors for use by the brain.

One of the most important characteristics of metabolic pathways is that they are highly exergonic – that is, they have large negative free energy changes, which provides them with a distinct direction to complete their reactions. Accordingly, if two metabolites are metabolically interconvertible, the pathway from the first to the second must differ from the pathway from the second back to the first. Also, in order to exert control on the flux of metabolites through a metabolic pathway, it is necessary to use enzymatic control to realize various regulations, such as regulating glycolysis, gluconeogenesis, the citric acid cycle [219], urea cycle, glycogen metabolism, fatty acids metabolism and pentose phosphate pathway [218].

A knowledge of metabolic pathways is indispensable for understanding a living system at the level of molecular networks. However, owing to the extreme complexity of the problem, it is both time-consuming and costly to determine metabolic pathways, and the network interactions therein, purely by means of biochemical experiments even for a very simple living system. Yet, even where the details of a metabolic pathways are known, our knowledge might still be incomplete, and the details of some network interactions between enzymes and substrates/products might be missing. In view of these problems, the development of an automated method, or a complementary tool, to provide rapid predictions of the network relationship between enzymes and substrates/products in a living system, would be highly desirable.

Interaction in metabolic pathways includes both enzymatic and hormone control. In one study [220], attention was focused on the enzyme control category, where the metabolic pathway is the network which links the various chemical reactions of compounds (substrates or products) catalyzed by enzymes. To cope with the problem, an approach combining GO [141], FunG (chemical functional group [221]) and PseAA composition [20] was adopted to represent the samples of enzyme–compound couples. For this, two basic identifiers were formulated: one was called 'GO-FunG', and the other 'PseAA-FunG'. The prediction was operated by hybridizing these two basic identifiers into one. As a showcase, the networking couples between enzymes and compounds in the 72 metabolic pathways of *Arabidopsis thaliana* (a small flowering plant widely used as a model organism for studies of the cellular and molecular biology of flowering plants) were investigated. The results thus obtained were quite encouraging, and suggested that the pioneer approach adopted in Ref. [220], although rather preliminary, might be used to study metabolic pathways as well as many other related problems in the cellular networking areas.

It is instructive to point out that the use of graphical approaches to study complicated biological systems can provide an intuitive picture and help gain useful insights. For example, a variety of graphical approaches have been used successfully to study enzyme-catalyzed systems (see Refs [222–231]), protein folding kinetics [232, 233], codon usage [234–237] and HIV reverse transcriptase inhibition mechanisms [169–171, 175]. Meanwhile, the cellular automaton images [238a,b, 239] have been used to represent biological sequences [240] for analyzing the fingerprint of SARS coronavirus [241], for predicting protein subcellular localization [40], transmembrane regions in proteins [242] and the effect on replication ratio by HBV virus gene missense mutation [243], as well as studying hepatitis B viral infections [244]. Recently, similar graphical approaches have also been used to represent DNA sequences (see Ref. [245]), to investigate p53 stress response networks [246], to analyze the network structure of amino acid metabolism [247], study cellular signaling network [248] and proteomics (see the recent review by González-Díaz *et al.* [249]), as well as to conduct a systems biology analysis of the *Drosophila* phagosome [250].

It is said that 'life is complicated', and in order to understand life at a deeper level, one must deal with an open system that comprises many complicated interactions, not only

Table 5.2 *List of the web servers introduced in this chapter and their web site addresses*

Web server predictor	Web site	Description
(1) **Cell-PLoc**	http://chou.med.harvard.edu/bioinf/Cell-PLoc/ or http://www.csbio.sjtu.edu.cn/bioinf/Cell-PLoc/	Predicting subcellular localization of proteins in various organisms [53]
(2) **Euk-OET-PLoc**	http://chou.med.harvard.edu/bioinf/euk-oet/	Predicting subcellular localization of Eukaryotic proteins [62]
(3) **Euk-mPLoc**	http://chou.med.harvard.edu/bioinf/euk-multi/ or http://www.csbio.sjtu.edu.cn/bioinf/euk-multi/	Predicting subcellular localization of Eukaryotic proteins with single or multiple location sites [59]
(4) **Hum**	http://chou.med.harvard.edu/bioinf/hum/	Predicting subcellular localization of human proteins [61]
(5) **Hum-mPLoc**	http://chou.med.harvard.edu/bioinf/hum-multi/ or http://www.csbio.sjtu.edu.cn/bioinf/hum-multi/	Predicting subcellular localization of human proteins with single or multiple location sites [46]
(6) **Plant-PLoc**	http://chou.med.harvard.edu/bioinf/plant/ or http://www.csbio.sjtu.edu.cn/bioinf/plant/	Predicting subcellular localization of plant proteins [45]
(7) **Gpos**	http://chou.med.harvard.edu/bioinf/Gpos/ or http://www.csbio.sjtu.edu.cn/bioinf/Gpos/	Predicting subcellular localization of Gram-positive proteins [56]
(8) **Gneg-PLoc**	http://chou.med.harvard.edu/bioinf/Gneg/ or http://www.csbio.sjtu.edu.cn/bioinf/Gneg/	Predicting subcellular localization of Gram-negative proteins [57]
(9) **Virus**	http://chou.med.harvard.edu/bioinf/virus/ or http://www.csbio.sjtu.edu.cn/bioinf/virus/	Predicting subcellular localization of virus proteins [58]
(10) **PseAAC**	http://chou.med.harvard.edu/bioinf/PseAAC/ or http://www.csbio.sjtu.edu.cn/bioinf/PseAAC/	Generating PseAA composition [114]
(11) **MemType-2L**	http://chou.med.harvard.edu/bioinf/MemType/ or http://www.csbio.sjtu.edu.cn/bioinf/MemType/	Predicting membrane protein type [115]
(12) **EzyPred**	http://chou.med.harvard.edu/bioinf/EzyPred/ or http://www.csbio.sjtu.edu.cn/bioinf/EzyPred/	Predicting enzyme functional class [123]
(13) **Signal-CF**	http://chou.med.harvard.edu/bioinf/Signal-CF/ or http://www.csbio.sjtu.edu.cn/bioinf/Signal-CF/	Predicting protein signal peptide [167]
(14) **Signal-3L**	http://chou.med.harvard.edu/bioinf/Signal-3L/ or http://www.csbio.sjtu.edu.cn/bioinf/Signal-3L/	Predicting protein signal peptide [168]
(15) **HIVcleave**	http://chou.med.harvard.edu/bioinf/HIV/ or http://www.csbio.sjtu.edu.cn/bioinf/HIV/	Predicting HIV protease cleavage sites [192]
(16) **Protease**	http://www.csbio.sjtu.edu.cn/bioinf/Protease/	Predicting protease type [254]

within the system itself but also with its external environment. Although the aim of systems biology is laudable, what has been achieved so far is clearly preliminary in nature.

5.5 List of Web Servers

Finally, for reader's convenience, a brief description of each of the web servers introduced in this chapter, as well as its web site address, is given in Table 5.2.

Recently, a web server, called "ProtIdent", was developed [254] for identifying proteases and their types by fusing functional domain and sequential evolution information. The web server is freely accessible to the public at http://www.csbio.sjtu.edu.cn/bioinf/Protease/.

References

1. Chou, K.C. (2002) *Gene Cloning and Expression Technologies*, Chapter 4 (eds P.W. Weinrer and Q. Lu), Eaton Publishing, Westborough, MA, pp. 57–70.
2. Shen, H.B., Yang, J. and Chou, K.C. (2007) Review: Methodology development for predicting subcellular localization and other attributes of proteins, *Expert Review of Proteomics*, **4**, 453–63.
3. Chou, K.C. (2004) Review: Structural bioinformatics and its impact to biomedical science. *Current Medicinal Chemistry*, **11**, 2105–34.
4. Lubec, G., Afjehi-Sadat, L., Yang, J.W. and John, J.P. (2005) Searching for hypothetical proteins: theory and practice based upon original data and literature. *Progress in Neurobiology*, **77**, 90–127.
5. Chou, K.C. (2006) *Frontiers in Medicinal Chemistry* (eds Atta-ur-Rahman and A.B. Reitz), Bentham Science Publishers, The Netherlands, pp. 455–502.
6. Alberts, B., Bray, D., Lewis, J. *et al.* (1994) *Molecular Biology of the Cell*, Chap. 1, Garland Publishing, New York and London.
7. Lodish, H., Baltimore, D., Berk, A. *et al.* (1995) *Molecular Cell Biology*, Chap. 3, Scientific American Books, New York.
8. Nakai, K. and Kanehisa, M. (1992) A knowledge base for predicting protein localization sites in eukaryotic cells. *Genomics*, **14**, 897–911.
9. Nakashima, H. and Nishikawa, K. (1994) Discrimination of intracellular and extracellular proteins using amino acid composition and residue-pair frequencies. *Journal of Molecular Biology*, **238**, 54–61.
10. Cedano, J., Aloy, P., P'erez-Pons, J.A. and Querol, E. (1997) Relation between amino acid composition and cellular location of proteins. *Journal of Molecular Biology*, **266**, 594–600.
11. Nakai, K. and Horton, P. (1999) PSORT: a program for detecting sorting signals in proteins and predicting their subcellular localization. *Trends in Biochemical Science*, **24**, 34–6.
12. Chou, K.C. and Elrod, D.W. (1998) Using discriminant function for prediction of subcellular location of prokaryotic proteins. *Biochemical and Biophysical Research Communications*, **252**, 63–8.
13. Reinhardt, A. and Hubbard, T. (1998) Using neural networks for prediction of the subcellular location of proteins. *Nucleic Acids Research*, **26**, 2230–36.
14. Chou, K.C. and Elrod, D.W. (1999) Protein subcellular location prediction. *Protein Engineering*, **12**, 107–18.
15. Yuan, Z. (1999) Prediction of protein subcellular locations using Markov chain models. *FEBS Letters*, **451**, 23–6.
16. Nakai, K. (2000) Protein sorting signals and prediction of subcellular localization. *Advances in Protein Chemistry*, **54**, 277–344.

17. Murphy, R.F., Boland, M.V. and Velliste, M. (2000) Towards a systematics for protein subcellular location: quantitative description of protein localization patterns and automated analysis of fluorescence microscope images. *Proceedings in International Conference on Intelligent Systems for Molecular Biology*, **8**, 251–9.
18. Chou, K.C. (2000) Review: Prediction of protein structural classes and subcellular locations. *Current Protein and Peptide Science*, **1**, 171–208.
19. Emanuelsson, O., Nielsen, H., Brunak, S. and von Heijne, G. (2000) Predicting subcellular localization of proteins based on their N-terminal amino acid sequence. *Journal of Molecular Biology*, **300**, 1005–16.
20. Chou, K.C. (2001) Prediction of protein cellular attributes using pseudo amino acid composition. *Proteins: Structure, Function, and Genetics*, **43**, 246–55. (Erratum: *Proteins: Structure, Function, and Genetics* (2001), **43**, 246)
21. Feng, Z.P. (2001) Prediction of the subcellular location of prokaryotic proteins based on a new representation of the amino acid composition. *Biopolymers*, **58**, 491–9.
22. Hua, S. and Sun, Z. (2001) Support vector machine approach for protein subcellular localization prediction. *Bioinformatics*, **17**, 721–8.
23. Feng, Z.P. and Zhang, C.T. (2001) Prediction of the subcellular location of prokaryotic proteins based on the hydrophobicity index of amino acids. *International Journal of Biological Macromolecules*, **28**. 255–61.
24. Feng, Z.P. (2002) An overview on predicting the subcellular location of a protein. *In Silico Biology*, **2**, 291–303.
25. Chou, K.C. and Cai, Y.D. (2002) Using functional domain composition and support vector machines for prediction of protein subcellular location. *Journal of Biological Chemistry*, **277**, 45765–9.
26. Zhou, G.P. and Doctor, K. (2003) Subcellular location prediction of apoptosis proteins. *PROTEINS: Structure, Function, and Genetics*, **50**, 44–8.
27. Pan, Y.X., Zhang, Z.Z., Guo, Z.M. *et al.* (2003) Application of pseudo amino acid composition for predicting protein subcellular location: stochastic signal processing approach. *Journal of Protein Chemistry*, **22**, 395–402.
28. Park, K.J. and Kanehisa, M. (2003) Prediction of protein subcellular locations by support vector machines using compositions of amino acid and amino acid pairs. *Bioinformatics*, **19**, 1656–63.
29. Gardy, J.L., Spencer, C., Wang, K. *et al.* (2003) PSORT-B: Improving protein subcellular localization prediction for Gram-negative bacteria. *Nucleic Acids Research*, **31**, 3613–17.
30. Huang, Y. and Li, Y. (2004) Prediction of protein subcellular locations using fuzzy k-NN method. *Bioinformatics*, **20**, 21–8.
31. Xiao, X., Shao, S., Ding, Y. *et al.* (2005) Using complexity measure factor to predict protein subcellular location. *Amino Acids*, **28**, 57–61.
32. Gao, Y., Shao, S.H., Xiao, X. *et al.* (2005) Using pseudo amino acid composition to predict protein subcellular location: approached with Lyapunov index, Bessel function, and Chebyshev filter. *Amino Acids*, **28**, 373–6.
33. Lei, Z. and Dai, Y. (2005) An SVM-based system for predicting protein subnuclear localizations. *BMC Bioinformatics* **6**, 291.
34. Shen, H.B. and Chou, K.C. (2005) Predicting protein subnuclear location with optimized evidence-theoretic K-nearest classifier and pseudo amino acid composition. *Biochemical and Biophysical Research Communications*, **337**, 752–6.
35. Garg, A., Bhasin, M. and Raghava, G.P. (2005) Support vector machine-based method for subcellular localization of human proteins using amino acid compositions, their order, and similarity search. *Journal of Biological Chemistry*, **280**, 14427–32.
36. Matsuda, S., Vert, J.P., Saigo, H. *et al.* (2005) A novel representation of protein sequences for prediction of subcellular location using support vector machines. *Protein Science*, **14**, 2804–13.

37. Gao, Q.B., Wang, Z.Z., Yan, C. and Du, Y.H. (2005) Prediction of protein subcellular location using a combined feature of sequence. *FEBS Letters*, **579**, 3444–8.
38. Chou, K.C. and Shen, H.B. (2006) Predicting protein subcellular location by fusing multiple classifiers. *Journal of Cellular Biochemistry*, **99**, 517–27.
39. Guo, J., Lin, Y. and Liu, X. (2006) GNBSL: A new integrative system to predict the subcellular location for Gram-negative bacteria proteins. *Proteomics*, **6**, 5099–105.
40. Xiao, X., Shao, S.H., Ding, Y.S. *et al.* (2006) Using cellular automata images and pseudo amino acid composition to predict protein subcellular location. *Amino Acids*, **30**, 49–54.
41. Hoglund, A., Donnes, P., Blum, T. *et al.* (2006) MultiLoc: prediction of protein subcellular localization using N-terminal targeting sequences, sequence motifs and amino acid composition. *Bioinformatics*, **22**, 1158–65.
42. Lee, K., Kim, D.W., Na, D. *et al.* (2006) PLPD: reliable protein localization prediction from imbalanced and overlapped datasets. *Nucleic Acids Research*, **34**, 4655–66.
43. Zhang, Z.H., Wang, Z.H., Zhang, Z.R. and Wang, Y.X. (2006) A novel method for apoptosis protein subcellular localization prediction combining encoding based on grouped weight and support vector machine. *FEBS Letters*, **580**, 6169–74.
44. Shi, J.Y., Zhang, S.W., Pan, Q. *et al.* (2007) Prediction of protein subcellular localization by support vector machines using multi-scale energy and pseudo amino acid composition. *Amino Acids*, **33**, 69–74.
45. Chou, K.C. and Shen, H.B. (2007) Large-scale plant protein subcellular location prediction. *Journal of Cellular Biochemistry*, **100**, 665–78.
46. Shen, H.B. and Chou, K.C. (2007) Hum-mPLoc: An ensemble classifier for large-scale human protein subcellular location prediction by incorporating samples with multiple sites. *Biochemical and Biophysical Research Communications*, **355**, 1006–11.
47. Shen, H.B., Yang, J. and Chou, K.C. (2007) Euk-PLoc: an ensemble classifier for large-scale eukaryotic protein subcellular location prediction. *Amino Acids*, **33**, 57–67.
48. Chen, Y.L. and Li, Q.Z. (2007) Prediction of apoptosis protein subcellular location using improved hybrid approach and pseudo amino acid composition. *Journal of Theoretical Biology*, **248**, 377–81.
49. Chen, Y.L. and Li, Q.Z. (2007) Prediction of the subcellular location of apoptosis proteins. *Journal of Theoretical Biology*, **245**, 775–83.
50. Mundra, P., Kumar, M., Kumar, K.K. *et al.* (2007) Using pseudo amino acid composition to predict protein subnuclear localization: Approached with PSSM. *Pattern Recognition Letters*, **28**, 1610–15.
51. Emanuelsson, O., Brunak, S., von Heijne, G. and Nielsen, H. (2007) Locating proteins in the cell using TargetP, SignalP and related tools. *Nature Protocols*, **2**, 953–71.
52. Chou, K.C. and Shen, H.B. (2007) Review: Recent progresses in protein subcellular location prediction. *Analytical Biochemistry*, **370**, 1–16.
53. Chou, K.C. and Shen, H.B. (2007) Cell-PLoc: A package of web-servers for predicting subcellular localization of proteins in various organisms. *Nature Protocols*, **3**, 153–62.
54. Jorgensen, R. (2006) Plant genomes. *Plant Cell*, **18**, 1099.
55. Jackson, S., Rounsley, S. and Purugganan, M. (2006) Comparative sequencing of plant genomes: choices to make. *Plant Cell*, **18**, 1100–4.
56. Shen, H.B. and Chou, K.C. (2007) Gpos-PLoc: an ensemble classifier for predicting subcellular localization of Gram-positive bacterial proteins. *Protein Engineering, Design, and Selection*, **20**, 39–46.
57. Chou, K.C. and Shen, H.B. (2006) Large-scale predictions of Gram-negative bacterial protein subcellular locations. *Journal of Proteome Research*, **5**, 3420–8.
58. Shen, H.B. and Chou, K.C. (2007) Virus-PLoc: A fusion classifier for predicting the subcellular localization of viral proteins within host and virus-infected cells. *Biopolymers*, **85**, 233–40.
59. Chou, K.C. and Shen, H.B. (2007) Euk-mPLoc: a fusion classifier for large-scale eukaryotic

protein subcellular location prediction by incorporating multiple sites. *Journal of Proteome Research*, **6**, 1728–34.

60. Glory, E. and Murphy, R.F. (2007) Automated subcellular location determination and high-throughput microscopy. *Developmental Cell*, **12**, 7–16.
61. Chou, K.C. and Shen, H.B. (2006) Hum-PLoc: A novel ensemble classifier for predicting human protein subcellular localization. *Biochemical and Biophysical Research Communications*, **347**, 150–7.
62. Chou, K.C. and Shen, H.B. (2006) Predicting eukaryotic protein subcellular location by fusing optimized evidence-theoretic K-nearest neighbor classifiers. *Journal of Proteome Research*, **5**, 1888–97.
63. Shen, H.B. and Chou, K.C. (2007) Using ensemble classifier to identify membrane protein types. *Amino Acids*, **32**, 483–8.
64. Shen, H.B. and Chou, K.C. (2005) Using optimized evidence-theoretic K-nearest neighbor classifier and pseudo amino acid composition to predict membrane protein types. *Biochemical and Biophysical Research Communications*, **334**, 288–92.
65. Shen, H.B. and Chou, K.C. (2006) Ensemble classifier for protein fold pattern recognition. *Bioinformatics*, **22**, 1717–22.

66a. Chou, K.C. (1993) Conformational change during photocycle of bacteriorhodopsin and its proton-pumping mechanism. *Journal of Protein Chemistry*, **12**, 337–50;

66b. Chou, K.C. (1994) Mini Review: A molecular piston mechanism of pumping protons by bacteriorhodopsin. *Amino Acids*, **7**, 1–17.

67. Schnell, J.R. and Chou, J.J. (2008) Structure and mechanism of the M2 proton channel of influenza A virus. *Nature*, **451**, 591–5.
68. Doyle, D.A., Morais, C.J., Pfuetzner, R.A. *et al.* (1998) The structure of the potassium channel: molecular basis of K+ conduction and selectivity. *Science*, **280**, 69–77.
69. Chou, K.C. (2004) Insights from modeling three-dimensional structures of the human potassium and sodium channels. *Journal of Proteome Research*, **3**, 856–61.
70. Oxenoid, K. and Chou, J.J. (2005) The structure of phospholamban pentamer reveals a channel-like architecture in membranes. *Proceedings of the National Academy of Sciences of the United States of America*, **102**, 10870–5.
71. Chou, K.C. (2004) Modelling extracellular domains of GABA-A receptors: subtypes 1, 2, 3, and 5. *Biochemical and Biophysical Research Communications*, **316**, 636–42.
72. Douglas, S.M., Chou, J.J. and Shih, W.M. (2007) DNA-nanotube-induced alignment of membrane proteins for NMR structure determination. *Proceedings of the National Academy of Sciences of the United States of America*, **104**, 6644–8.
73. Nakashima, H., Nishikawa, K. and Ooi, T. (1986) The folding type of a protein is relevant to the amino acid composition. *Journal of Biochemistry*, **99**, 152–62.
74. Klein, P. and Delisi, C. (1986) Prediction of protein structural class from amino acid sequence. *Biopolymers*, **25**, 1659–72.
75. Klein, P. (1986) Prediction of protein structural class by discriminant analysis. *Biochimica et Biophysica Acta*, **874**, 205–15.
76. Chou, K.C. and Zhang, C.T. (1994) Predicting protein folding types by distance functions that make allowances for amino acid interactions. *Journal of Biological Chemistry*, **269**, 22014–20.
77. Chou, K.C. (1995) A novel approach to predicting protein structural classes in a (20-1)-D amino acid composition space. *Proteins: Structure, Function and Genetics*, **21**, 319–44.
78. Liu, W. and Chou, K.C. (1998) Prediction of protein structural classes by modified Mahalanobis discriminant algorithm. *Journal of Protein Chemistry*, **17**, 209–17.
79. Chou, K.C., Liu, W., Maggiora, G.M. and Zhang, C.T. (1998) Prediction and classification of domain structural classes. *PROTEINS: Structure, Function, and Genetics*, **31**, 97–103.
80. Chou, K.C. and Maggiora, G.M. (1998) Domain structural class prediction. *Protein Engineering*, **11**, 523–38.

81. Chou, K.C. (1999) A key driving force in determination of protein structural classes. *Biochemical and Biophysical Research Communications*, **264**, 216–24.
82. Chou, K.C. and Elrod, D.W. (1999) Prediction of membrane protein types and subcellular locations. *PROTEINS: Structure, Function, and Genetics*, **34**, 137–53.
83. Cai, Y.D., Liu, X.J. and Chou, K.C. (2001) Artificial neural network model for predicting membrane protein types. *Journal of Biomolecular Structure and Dynamics*, **18**, 607–10.
84. Guo, Z.M. (2002) Prediction of membrane protein types by using pattern recognition method based on pseudo amino acid composition. Masters Thesis, Bio-X Life Science Research Center, Shanghai Jiaotong University.
85. Cai, Y.D., Zhou, G.P. and Chou, K.C. (2003) Support vector machines for predicting membrane protein types by using functional domain composition. *Biophysical Journal*, **84**, 3257–63.
86. Cai, Y.D., Pong-Wong, R., Feng, K. *et al.* (2004) Application of SVM to predict membrane protein types. *Journal of Theoretical Biology*, **226**, 373–6.
87. Wang, M., Yang, J., Liu, G.P. *et al.* (2004) Weighted-support vector machines for predicting membrane protein types based on pseudo amino acid composition. *Protein Engineering, Design, and Selection*, **17**, 509–16.
88. Chou, K.C. (2005) Using amphiphilic pseudo amino acid composition to predict enzyme subfamily classes. *Bioinformatics*, **21**, 10–19.
89. Chou, K.C. and Cai, Y.D. (2005) Prediction of membrane protein types by incorporating amphipathic effects. *Journal of Chemical Information and Modeling*, **45**, 407–13.
90. Liu, H., Wang, M. and Chou, K.C. (2005) Low-frequency Fourier spectrum for predicting membrane protein types. *Biochemical and Biophysical Research Communications*, **336**, 737–9.
91. Wang, M., Yang, J., Xu, Z.J. *et al.* (2005) SLLE for predicting membrane protein types. *Journal of Theoretical Biology*, **232**, 7–15.
92. Shen, H.B., Yang, J. and Chou, K.C. (2006) Fuzzy KNN for predicting membrane protein types from pseudo amino acid composition. *Journal of Theoretical Biology*, **240**, 9–13.
93. Wang, S.Q., Yang, J. and Chou, K.C. (2006) Using stacked generalization to predict membrane protein types based on pseudo amino acid composition. *Journal of Theoretical Biology*, **242**, 941–6.
94. Yang, X.G., Luo, R.Y. and Feng, Z.P. (2007) Using amino acid and peptide composition to predict membrane protein types. *Biochemical and Biophysical Research Communications*, **353**, 164–9.
95. Pu, X., Guo, J., Leung, H. and Lin, Y. (2007) Prediction of membrane protein types from sequences and position-specific scoring matrices. *Journal of Theoretical Biology*, **247**, 259–65.
96. Liu, H., Yang, J., Wang, M. *et al.* (2005) Using Fourier spectrum analysis and pseudo amino acid composition for prediction of membrane protein types. *The Protein Journal*, **24**, 385–9.
97. Xiao, X., Shao, S.H., Huang, Z.D. and Chou, K.C. (2006) Using pseudo amino acid composition to predict protein structural classes: approached with complexity measure factor. *Journal of Computational Chemistry*, **27**, 478–82.
98. Zhang, T., Ding, Y. and Chou, K.C. (2006) Prediction of protein subcellular location using hydrophobic patterns of amino acid sequence. *Computational Biology and Chemistry*, **30**, 367–71.
99. Chen, C., Zhou, X., Tian, Y. *et al.* (2006) Predicting protein structural class with pseudoamino acid composition and support vector machine fusion network. *Analytical Biochemistry*, **357**, 116–21.
100. Chen, C., Tian, Y.X., Zou, X.Y. *et al.* (2006) Using pseudoamino acid composition and support vector machine to predict protein structural class. *Journal of Theoretical Biology*, **243**, 444–8.
101. Zhang, S.W., Pan, Q., Zhang, H.C. *et al.* (2006) Prediction protein homo-oligomer types by pseudo amino acid composition: Approached with an improved feature extraction and naive Bayes feature fusion. *Amino Acids*, **30**, 461–8.
102. Du, P. and Li, Y. (2006) Prediction of protein submitochondria locations by hybridizing

pseudoamino acid composition with various physicochemical features of segmented sequence. *BMC Bioinformatics*, **7**, 518.

103. Mondal, S., Bhavna, R., Mohan Babu, R. and Ramakumar, S. (2006) Pseudo amino acid composition and multi-class support vector machines approach for conotoxin superfamily classification. *Journal of Theoretical Biology*, **243**, 252–60.
104. Lin, H. and Li, Q.Z. (2007) Predicting conotoxin superfamily and family by using pseudo amino acid composition and modified Mahalanobis discriminant. *Biochemical and Biophysical Research Communications*, **354**, 548–51.
105. Lin, H. and Li, Q.Z. (2007) Using pseudo amino acid composition to predict protein structural class: Approached by incorporating 400 dipeptide components. *Journal of Computational Chemistry*, **28**, 1463–6.
106. Kurgan, L.A., Stach, W. and Ruan, J. (2007) Novel scales based on hydrophobicity indices for secondary protein structure. *Journal of Theoretical Biology*, **248**, 354–66.
107. Zhou, X.B., Chen, C., Li, Z.C. and Zou, X.Y. (2007) Using Chou's amphiphilic pseudoamino acid composition and support vector machine for prediction of enzyme subfamily classes. *Journal of Theoretical Biology*, **248**, 546–51.
108. Zhang, T.L. and Ding, Y.S. (2007) Using pseudo amino acid composition and binary-tree support vector machines to predict protein structural classes. *Amino Acids*, **33**, 623–29.
109. Ding, Y.S., Zhang, T.L. and Chou, K.C. (2007) Prediction of protein structure classes with pseudo amino acid composition and fuzzy support vector machine network. *Protein and Peptide Letters*, **14**, 811–15.
110. Fang, Y., Guo, Y., Feng, Y. and Li, M. (2007) Predicting DNA-binding proteins: approached from Chou's pseudo amino acid composition and other specific sequence features. *Amino Acids*, **34**, 103–9.
111. Zhang, S.W., Zhang, Y.L., Yang, H.F. *et al.* (2007) Using the concept of Chou's pseudo amino acid composition to predict protein subcellular localization: an approach by incorporating evolutionary information and von Neumann entropies. *Amino Acids*, **34**, 565–72.
112. Shi, J.Y., Zhang, S.W., Pan, Q. and Zhou, G.P. (2008) Using pseudo amino acid composition to predict protein subcellular location: Approached with amino acid composition distribution. *Amino Acids*, DOI 10.1007/s00726-00007-00623-z.
113. Xiao, X. and Chou, K.C. (2007) Digital coding of amino acids based on hydrophobic index. *Protein and Peptide Letters*, **14**, 871–5.
114. Shen, H.B. and Chou, K.C. (2007) PseAAC: a flexible web-server for generating various kinds of protein pseudo amino acid composition. *Analytical Biochemistry*, **373**, 386–8.
115. Chou, K.C. and Shen, H.B. (2007) MemType-2L: A Web server for predicting membrane proteins and their types by incorporating evolution information through Pse-PSSM. *Biochemical and Biophysical Research Communications*, **360**, 339–45.
116. Schaffer, A.A., Aravind, L., Madden, T.L. *et al.* (2001) Altschul, Improving the accuracy of PSI-BLAST protein database searches with composition-based statistics and other refinements. *Nucleic Acids Research*, **29**, 2994–3005.
117. Afjehi-Sadat, L. and Lubec, G. (2007) Identification of enzymes and activity from two-dimensional gel electrophoresis. *Nature Protocols*, **2**, 2318–24.
118. Chou, K.C. and Elrod, D.W. (2003) Prediction of enzyme family classes. *Journal of Proteome Research*, **2**, 183–90.
119. Chou, K.C. and Cai, Y.D. (2004) Predicting enzyme family class in a hybridization space. *Protein Science*, **13**, 2857–63.
120. Cai, C.Z., Han, L.Y., Ji, Z.L. and Chen, Y.Z. (2004) Enzyme family classification by support vector machines. *PROTEINS: Structure, Function, and Bioinformatics*, **55**, 66–76.
121. Cai, Y.D. and Chou, K.C. (2005) Predicting enzyme subclass by functional domain composition and pseudo amino acid composition. *Journal of Proteome Research*, **4**, 967–71.

122. Huang, W.L., Chen, H.M., Hwang, S.F. and Ho, S.Y. (2006) Accurate prediction of enzyme subfamily class using an adaptive fuzzy k-nearest neighbor method. *Biosystems*, **90**, 405–13.
123. Shen, H.B. and Chou, K.C. (2007) EzyPred: A top-down approach for predicting enzyme functional classes and subclasses. *Biochemical and Biophysical Research Communication*, **364**, 53–9.
124. Bairoch, A. (2000) The ENZYME Database in 2000. *Nucleic Acids Research*, **28**, 304–5.
125. Finn, R.D., Mistry, J., Schuster-Bockler, B. *et al.* (2006) Pfam: clans, web tools and services. *Nucleic Acids Research*, **34**, D247–51.
126. Barrett, A.J. and McDonald, J.K. (1986) Nomenclature: protease, proteinase and peptidase. *Biochemical Journal*, **237**, 935.
127. Puente, X.S., Sanchez, L.M., Overall, C.M. and Lopez-Otin, C. (2003) Human and mouse proteases: a comparative genomic approach. *Nature Review Genetics*, **4**, 544–8.
128. Rawlings, N.D., Tolle, D.P. and Barrett, A.J. (2004) MEROPS: the peptidase database. *Nucleic Acids Research*, **32**, D160–4.
129. Chou, J.J., Matsuo, H., Duan, H. and Wagner, G. (1998) Solution structure of the RAIDD CARD and model for CARD/CARD interaction in caspase-2 and caspase-9 recruitment. *Cell*, **94**, 171–80.
130. Qin, H., Srinvasula, S.M., Wu, G. *et al.* (1999) Structural basis of procaspase-9 recruitment by the apoptotic protease-activating factor 1. *Nature*, **399**, 549–57.
131. Chou, J.J., Li, H., Salvessen, G.S. *et al.* (1999) Solution structure of BID, an intracellular amplifier of apoptotic signalling. *Cell*, **96**, 615–24.
132. Watt, W., Koeplinger, K.A., Mildner, A.M. *et al.* (1999) The atomic resolution structure of human caspase-8, a key activator of apoptosis. *Structure*, **7**, 1135–43.
133. Chou, K.C., Tomasselli, A.G. and Heinrikson, R.L. (2000) Prediction of the tertiary structure of a caspase-9/inhibitor complex. *FEBS Letters*, **470**, 249–56.
134. Chou, K.C. and Howe, W.J. (2002) Prediction of the tertiary structure of the beta-secretase zymogen. *Biochemical and Biophysical Research Communications*, **292**, 702–8.
135. Chou, K.C., Wei, D.Q. and Zhong, W.Z. (2003) Binding mechanism of coronavirus main proteinase with ligands and its implication to drug design against SARS. *Biochemical and Biophysical Research Communications*, **308**, 148–151. (Erratum: *Biochemical and Biophysical Research Communications* (2003), **310**, 675.)
136. Chou, K.C., Wei, D.Q., Du, Q.S. *et al.* (2006) Review: Progress in computational approach to drug development against SARS. *Current Medicinal Chemistry*, **13**, 3263–70.
137. Chou, K.C. (1993) A vectorized sequence-coupling model for predicting HIV protease cleavage sites in proteins. *Journal of Biological Chemistry*, **268**, 16938–48.
138. Chou, K.C. (1996) Review: Prediction of HIV protease cleavage sites in proteins. *Analytical Biochemistry*, **233**, 1–14.
139. Chou, K.C. and Cai, Y.D. (2006) Prediction of protease types in a hybridization space. *Biochemical and Biophysical Research Communications*, **339**, 1015–20.
140. Zhou, G.P. and Cai, Y.D. (2006) Predicting protease types by hybridizing gene ontology and pseudo amino acid composition. *PROTEINS: Structure, Function, and Bioinformatics*, **63**, 681–4.
141. Ashburner, M., Ball, C.A., Blake, J.A. *et al.* (2000) Gene ontology: tool for the unification of biology. *Nature Genetics*, **25**, 25–9.
142. Chou, K.C. and Cai, Y.D. (2005) Using GO-PseAA predictor to identify membrane proteins and their types. *Biochemical and Biophysical Research Communications*, **327**, 845–7.
143. Roth, B.L., Willins, D.L. and Kroeze, W.K. (1998) G protein-coupled receptor (GPCR) trafficking in the central nervous system: relevance for drugs of abuse. *Drug and Alcohol Dependence*, **51**, 73–85.
144. Elrod, D.W. and Chou, K.C. (2002) A study on the correlation of G-protein-coupled receptor types with amino acid composition. *Protein Engineering*, **15**, 713–15.

145. Chou, K.C. and Elrod, D.W. (2002) Bioinformatical analysis of G-protein-coupled receptors. *Journal of Proteome Research*, **1**, 429–33.
146. Bhasin, M. and Raghava, G.P. (2005) GPCRsclass: a web tool for the classification of amine type of G-protein-coupled receptors. *Nucleic Acids Research*, **33**, W143–7.
147. Chou, K.C. (2005) Prediction of G-protein-coupled receptor classes. *Journal of Proteome Research*, **4**, 1413–18.
148. Wen, Z., Li, M., Li, Y. *et al.* (2006) Delaunay triangulation with partial least squares projection to latent structures: a model for G-protein coupled receptors classification and fast structure recognition. *Amino Acids*, **32**, 277–83.
149. Gao, Q.B. and Wang, Z.Z. (2006) Classification of G-protein coupled receptors at four levels. *Protein Engineering Design and Selection*, **19**, 511–16.
150. Horn, F., Weare, J., Beukers, M.W. *et al.* (1998) GPCRDB: an information system for G protein-coupled receptors. *Nucleic Acids Research*, **26**, 275–9.
151. Horn, F., Vriend, G. and Cohen, F.E. (2001) Collecting and harvesting biological data: the GPCRDB and NucleaRDB information systems. *Nucleic Acids Research*, **29**, 346–9.
152. Kutay, U., Ahnert-Hilger, G., Hartmann, E. *et al.* (1995) Transport route for synaptobrevin via a novel pathway of insertion into the endoplasmic reticulum membrane. *The EMBO Journal*, **14**, 217–23.
153. Chou, K.C. (2002) Review: Prediction of protein signal sequences. *Current Protein and Peptide Science*, **3**, 615–22.
154. Rapoport, T.A. (1992) Transport of proteins across the endoplasmic reticulum membrane. *Science*, **258**, 931–6.
155. Zheng, N. and Gierasch, L.M. (1996) Signal sequences: the same yet different. *Cell*, **86**, 849–52.
156. McGeoch, D.J. (1985) On the predictive recognition of signal peptide sequences. *Virus Research*, **3**, 271–86.
157. von Heijne, G. (1986) A new method for predicting signal sequence cleavage sites. *Nucleic Acids Research*, **14**, 4683–90.
158. Folz, R.J. and Gordon, J.I. (1987) Computer-assisted predictions of signal peptidase processing sites. *Biochemical and Biophysical Research Communications*, **146**, 870–7.
159. Ladunga, I., Czako, F., Csabai, I. and Geszti, T. (1991) Improving signal peptide prediction accuracy by simulated neural network. *Computer Applied Bioscience*, **7**, 485–7.
160. Arrigo, P., Giuliano, F., Scalia, F. *et al.* (1991) Identification of a new motif on nucleic acid sequence data using Kohonen's self-organizing map. *Computer Applied Bioscience*, **7**, 353–7.
161. Schneider, G. and Wrede, P. (1993) Signal analysis of protein targeting sequences. *Protein Sequence Data Analysis*, **5**, 227–36.
162. Nielsen, H., Engelbrecht, J., Brunak, S. and von Heijne, G. (1997) Identification of prokaryotic and eukaryotic signal peptides and prediction of their cleavage sites. *Protein Engineering*, **10**, 1–6.
163. Emanuelsson, O., Nielsen, H. and von Heijne, G. (1999) ChloroP, a neural network-based method for predicting chloroplast transit peptides and their cleavage sites. *Protein Science*, **8**, 978–84.
164. Chou, K.C. (2001) Using subsite coupling to predict signal peptides. *Protein Engineering*, **14**, 75–9.
165. Bendtsen, J.D., Nielsen, H., von Heijne, G. and Brunak, S. (2004) Improved prediction of signal peptides: SignalP 3.0. *Journal of Molecular Biology*, **340**, 783–95.
166. Hiller, K., Grote, A., Scheer, M. *et al.* (2004) PrediSi: prediction of signal peptides and their cleavage positions. *Nucleic Acids Research*, **32**, W375–9.
167. Chou, K.C. and Shen, H.B. (2007) Signal-CF: a subsite-coupled and window-fusing approach for predicting signal peptides. *Biochemical and Biophysical Research Communications*, **357**, 633–40.

168. Shen, H.B. and Chou, K.C. (2007) Signal-3L: a 3-layer approach for predicting signal peptide. *Biochemical and Biophysical Research Communications*, **363**, 297–303.
169. Althaus, I.W., Chou, J.J., Gonzales, A.J. *et al.* (1993) Steady-state kinetic studies with the non-nucleoside HIV-1 reverse transcriptase inhibitor U-87201E. *Journal of Biological Chemistry*, **268**, 6119–24.
170. Althaus, I.W., Gonzales, A.J., Chou, J.J. *et al.* (1993) The quinoline U-78036 is a potent inhibitor of HIV-1 reverse transcriptase. *Journal of Biological Chemistry*, **268**, 14875–80.
171. Althaus, I.W., Chou, J.J., Gonzales, A.J. *et al.* (1993) Kinetic studies with the nonnucleoside HIV-1 reverse transcriptase inhibitor U-88204E. *Biochemistry*, **32**, 6548–54.
172. Althaus, I.W., Chou, J.J., Gonzales, A.J. *et al.* (1994) Steady-state kinetic studies with the polysulfonate U-9843, a HIV reverse transcriptase inhibitor. *Experientia*, **50**, 23–8.
173. Althaus, I.W., Chou, J.J., Gonzales, A.J. *et al.* (1994) Kinetic studies with the non-nucleoside HIV-1 reverse transcriptase inhibitor U-90152E. *Biochemical Pharmacology*, **47**, 2017–28.
174. Althaus, I.W., Chou, K.C., Franks, K.M. *et al.* (1996) The benzylthio-pyrididine U-31,355 is a potent inhibitor of HIV-1 reverse transcriptase. *Biochemical Pharmacology*, **51**, 743–50.
175. Chou, K.C., Kezdy, F.J. and Reusser, F. (1994) Review: Steady-state inhibition kinetics of processive nucleic acid polymerases and nucleases. *Analytical Biochemistry*, **221**, 217–30.
176. McQuade, T.J., Tomasselli, A.G., Liu, L. *et al.* (1990) A synthetic HIV-1 protease inhibitor with antiviral activity arrests HIV-like particle maturation. *Science*, **247**, 454–6.
177. Meek, T.D., Lambert, D.M., Dreyer, G.B. *et al.* (1990) Inhibition of HIV-1 protease in infected T-lymphocytes by synthetic peptide analogues. *Nature*, **343**, 90–2.
178. Poorman, R.A., Tomasselli, A.G., Heinrikson, R.L. and Kezdy, F.J. (1991) A cumulative specificity model for proteases from human immunodeficiency virus types 1 and 2, inferred from statistical analysis of an extended substrate data base. *Journal of Biological Chemistry*, **266**, 14554–61.
179. Wlodawer, A. and Erickson, J.W. (1993) Structure-based inhibitors of HIV-1 protease. *Annual Review Biochemistry*, **62**, 543–85.
180. Rognvaldsson, T., You, L. and Garwicz, D. (2007) Bioinformatic approaches for modeling the substrate specificity of HIV-1 protease: an overview. *Expert Review of Molecular Diagnostics*, **7**, 435–51.
181. Liang, G.Z. and Li, S.Z. (2007) A new sequence representation as applied in better specificity elucidation for human immunodeficiency virus type 1 protease. *Biopolymers*, **88**, 401–12.
182. Barre-Sinoussi, F., Chermann, J.C., Rey, F. *et al.* (1983) Isolation of a T-lymphotropic retrovirus from a patient at risk for acquired immune deficiency syndrome (AIDS). *Science*, **220**, 868–71.
183. Gallo, R.C., Salahuddin, S.Z., Popovic, M. *et al.* (1984) Frequent detection and isolation of cytopathic retroviruses (HTLV-III) from patients with AIDS and at risk for AIDS. *Science*, **224**, 500–3.
184. Miller, M., Schneider, J., Sathyanarayana, B.K. *et al.* (1989) Structure of complex of synthetic HIV-1 protease with a substrate-based inhibitor at 2.3 A resolution. *Science*, **246**, 1149–52.
185. Schechter, I. and Berger, A. (1967) On the size of the active site in protease. I. Papain. *Biochemical and Biophysical Research Communications*, **27**, 157–62.
186. Chou, K.C., Zhang, C.T. and Kezdy, F.J. (1993) A vector approach to predicting HIV protease cleavage sites in proteins. *Proteins: Structure, Function, and Genetics*, **16**, 195–204.
187. Chou, J.J. (1993) Predicting cleavability of peptide sequences by HIV protease via correlation-angle approach. *Journal of Protein Chemistry*, **12**, 291–302.
188. Chou, K.C. and Zhang, C.T. (1993) Studies on the specificity of HIV protease: an application of Markov chain theory. *Journal of Protein Chemistry*, **12**, 709–24.
189. Zhang, C.T. and Chou, K.C. (1993) An alternate-subsite-coupled model for predicting HIV protease cleavage sites in proteins. *Protein Engineering*, **7**, 65–73.
190. Thompson, T.B., Chou, K.C., Zheng, C. (1995) Neural network prediction of the HIV-1 protease cleavage sites. *Journal of Theoretical Biology*, **177**, 369–79.

191. Chou, K.C., Tomasselli, A.L., Reardon, I.M. and Heinrikson, R.L. (1996) Predicting HIV protease cleavage sites in proteins by a discriminant function method. *PROTEINS: Structure, Function, and Genetics*, **24**, 51–72.
192. Shen, H.B. and Chou, K.C. (2008) HIVcleave: a web server for predicting HIV protease cleavage sites in proteins. *Analytical Biochemistry*, **375**, 388–90.
193. Anand, K., Ziebuhr, J., Wadhwani, P. *et al.* (2003) Coronavirus main proteinase (3CLpro) structure: basis for design of anti-SARS drugs. *Science*, **300**, 1763–7.
194. Gao, F., Ou, H.Y., Chen, L.L. *et al.* (2003) Prediction for proteinase cleavage sites in polyproteins of coronaviruses and its applications in analyzing SARS-CoV genomes. *FEBS Letters*, **553**, 451–6.
195. Sirois, S., Tsoukas, C.M., Chou, K.C. *et al.* (2005) Selection of molecular descriptors with artificial intelligence for the understanding of HIV-1 protease peptidomimetic inhibitors-activity. *Medicinal Chemistry*, **1**, 173–84.
196. Gao, W.N., Wei, D.Q., Li, Y. *et al.* (2007) Agaritine and its derivatives are potential inhibitors against HIV proteases. *Medicinal Chemistry*, **3**, 221–6.
197. Du, Q.S., Wang, S.Q., Wei, D.Q. *et al.* (2004) Polyprotein cleavage mechanism of SARS CoV Mpro and chemical modification of octapeptide. *Peptides*, **25**, 1857–64.
198. Sirois, S., Wei, D.Q., Du, Q.S. and Chou, K.C. (2004) Virtual screening for SARS-CoV protease based on KZ7088 pharmacophore points. *Journal of Chemical Information and Computer Science*, **44**, 1111–22.
199. Du, Q.S., Wang, S., Wei, D.Q. *et al.* (2005) Molecular modelling and chemical modification for finding peptide inhibitor against SARS CoV Mpro. *Analytical Biochemistry*, **337**, 262–70.
200. Du, Q.S., Wang, S.Q., Jiang, Z.Q. *et al.* (2005) Application of bioinformatics in search for cleavable peptides of SARS-CoV Mpro and chemical modification of octapeptides. *Medicinal Chemistry*, **1**, 209–13.
201 Wei, D.Q., Zhang, R., Du, Q.S. *et al.* (2006) Anti-SARS drug screening by molecular docking. *Amino Acids*, **31**, 73–80.
202. Gan, Y.R., Huang, H., Huang, Y.D. *et al.* (2006) Synthesis and activity assess of an octapeptide inhibitor designed for SARS coronavirus main proteinase. *Peptides*, **27**, 622–5.
203. Wei, D.Q., Chou, K.C., Gan, Y.R. and Du, Q.S. (2005) Patent Application No: CN 1560074A, China.
204. Zhang, R., Wei, D.Q., Du, Q.S. and Chou, K.C. (2006) Molecular modeling studies of peptide drug candidates against SARS. *Medicinal Chemistry*, **2**, 309–14.
205. Wang, S.Q., Du, Q.S., Zhao, K. *et al.* (2007) Virtual screening for finding natural inhibitor against cathepsin-L for SARS therapy. *Amino Acids*, **33**, 129–35.
206. Vollert, C.S. and Uetz, P. (2004) The phox homology (PX) domain protein interaction network in yeast. *Molecular and Cellular Proteomics*, **3**, 1053–64.
207. Chou, K.C. (2005) Coupling interaction between thromboxane A2 receptor and alpha-13 subunit of guanine nucleotide-binding protein. *Journal of Proteome Research*, **4**, 1681–6.
208. Chou, K.C., Watenpaugh, K.D. and Heinrikson, R.L. (1999) A Model of the complex between cyclin-dependent kinase 5(Cdk5) and the activation domain of neuronal Cdk5 activator. *Biochemical and Biophysical Research Communications*, **259**, 420–8.
209. Kanehisa, M., Goto, S., Kawashima, S. *et al.* (2004) The KEGG resources for deciphering the genome. *Nucleic Acids Research*, **32**, D277–80.
210. Yan, C., Dobbs, D. and Honavar, V. (2004) A two-stage classifier for identification of protein-protein interface residues. *Bioinformatics*, **20** (Suppl 1), I371–8.
211. Sprinzak, E. and Margalit, H. (2001) Correlated sequence-signatures as markers of protein-protein interaction. *Journal of Molecular Biology*, **311**, 681–92.
212. Bock, J.R. and Gough, D.A. (2001) Predicting protein-protein interactions from primary structure. *Bioinformatics*, **17**, 455–60.

213. Jansen, R., Yu, H., Greenbaum, D. *et al.* (2003) A Bayesian networks approach for predicting protein-protein interactions from genomic data. *Science*, **302**, 449–53.
214. Martin, S., Roe, D. and Faulon, J.L. (2005) Predicting protein-protein interactions using signature products. *Bioinformatics*, **21**, 218–26.
215. Ben-Hur, A. and Noble, W.S. (2005) Kernel methods for predicting protein-protein interactions. *Bioinformatics*, **21** (Suppl. 1), i38–46.
216. Chou, K.C. and Cai, Y.D. (2006) Predicting protein-protein interactions from sequences in a hybridization space. *Journal of Proteome Research*, **5**, 316–22.
217. Chou, K.C. and Cai, Y.D. (2004) Using GO-PseAA predictor to predict enzyme subclass. *Biochemical and Biophysical Research Communications*, **325**, 506–9.
218. Voet, D., Voet, J.G. and Pratt, C.W. (2002) *Fundamentals of Biochemistry*, Chap. 13, John Wiley & Sons, Inc., New York.
219. Krebs, H.A. and Johnson, W.A. (1937) The role of citric acid in intermediate metabolism in animal tissues. *Enzymologia*, **4**, 148–56.
220. Chou, K.C., Cai, Y.D. and Zhong, W.Z. (2006) Predicting networking couples for metabolic pathways of *Arabidopsis*. *EXCLI Journal*, **5**, 55–65.
221. Marchand-Geneste, N., Watson, K.A., Alsberg, B.K. and King, R.D. (2002) New approach to pharmacophore mapping and QSAR analysis using inductive logic programming. Application to thermolysin inhibitors and glycogen phosphorylase B inhibitors. *Journal of Medicinal Chemistry*, **45**, 399–409.
222. Chou, K.C., Jiang, S.P., Liu, W.M. and Fee, C.H. (1979) Graph theory of enzyme kinetics: 1. Steady-state reaction system. *Scientia Sinica*, **22**, 341–58.
223. Chou, K.C. and Forsen, S. (1980) Graphical rules for enzyme-catalyzed rate laws. *Biochemical Journal*, **187**, 829–35.
224. Chou, K.C. (1981) Two new schematic rules for rate laws of enzyme-catalyzed reactions. *Journal of Theoretical Biology*, **89**, 581–92.
225. Chou, K.C. and Forsen, S. (1981) Graphical rules of steady-state reaction systems. *Canadian Journal of Chemistry*, **59**, 737–55.
226. Chou, K.C. and Liu, W.M. (1981) Graphical rules for nonsteady state enzyme kinetics. *Journal of Theoretical Biology*, **91**, 637–54.
227. Zhou, G.P. and Deng, M.H. (1984) An extension of Chou's graphical rules for deriving enzyme kinetic equations to system involving parallel reaction pathways. *Biochemical Journal*, **222**, 169–76.
228. Chou, K.C. (1989) Graphical rules in steady and nonsteady enzyme kinetics. *Journal of Biological Chemistry*, **264**, 12074–9.
229. Kuzmic, P., Ng, K.Y. and Heath, T.D. (1992) Mixtures of tight-binding enzyme inhibitors. Kinetic analysis by a recursive rate equation. *Analytical Biochemistry*, **200**, 68–73.
230. Lin, S.X. and Neet, K.E. (1990) Demonstration of a slow conformational change in liver glucokinase by fluorescence spectroscopy. *Journal of Biological Chemistry*, **265**, 9670–5.
231. Andraos, J. (2008) Kinetic plasticity and the determination of product ratios for kinetic schemes leading to multiple products without rate laws: new methods based on directed graphs. *Canadian Journal of Chemistry*, **86**, 342–57.
232. Chou, K.C. (1990) Review: Applications of graph theory to enzyme kinetics and protein folding kinetics. Steady and nonsteady state systems. *Biophysical Chemistry*, **35**, 1–24.
233. Chou, K.C. (1993) Graphic rule for nonsteady-state enzyme kinetics and protein folding kinetics. *Journal of Mathematical Chemistry*, **12**, 97–108.
234. Chou, K.C. and Zhang, C.T. (1992) Diagrammatization of codon usage in 339 HIV proteins and its biological implication. *AIDS Research and Human Retroviruses*, **8**, 1967–6.
235. Zhang, C.T. and Chou, K.C. (1994) Analysis of codon usage in 1562 E. Coli protein coding sequences. *Journal of Molecular Biology*, **238**, 1–8.
236. Sorimachi, K. and Okayasu, T. (2008) Codon evolution is governed by linear formulas. *Amino Acids*, DOI 10.1007/s00726-00007-00024-00723.

237. Okayasu, T. and Sorimachi, K. (2008) Organisms can essentially be classified according to two codon patterns. *Amino Acids*, DOI 10.1007/s00726-00008-00059-00720.
238a. Wolfram, S. (1984) Cellular automation as models of complexity. *Nature*, **311**, 419–24.
238b. Wolfram, S. (2002) *A New Kind of Science*, Wolfram Media Inc., Champaign, IL.
239. Xiao, X., Wang, P. and Chou, K.C. (2008) GPCR-CA: A cellular automaton image approach for predicting G-protein-coupled receptor functional classes. *Journal of Computational Chemistry*, DOI 10.1002/jcc.21163.
240. Xiao, X., Shao, S., Ding, Y. *et al.* (2005) Using cellular automata to generate image representation for biological sequences. *Amino Acids*, **28**, 29–35.
241. Wang, M., Yao, J.S., Huang, Z.D. *et al.* (2005) A new nucleotide-composition based fingerprint of SARS-CoV with visualization analysis. *Medicinal Chemistry*, **1**, 39–47.
242. Diao, Y., Ma, D., Wen, Z. *et al.* (2007) Using pseudo amino acid composition to predict transmembrane regions in protein: cellular automata and Lempel-Ziv complexity. *Amino Acids*, **34**, 111–17.
243. Xiao, X., Shao, S., Ding, Y. *et al.* (2005) An application of gene comparative image for predicting the effect on replication ratio by HBV virus gene missense mutation. *Journal of Theoretical Biology*, **235**, 555–65.
244. Xiao, X., Shao, S.H. and Chou, K.C. (2006) A probability cellular automaton model for hepatitis B viral infections. *Biochemical and Biophysical Research Communications*, **342**, 605–10.
245. Qi, X.Q., Wen, J. and Qi, Z.H. (2007) New 3D graphical representation of DNA sequence based on dual nucleotides. *Journal of Theoretical Biology*, **249**, 681–90.
246. Qi, J.P., Shao, S.H., Li, D.D. and Zhou, G.P. (2007) A dynamic model for the p53 stress response networks under ion radiation. *Amino Acids*, **33**, 75–83.
247. Shikata, N., Maki, Y., Noguchi, Y. *et al.* (2007) Multi-layered network structure of amino acid (AA) metabolism characterized by each essential AA-deficient condition. *Amino Acids*, **33**, 113–21.
248. Diao, Y., Li, M., Feng, Z. *et al.* (2007) The community structure of human cellular signaling network. *Journal of Theoretical Biology*, **247**, 608–15.
249. González-Díaz, H., González-Díaz, Y., Santana, L. *et al.* (2008) Proteomics, networks, and connectivity indices. *Proteomics*, **8**, 750–78.
250. Stuart, L.M., Boulais, J., Charriere, G.M. *et al.* (2007) A systems biology analysis of the Drosophila phagosome. *Nature*, **445**, 95–101.
251. Zhang, Z. and Henzel, W.J. (2004) Signal peptide prediction based on analysis of experimentally verified cleavage sites. *Protein Science*, **13**, 2819–24.
252. Spiess, M. (1995) Heads or tails - what determines the orientation of proteins in the membrane. *FEBS Letters*, **369**, 76–9.
253. Schulz, G.E. and Schirmer, R.H. (1985) *Principles of Protein Structure*, Chapter 2, Springer-Verlag, New York, pp. 17–18.
254. Chou, K.C. and Shen, H.B. (2008) ProtIdent: A web server for identifying proteases and their types by fusing functional domain and sequential evolution information. *Biochem Biophys Res Comm*, **376**, 321–325.

5.6 Appendix A

A comparison between the predicted results by Euk-PLoc and the experimental results reported latter. Listed in column 4 are the predicted results (marked in blue); those in column 5 are the experimental results. The comments in column 6 indicate whether the proteins concerned are with single location or multiple locations; the comment content is colored in red when the prediction is inconsistent or partly inconsistent with observation. See the text for further explanation.

No.	Accession number	Subcellular location annotated in Swiss-Prot 50.7 released on 19 September 2006	Subcellular location predicted prior to experimental reports by **Euk-mPLoc** before November 2006	Subcellular location observed by experiments later and annotated in Swiss-Prot 53.2 released on 26 June 2007	Comment
1	O13674	Unknown	Cytoplasm	Cytoplasm	Single
2	O13699	Unknown	Mitochondrion	Mitochondrion	Single
3	O13715	Unknown	Cytoplasm	Cytoplasm	Single
4	O13795	Unknown	Cytoplasm; Nucleus	Cytoplasm; Nucleus	Multiple
5	O13826	Unknown	Nucleus	Nucleus	Single
6	O13859	Unknown	Cytoplasm; Nucleus	Cytoplasm; Nucleus	Multiple
7	O13894	Unknown	Nucleus	Nucleus (nucleolus)	Single
8	O14013	Unknown	Nucleus	Nucleus (nucleolus)	Single
9	O14015	Unknown	Cytoplasm; Nucleus.	Cytoplasm; Nucleus	Multiple
10	O14019	Unknown	Cytoplasm	Cytoplasm	Single
11	O14077	Unknown	Cytoplasm	Cytoplasm	Single
12	O14140	Unknown	Cytoplasm; Nucleus	Cytoplasm; Nucleus	Multiple
13	O14183	Unknown	Cytoplasm	Cytoplasm	Single
14	O14185	Unknown	Cytoplasm	Cytoplasm (localizes to the barrier septum)	Single
15	O14202	Unknown	Mitochondrion	Mitochondrion (mitochondrial inner membrane; single-pass membrane protein)	Single
16	O14216	Unknown	Cytoplasm; Nucleus	Cytoplasm; Nucleus	Multiple
17	O14235	Unknown	Mitochondrion	Mitochondrion	Single
18	O14455	Unknown	Cytoplasm	Cytoplasm	Single
19	O42654	Unknown	Cytoplasm	Cytoplasm (cell cortex)	Single
20	O42980	Unknown	Cytoplasm	Cytoplasm	Single
21	O43541	Unknown	Nucleus	Nucleus	Single
22	O47950	Unknown	Mitochondrion	Mitochondrion	Single
23	O60094	Unknown	Nucleus	Nucleus	Single
24	O74317	Unknown	Mitochondrion	Mitochondrion	Single
25	O74381	Unknown	Cytoplasm; Nucleus	Cytoplasm; Nucleus	Multiple
26	O74405	Unknown	Cytoplasm; Nucleus	Cytoplasm; Nucleus	Multiple
27	O74531	Unknown	Cytoplasm	Cytoplasm	Single
28	O74783	Unknown	Mitochondrion	Mitochondrion	Single
29	O74854	Unknown	Cytoplasm; Nucleus	Cytoplasm; Nucleus (constantly expressed throughout the cell cycle; expressed in nucleus except the nucleolus and is localized at cell tips on both sides of the septum in septated cells)	Multiple
30	O74910	Unknown	Cytoplasm; Nucleus	Cytoplasm; Nucleus	Multiple
31	O75251	Unknown	Mitochondrion	Mitochondrion	Single
32	O80448	Unknown	Cytoplasm	Cytoplasm	Single
33	O94334	Unknown	Cytoplasm	Cytoplasm	Single
34	O94435	Unknown	Cytoplasm	Cytoplasm	Single

No.	Accession number	Subcellular location annotated in Swiss-Prot 50.7 released on 19 September 2006	Subcellular location predicted prior to experimental reports by **Euk-mPLoc** before November 2006	Subcellular location observed by experiments later and annotated in Swiss-Prot 53.2 released on 26 June 2007	Comment
35	O94661	Unknown	Endoplasmic reticulum	Endoplasmic reticulum (endoplasmic reticulum membrane; single-pass membrane protein)	Single
36	O94665	Unknown	Cytoplasm; Nucleus	Cytoplasm; Nucleus	Multiple
37	O94668	Unknown	Cytoplasm; Nucleus	Cytoplasm; Nucleus	Multiple
38	P01014	Unknown	Secreted protein	Secreted protein	Single
39	P01023	Unknown	Secreted protein.	Secreted protein	Single
40	P01025	Unknown	Secreted protein	Secreted protein	Single
41	P01026	Unknown	Secreted protein	Secreted protein	Single
42	P01029	Unknown	Secreted protein	Secreted protein	Single
43	P01031	Unknown	Secreted protein	Secreted protein	Single
44	P01032	Unknown	Secreted protein	Secreted protein	Single
45	P01034	Unknown	Secreted protein	Secreted protein	Single
46	P01035	Unknown	Secreted protein	Secreted protein	Single
47	P01036	Unknown	Secreted protein	Secreted protein	Single
48	P01037	Unknown	Secreted protein	Secreted protein	Single
49	P01038	Unknown	Secreted protein	Secreted protein	Single
50	P01127	Unknown	Secreted protein	Secreted protein	Single
51	P01356	Unknown	Secreted protein	Secreted protein	Single
52	P02400	Unknown	Cytoplasm	Cytoplasm	Single
53	P02405	Unknown	Cytoplasm	Cytoplasm	Single
54	P02407	Unknown	Cytoplasm	Cytoplasm	Single
55	P02735	Unknown	Secreted protein	Secreted protein	Single
56	P02738	Unknown	Secreted protein	Secreted protein	Single
57	P02739	Unknown	Secreted protein	Secreted protein	Single
58	P02740	Unknown	Secreted protein	Secreted protein	Single
59	P03952	Unknown	Secreted protein	Secreted protein	Single
60	P04003	Unknown	Secreted protein	Secreted protein	Single
61	P04085	Unknown	Secreted protein	Secreted protein	Single
62	P04449	Unknown	Cytoplasm	Cytoplasm	Single
63	P04551	Unknown	Cytoplasm	Cytoplasm	Single
64	P05318	Unknown	Cytoplasm	Cytoplasm	Single
65	P05319	Unknown	Cytoplasm	Cytoplasm	Single
66	P05735	Unknown	Cytoplasm	Cytoplasm	Single
67	P05736	Unknown	Cytoplasm	Cytoplasm	Single
68	P05737	Unknown	Cytoplasm	Cytoplasm	Single
69	P05738	Unknown	Cytoplasm	Cytoplasm	Single
70	P05745	Unknown	Cytoplasm	Cytoplasm	Single
71	P05747	Unknown	Cytoplasm	Cytoplasm	Single
72	P05749	Unknown	Cytoplasm	Cytoplasm	Single
73	P05753	Unknown	Cytoplasm	Cytoplasm	Single
74	P06307	Unknown	Secreted protein	Secreted protein	Single
75	P06684	Unknown	Secreted protein	Secreted protein	Single
76	P06911	Unknown	Secreted protein	Secreted protein	Single
77	P07279	Unknown	Cytoplasm	Cytoplasm	Single

No.	Accession number	Subcellular location annotated in Swiss-Prot 50.7 released on 19 September 2006	Subcellular location predicted prior to experimental reports by **Euk-mPLoc** before November 2006	Subcellular location observed by experiments later and annotated in Swiss-Prot 53.2 released on 26 June 2007	Comment
78	P07280	Unknown	Cytoplasm	Cytoplasm	Single
79	P07281	Unknown	Cytoplasm	Cytoplasm	Single
80	P08607	Unknown	Secreted protein	Secreted protein	Single
81	P08621	Unknown	Nucleus	Nucleus	Single
82	P08649	Unknown	Secreted protein	Secreted protein	Single
83	P09040	Unknown	Secreted protein	Secreted protein	Single
84	P09240	Unknown	Secreted protein	Secreted protein	Single
85	P09859	Unknown	Secreted protein	Secreted protein	Single
86	P09932	Unknown	Nucleus	Nucleus	Single
87	P0C0L4	Unknown	Secreted protein	Secreted protein	Single
88	P0C0L5	Unknown	Secreted protein	Secreted protein	Single
89	P0C0T4	Unknown	Cytoplasm; Mitochondrion	Cytoplasm	Single
90	P0C0V8	Unknown	Cytoplasm	Cytoplasm	Single
91	P0C0W9	Unknown	Cytoplasm	Cytoplasm	Single
92	P0C0X0	Unknown	Cytoplasm	Cytoplasm	Single
93	P10622	Unknown	Cytoplasm	Cytoplasm	Single
94	P10664	Unknown	Cytoplasm	Cytoplasm	Single
95	P12082	Unknown	Secreted protein	Secreted protein	Single
96	P14127	Unknown	Cytoplasm	Cytoplasm	Single
97	P14272	Unknown	Secreted protein	Secreted protein	Single
98	P14605	Unknown	Cytoplasm; Membrane	Cytoplasm	Single
99	P14796	Unknown	Cytoplasm	Cytoplasm	Single
100	P14841	Unknown	Secreted protein	Secreted protein	Single
101	P15638	Unknown	Secreted protein	Secreted protein	Single
102	P17076	Unknown	Cytoplasm	Cytoplasm	Single
103	P17079	Unknown	Chloroplast; Cytoplasm	Cytoplasm	Single
104	P17157	Unknown	Centriole; Nucleus	Cytoplasm; Nucleus	Multiple
105	P17248	Unknown	Cytoplasm	Cytoplasm	Single
106	P17629	Unknown	Nucleus	Nucleus	Single
107	P19313	Unknown	Secreted protein	Secreted protein	Single
108	P19707	Unknown	Secreted protein	Secreted protein	Single
109	P19708	Unknown	Secreted protein	Secreted protein	Single
110	P19823	Unknown	Secreted protein	Secreted protein	Single
111	P19827	Unknown	Secreted protein	Secreted protein	Single
112	P20033	Unknown	Secreted protein	Secreted protein	Single
113	P20851	Unknown	Secreted protein	Secreted protein	Single
114	P21651	Unknown	Nucleus	Nucleus (localizes to chromosomes	Single
115	P22227	Unknown	Nucleus	Nucleus	Single
116	P22298	Unknown	Secreted protein	Secreted protein	Single
117	P23023	Unknown	Nucleus	Nucleus	Single
118	P23248	Unknown	Cytoplasm	Cytoplasm	Single
119	P23362	Unknown	Secreted protein	Secreted protein	Single
120	P23381	Unknown	Cytoplasm	Cytoplasm	Single

No.	Accession number	Subcellular location annotated in Swiss-Prot 50.7 released on 19 September 2006	Subcellular location predicted prior to experimental reports by **Euk-mPLoc** before November 2006	Subcellular location observed by experiments later and annotated in Swiss-Prot 53.2 released on 26 June 2007	Comment
121	P23699	Unknown	Acrosome	Secreted protein	Single
122	P24000	Unknown	Cytoplasm	Cytoplasm	Single
123	P25328	Unknown	Cytoplasm	Cytoplasm (the virus has no extracellular transmission pathway; it exists as a ribonucleoprotein viral particle in the host cytoplasm and can be transmitted through mating or cytoplasmic mixing, i.e. cytoduction)	Single
124	P25355	Unknown	Cytoplasm	Cytoplasm	Single
125	P25454	Unknown	Nucleus	Nucleus (localizes as foci on meiotic chromosomes)	Single
126	P25574	Unknown	Endoplasmic reticulum	Endoplasmic reticulum (Endoplasmic reticulum membrane; single-pass type 1 membrane protein)	Single
127	P25586	Unknown	Nucleus	Nucleus (nucleolus)	Single
128	P26262	Unknown	Secreted protein	Secreted protein	Single
129	P26781	Unknown	Cytoplasm	Cytoplasm	Single
130	P26782	Unknown	Cytoplasm; Mitochondrion	Cytoplasm	Single
131	P28325	Unknown	Secreted protein	Secreted protein	Single
132	P28576	Unknown	Secreted protein	Secreted protein	Single
133	P29453	Unknown	Cytoplasm	Cytoplasm	Single
134	P30183	Unknown	Centriole; Nucleus	Nucleus	Single
135	P31532	Unknown	Secreted protein	Secreted protein	Single
136	P32344	Unknown	Mitochondrion	Mitochondrion	Single
137	P32452	Unknown	Chloroplast; Cytoplasm	Cytoplasm	Single
138	P32769	Unknown	Cytoplasm	Cytoplasm	Single
139	P32827	Unknown	Cytoplasm	Cytoplasm	Single
140	P32841	Unknown	Cytoplasm; Nucleus	Nucleus (localizes to chromosomes)	Single
141	P32921	Unknown	Cytoplasm	Cytoplasm	Single
142	P33420	Unknown	Cytoplasm	Cytoplasm (localizes to spindle poles throughout the cell cycle)	Single
143	P33442	Unknown	Cytoplasm	Cytoplasm	Single
144	P34007	Unknown	Secreted protein	Secreted protein	Single
145	P34217	Unknown	Cytoplasm	Cytoplasm	Single

No.	Accession number	Subcellular location annotated in Swiss-Prot 50.7 released on 19 September 2006	Subcellular location predicted prior to experimental reports by **Euk-mPLoc** before November 2006	Subcellular location observed by experiments later and annotated in Swiss-Prot 53.2 released on 26 June 2007	Comment
146	P34241	Unknown	Nucleus	Nucleus (accumulates in the immediate vicinity of the dense fibrillar component of the nucleolus)	Single
147	P34544	Unknown	Nucleus	Nucleus	Single
148	P35481	Unknown	Secreted protein	Secreted protein	Single
149	P35541	Unknown	Secreted protein	Secreted protein	Single
150	P35542	Unknown	Secreted protein	Secreted protein	Single
151	P35735	Unknown	Membrane	Cell membrane (multi-pass membrane protein)	Single
152	P35997	Unknown	Cytoplasm	Cytoplasm	Single
153	P36013	Unknown	Mitochondrion	Mitochondrion (mitochondrial matrix)	Single
154	P36038	Unknown	Mitochondrion	Mitochondrion	Single
155	P36056	Unknown	Mitochondrion	Mitochondrion	Single
156	P36105	Unknown	Cytoplasm	Cytoplasm	Single
157	P36138	Unknown	Cytoplasm	Cytoplasm	Single
158	P36141	Unknown	Mitochondrion	Mitochondrion	Single
159	P38175	Unknown	Mitochondrion	Mitochondrion	Single
160	P38212	Unknown	Endoplasmic reticulum; Golgi	Endoplasmic reticulum (endoplasmic reticulum membrane; single-pass type 1 membrane protein)	Single
161	P38260	Unknown	Cytoplasm	Cytoplasm	Single
162	P38289	Unknown	Mitochondrion	Mitochondrion	Single
163	P38324	Unknown	Nucleus	Nucleus	Single
164	P38334	Unknown	Golgi	Golgi apparatus (*cis*-Golgi network)	Single
165	P38339	Unknown	Cytoplasm	Cytoplasm (bud and bud neck)	Single
166	P38344	Unknown	Cytoplasm	Cytoplasm	Single
167	P38711	Unknown	Cytoplasm	Cytoplasm	Single
168	P38754	Unknown	Cytoplasm	Cytoplasm	Single
169	P38779	Unknown	Nucleus	Nucleus (nucleolus)	Single
170	P38783	Unknown	Mitochondrion	Mitochondrion	Single
171	P38813	Unknown	Endoplasmic reticulum; Golgi	Endoplasmic reticulum (endoplasmic reticulum membrane; single-pass type 1 membrane protein)	Single
172	P38844	Unknown	Cell wall; Secreted protein	Cell wall (lipid-anchor; GPI-anchored cell wall protein	Single
173	P38961	Unknown	Nucleus	Nucleus (nucleolus)	Single

No.	Accession number	Subcellular location annotated in Swiss-Prot 50.7 released on 19 September 2006	Subcellular location predicted prior to experimental reports by **Euk-mPLoc** before November 2006	Subcellular location observed by experiments later and annotated in Swiss-Prot 53.2 released on 26 June 2007	Comment
174	P39016	Unknown	Cytoplasm	Cytoplasm	Single
175	P39729	Unknown	Cytoplasm	Cytoplasm	Single
176	P39732	Unknown	Cytoplasm	Cytoplasm	Single
177	P39741	Unknown	Cytoplasm	Cytoplasm	Single
178	P39939	Unknown	Cytoplasm	Cytoplasm	Single
179	P40005	Unknown	Cytoplasm	Cytoplasm	Single
180	P40033	Unknown	Mitochondrion	Mitochondrion	Single
181	P40048	Unknown	Cytoplasm	Cytoplasm	Single
182	P40096	Unknown	Nucleus	Nucleus	Single
183	P40186	Unknown	Cytoplasm	Cytoplasm	Single
184	P40212	Unknown	Cytoplasm	Cytoplasm	Single
185	P40213	Unknown	Cytoplasm	Cytoplasm	Single
186	P40215	Unknown	Mitochondrion	Mitochondrion (mitochondrial intermembrane space)	Single
187	P40453	Unknown	Cytoplasm; Nucleus	Cytoplasm	Single
188	P40496	Unknown	Mitochondrion	Mitochondrion	Single
189	P40525	Unknown	Cytoplasm	Cytoplasm	Single
190	P40530	Unknown	Mitochondrion	Mitochondrion (mitochondrial matrix)	Single
191	P40558	Unknown	Cytoplasm	Cytoplasm	Single
192	P40976	Unknown	Chloroplast; Cytoplasm	Cytoplasm	Single
193	P41056	Unknown	Cytoplasm	Cytoplasm	Single
194	P41057	Unknown	Cytoplasm	Cytoplasm	Single
195	P41058	Unknown	Cytoplasm	Cytoplasm	Single
196	P41229	Unknown	Nucleus	Nucleus	Single
197	P41520	Unknown	Secreted protein	Secreted protein	Single
198	P42027	Unknown	Mitochondrion	Mitochondrion	Single
199	P42028	Unknown	Mitochondrion	Mitochondrion	Single
200	P42819	Unknown	Secreted protein	Secreted protein	Single
201	P42846	Unknown	Nucleus	Nucleus; nucleolus	Single
202	P43565	Unknown	Cytoplasm; Nucleus	Cytoplasm; Nucleus	Multiple
203	P46784	Unknown	Cytoplasm; Mitochondrion	Cytoplasm	Single
204	P46990	Unknown	Chloroplast; Cytoplasm	Cytoplasm	Single
205	P46995	Unknown	Nucleus	Nucleus	Single
206	P47006	Unknown	Nucleus	Nucleus (nucleolus)	Single
207	P47025	Unknown	Mitochondrion	Mitochondrion (mitochondrial outer membrane; cytoplasmic side)	Single
208	P47076	Unknown	Nucleus	Nucleus	Single
209	P47108	Unknown	Nucleus	Nucleus (nucleolus)	Single
210	P47122	Unknown	Cytoplasm	Cytoplasm	Single
211	P47141	Unknown	Mitochondrion	Mitochondrion	Single

No.	Accession number	Subcellular location annotated in Swiss-Prot 50.7 released on 19 September 2006	Subcellular location predicted prior to experimental reports by **Euk-mPLoc** before November 2006	Subcellular location observed by experiments later and annotated in Swiss-Prot 53.2 released on 26 June 2007	Comment
212	P47150	Unknown	Mitochondrion	Mitochondrion	Single
213	P48524	Unknown	Cytoplasm	Cytoplasm	Single
214	P49166	Unknown	Cytoplasm	Cytoplasm	Single
215	P49167	Unknown	Cytoplasm	Cytoplasm	Single
216	P49591	Unknown	Cytoplasm	Cytoplasm	Single
217	P49626	Unknown	Cytoplasm	Cytoplasm	Single
218	P49631	Unknown	Cytoplasm	Cytoplasm	Single
219	P50109	Unknown	Cytoplasm	Cytoplasm	Single
220	P51401	Unknown	Cytoplasm	Cytoplasm	Single
221	P51402	Unknown	Cytoplasm	Cytoplasm	Single
222	P53030	Unknown	Cytoplasm	Cytoplasm	Single
223	P53080	Unknown	Cytoplasm	Cytoplasm	Single
224	P53088	Unknown	Mitochondrion	Mitochondrion	Single
225	P53124	Unknown	Cytoplasm	Cytoplasm	Single
226	P53188	Unknown	Nucleus	Nucleus (nucleolus)	Single
227	P53292	Unknown	Mitochondrion	Mitochondrion	Single
228	P53305	Unknown	Mitochondrion	Mitochondrion	Single
229	P53552	Unknown	Cytoplasm; Nucleus	Nucleus	Single
230	P53743	Unknown	Cytoplasm; Nucleus	Nucleus (nucleolus)	Single
231	P53890	Unknown	Cytoplasm	Cytoplasm (arrives at the bud site approximately coincident with bud emergence and dissociates from the septin scaffold before cytokinesis)	Single
232	P53908	Unknown	Chloroplast; Membrane	Membrane (multi-pass membrane protein)	Single
233	P53964	Unknown	Cytoplasm; membrane	Membrane (single-pass membrane protein)	Single
234	P54005	Unknown	Cytoplasm	Cytoplasm	Single
235	P54867	Unknown	Membrane	Cell membrane (single-pass type 1 membrane protein)	Single
236	P56628	Unknown	Cytoplasm	Cytoplasm	Single
237	P79263	Unknown	Secreted protein	Secreted protein	Single
238	P80110	Unknown	Secreted protein	Secreted protein	Single
239	P80111	Unknown	Secreted protein	Secreted protein	Single
240	P80344	Unknown	Secreted protein	Secreted protein	Single
241	P81061	Unknown	Secreted protein	Secreted protein	Single
242	P87054	Unknown	Cytoplasm; Nucleus	Cytoplasm; Nucleus	Multiple
243	P87133	Unknown	Mitochondrion	Mitochondrion	Single
244	P87262	Unknown	Cytoplasm	Cytoplasm	Single
245	P87299	Unknown	Cytoplasm	Cytoplasm	Single
246	P97278	Unknown	Secreted protein	Secreted protein	Single
247	P97279	Unknown	Secreted protein	Secreted protein	Single

No.	Accession number	Subcellular location annotated in Swiss-Prot 50.7 released on 19 September 2006	Subcellular location predicted prior to experimental reports by **Euk-mPLoc** before November 2006	Subcellular location observed by experiments later and annotated in Swiss-Prot 53.2 released on 26 June 2007	Comment
248	P97280	Unknown	Secreted protein	Secreted protein	Single
249	P97430	Unknown	Secreted protein	Secreted protein	Single
250	P98119	Unknown	Secreted protein	Secreted protein	Single
251	P98121	Unknown	Secreted protein	Secreted protein	Single
252	Q00420	Unknown	Nucleus	Nucleus	Single
253	Q01163	Unknown	Mitochondrion	Mitochondrion	Single
254	Q01448	Unknown	Nucleus	Nucleus	Single
255	Q02326	Unknown	Chloroplast; Cytoplasm	Cytoplasm	Single
256	Q02753	Unknown	Cytoplasm	Cytoplasm	Single
257	Q03213	Unknown	Nucleus	Nucleus	Single
258	Q03337	Unknown	Golgi	Golgi apparatus (*cis*-Golgi network)	Single
259	Q03758	Unknown	Cytoplasm	Cytoplasm	Single
260	Q03784	Unknown	Golgi	Golgi apparatus (*cis*-Golgi network)	Single
261	Q04231	Unknown	Centriole; Cytoplasm	Nucleus	Single
262	Q04235	Unknown	Cytoplasm	Cytoplasm	Single
263	Q04264	Unknown	Nucleus	Nucleus	Single
264	Q04806	Unknown	Cytoplasm	Cytoplasm	Single
265	Q04949	Unknown	Cytoplasm	Cytoplasm (concentrates at motile dots in the cytoplasm corresponding to the plus ends of cytoplasmic microtubules)	Single
266	Q06033	Unknown	Secreted protein	Secreted protein	Single
267	Q06078	Unknown	Nucleus	Nucleus (nucleolus)	Single
268	Q06547	Unknown	Nucleus	Nucleus	Single
269	Q07092	Unknown	Secreted protein	Secreted protein (extracellular space; extracellular matrix)	Single
270	Q09094	Unknown	Centriole; Nucleus	Nucleus	Single
271	Q09792	Unknown	Cytoplasm; Nucleus	Cytoplasm; Nucleus	Multiple
272	Q09796	Unknown	Nucleus	Nucleus (nuclear rim)	Single
273	Q09815	Unknown	Cytoplasm	Cytoplasm (septum)	Single
274	Q09855	Unknown	Cytoplasm	Cytoplasm	Single
275	Q09868	Unknown	Cytoplasm	Cytoplasm	Single
276	Q09884	Unknown	Chloroplast; Cytoplasm	Cytoplasm; Nucleus	Multiple
277	Q09902	Unknown	Cytoplasm; Nucleus.	Cytoplasm; Nucleus	Multiple
278	Q10168	Unknown	Nucleus	Nucleus (nuclear pore complex; cytoplasmic side. Nucleus; nuclear pore complex; nucleoplasmic side)	Single

No.	Accession number	Subcellular location annotated in Swiss-Prot 50.7 released on 19 September 2006	Subcellular location predicted prior to experimental reports by **Euk-mPLoc** before November 2006	Subcellular location observed by experiments later and annotated in Swiss-Prot 53.2 released on 26 June 2007	Comment
279	Q10180	Unknown	Cytoplasm	Cytoplasm (localizes to the barrier septum and cell tip)	Single
280	Q10223	Unknown	Cytoplasm; Nucleus	Cytoplasm; Nucleus (associated with vesicle-like and endoplasmic reticulum structures)	Multiple
281	Q10225	Unknown	Cytoplasm	Cytoplasm	Single
282	Q10253	Unknown	Cytoplasm	Cytoplasm	Single
283	Q10257	Unknown	Nucleus	Nucleus (nucleolus)	Single
284	Q10271	Unknown	Cytoplasm; Nucleus	Nucleus; Cytoplasm (localizes to a large number of foci in both the nucleus and cytoplasm)	Multiple
285	Q10274	Unknown	Cytoplasm; Nucleus	Nucleus	Single
286	Q10308	Unknown	Mitochondrion	Mitochondrion	Single
287	Q10326	Unknown	Cytoplasm	Cytoplasm (localizes to the barrier septum)	Single
288	Q10432	Unknown	Nucleus	Nucleus; Nucleoplasm	Single
289	Q10434	Unknown	Cytoplasm	Cytoplasm (localizes to cell tips during interphase)	Single
290	Q10447	Unknown	Cytoplasm	Cytoplasm (located at the cell tip)	Single
291	Q10474	Unknown	Cytoplasm; Nucleus	Cytoplasm; Nucleus	Multiple
292	Q12087	Unknown	Cytoplasm	Cytoplasm	Single
293	Q12213	Unknown	Chloroplast; Cytoplasm	Cytoplasm	Single
294	Q12215	Unknown	Membrane	Membrane (multipass membrane protein)	Single
294	Q12263	Unknown	Cytoplasm	Cytoplasm (bud neck)	Single
295	Q12690	Unknown	Cytoplasm	Cytoplasm	Single
296	Q14624	Unknown	Secreted protein	Secreted protein	Single
297	Q20347	Unknown	Cytoplasm	Cytoplasm	Single
298	Q22866	Unknown	Cytoplasm	Cytoplasm	Single
299	Q28065	Unknown	Secreted protein	Secreted protein	Single
300	Q28066	Unknown	Secreted protein	Secreted protein	Single
301	Q3E754	Unknown	Cytoplasm	Cytoplasm	Single
302	Q3E757	Unknown	Cytoplasm	Cytoplasm	Single
303	Q3E792	Unknown	Cytoplasm; Mitochondrion	Cytoplasm	Single
304	Q3E7X9	Unknown	Cytoplasm	Cytoplasm	Single
305	Q42577	Unknown	Mitochondrion	Mitochondrion	Single
306	Q43844	Unknown	Mitochondrion	Mitochondrion	Single
307	Q61702	Unknown	Secreted protein	Secreted protein	Single

No.	Accession number	Subcellular location annotated in Swiss-Prot 50.7 released on 19 September 2006	Subcellular location predicted prior to experimental reports by **Euk-mPLoc** before November 2006	Subcellular location observed by experiments later and annotated in Swiss-Prot 53.2 released on 26 June 2007	Comment
308	Q61703	Unknown	Secreted protein	Secreted protein	Single
309	Q61704	Unknown	Secreted protein	Secreted protein	Single
310	Q62261	Unknown	Membrane	Cell membrane (peripheral membrane protein; cytoplasmic side)	Single
311	Q63514	Unknown	Secreted protein	Secreted protein	Single
312	Q63515	Unknown	Secreted protein	Secreted protein	Single
313	Q6NS38	Unknown	Cytoplasm; Nucleus	Nucleus (detected in replication foci during s-phase)	Single
314	Q86XK2	Unknown	Nucleus	Nucleus	Single
315	Q8TCJ0	Unknown	Cytoplasm; Nucleus	Nucleus	Single
316	Q96Q83	Unknown	Cytoplasm; Nucleus	Cytoplasm; Nucleus	Multiple
317	Q96RK4	Unknown	Centriole; Cytoplasm	Centrosome (localizes to the pericentriolar region throughout the cell cycle)	Single
318	Q9C0U3	Unknown	Mitochondrion	Mitochondrion	Single
319	Q9C0W0	Unknown	Nucleus	Nucleus	Single
320	Q9C104	Unknown	Cytoplasm	Cytoplasm	Single
321	Q9C110	Unknown	Cytoplasm	Cytoplasm	Single
322	Q9DB96	Unknown	Centriole; Cytoplasm; Nucleus	Nucleus; Cytoplasm (detected in axons, dendrites and filopodia)	Multiple
323	Q9H7D7	Unknown	Cytoplasm	Cytoplasm	Single
324	Q9NR20	Unknown	Cytoplasm	Cytoplasm	Single
325	Q9P7N0	Unknown	Cytoplasm; Nucleus	Cytoplasm; Nucleus	Multiple
326	Q9UPN7	Unknown	Cytoplasm	Cytoplasm	Single
327	Q9US49	Unknown	Cytoplasm; Nucleus	Cytoplasm; Nucleus	Multiple
328	Q9USR9	Unknown	Nucleus	Nucleus (nucleoplasm)	Single
329	Q9USV4	Unknown	Cytoplasm	Cytoplasm	Single
330	Q9UT31	Unknown	Mitochondrion	Mitochondrion	Single
331	Q9UTR7	Unknown	Cytoplasm	Cytoplasm	Single
332	Q9UU87	Unknown	Cytoplasm; Nucleus	Cytoplasm; Nucleus	Multiple
333	Q9Y7V0	Unknown	Endoplasmic reticulum	Endoplasmic reticulum (endoplasmic reticulum membrane; single-pass type 2 membrane protein)	Single
334	Q9ZNR6	Unknown	Cytoplasm	Cytoplasm	Single

6

Molecular Interaction Networks: Topological and Functional Characterizations

Xiaogang Wu and Jake Y. Chen

Indiana University School of Informatics/Purdue University School of Science/ Indiana Center for Systems Biology and Personalized Medicine, Indianapolis, IN 46202, USA

In this chapter, both the basic concepts and current research trends in network biology – an emerging study of molecular interaction networks in cells [1] – are introduced. Recent breakthroughs in 'Omics' technologies [2], such as genomics, transcriptomics, metabolomics, proteomics and glycomics in biological sciences have created new computational opportunities to help researchers understand how genes, messenger RNAs (mRNAs), microRNAs (miRNAs), proteins, metabolites and chemical compounds function in the context of one another, as well as together as a whole through biological pathways. 'Omics' technologies also create massive opportunities for engineering professionals to automate the sifting and interpretation of Omics data and, therefore, to participate in postgenome biological discoveries and applications. In contrast to conventional Omics studies, which concentrate on the parallel or groupwise analysis of biomolecular structures and functions, network biology concentrates on the study of structural and functional relationships between biological molecules – for example, 'protein X binds to protein Y', 'transcription factor X activates the expressions of a group of genes, A, B, C . . .' or 'chemical compounds with a substructure feature of *f* can inhibit a subclass of protein kinases'. In spite of these important differences, network biology studies have also been provided with an 'Omics' name – Interactomics – due primarily to its large genome-scale characteristics that are similar to those of conventional 'Omics', and the high volumes of data being collected from

Automation in Proteomics and Genomics: An Engineering Case-Based Approach Edited by Gil Alterovitz, Roseann Benson and Marco Ramoni

both experimental and literature-based sources [3–5]. Whether it is referred to as network biology or as Interactomics, the study of molecular interaction networks has been crucial in determining relationships between molecular entities, understanding molecular signaling events in cells, and finding new functional insights of complex biological processes. Network biology is also an essential component of systems biology [6, 7], which aims to integrate our Omics knowledge of cells and develop coherent computational models that may be used for simulation and engineering purposes in future biomedical applications, such as diagnostics and the treatment of complex human diseases.

In the study of molecular interaction networks (or network biology), many concepts – as well as computational methods recently applied to the field – will be introduced. In this chapter, we first describe knowledge representations for biomolecular interaction networks with computer representation formats and methods, mathematical abstractions and visual layout methods, after which commonly used network properties from topological, functional and dynamical characteristics – which collectively lay the foundations for network biology research – are introduced. Finally, computational methods – from both topological and functional perspectives – that describe and predict network modules are presented. At this point, related biological pathway analysis methods will not be explored, as that would be a separate study in and of itself.

6.1 Network Representations

There are many ways to represent the molecular interaction networks and its most basic components – biological molecules and molecular interactions. Over the past two decades, the representation of biological molecules such as DNAs, genes, genomic sequences, mRNAs, small nuclear RNAs (snRNA)s), microRNAs, proteins, structures, metabolites and chemical compounds has been the main subject of bioinformatic study. During this time, bioinformatic studies have evolved and matured. For example, the computer representation of a DNA/RNA/protein molecule is also achieved with a one-dimensional sequence of strings, with each character of the string representation being a basic biochemical unit of the molecule. The structures of macromolecules and chemical compound are represented as more complex three-dimensional (3D) coordinates of each atom's position and the chemical bonds that link them. Many international databases, including GeneBank for DNA/RNA sequences, UniProt for protein sequences, PDB for protein structures and PubChem for chemical compounds, have been developed and are currently widely used. Biomolecular function annotation has been aided by the development of gene ontology (for gene functions) and, most recently, of sequence ontology (for all biomolecular functions). In this section, it is assumed that the reader is familiar with basic biological data representation schemes and, therefore, the discussion will concentrate on the representation of molecular interactions in the following sequence. First, the ontological representation, data exchange formats and current database developed for biomolecular interaction network data will be described. Second, both graph adjacency list and graph matrix abstraction of biomolecular interaction network data will be explained. Finally, visual network layout methods such as radial layout, hierarchical layout and force-directed layout and visualization software tools will be illustrated.

6.1.1 Computer Representation

6.1.1.1 Ontological Representation

To biological researchers, any molecular interaction data that merely capture a pair of interaction relationship as 'A–B' – that is, molecule A interacting with molecule B – is not useful for practical purposes. The experimental context of 'A–B' interactions is as important as the collected interaction data itself in network biology. The requirement for capturing additional experimental contexts is necessary, because molecular interaction data are often generated from multiple sources and are of different types (e.g. high-throughput yeast two-hybrid screening for binary protein interactions; immunoprecipitation-coupled mass spectrometry (MS) for protein interactions; ChIP-Chip experiments for protein–DNA interactions; text mining for molecular associations). These experimental platforms generate data of highly varying quality and coverage, making data preprocessing and integration necessary. In order to assess whether or not a particular interaction collected is biologically meaningful, one must examine all experimental details associated with the biomolecular interaction before determining the reliability of the data. This examination reduces the risk of making erroneous inference from data noise. Towards this purpose, molecular interaction (MI) ontology was created as an international effort to integrate publicly available biomedical ontology into a standardized collection of vocabulary and definitions to characterize the terms and features used in molecular interactions [8]. An example is the Ontology Lookup Service (OLS), that helps search current MI database (see Figure 6.1a). Although MI provides a good framework that describes which information must be captured for each molecular interaction data, the challenge remains as to how this information could be applied to annotate the growing accumulation of the molecular interaction data, which may be several orders of magnitude larger than conventional sequence data for the same organism.

6.1.1.2 Data Exchange Format

With the rapid accumulation of Omics data in public databases, and the accelerated need for interpreting experimental data from heterogeneous sources, there is a rising demand for developing standardizing data exchange formats in network biology. The current systems that provide integrated analyses of molecular interaction networks are still in their infancy, and to facilitate bioinformatics software programs that exchange molecular interaction data two primary data exchange format standards are proposed – the PSI-MI XML format and the MIMIx format. The PSI-MI XML, with its current XML schema (see Figure 6.1b) was developed by members of the Human Proteome Organization (HUPO) Proteomics Standards Initiatives (PSI) to describe molecular interaction information in XML formats that are compliant with the MI ontology. The members of the PSI-MI XML include major molecular interaction database developers and researchers primarily from academia – that is BIND, Cellzome, Dana Faber Cancer Institute, DIP, HPRD, IntAct, MINT and a few other protein interaction data providers such as Hybrigenics [4]. As a complementary standard, MIMIx is a structured data exchange format with the intent of providing minimal, but essential, experimental information for molecular interactions [5]. Due to the general similarity between molecular interaction data and the domain-neutral entity relationship data found in many other domains (such as social networks and computer networks), many applications (shown in Table 6.1) also support domain-neutral interaction data formats,

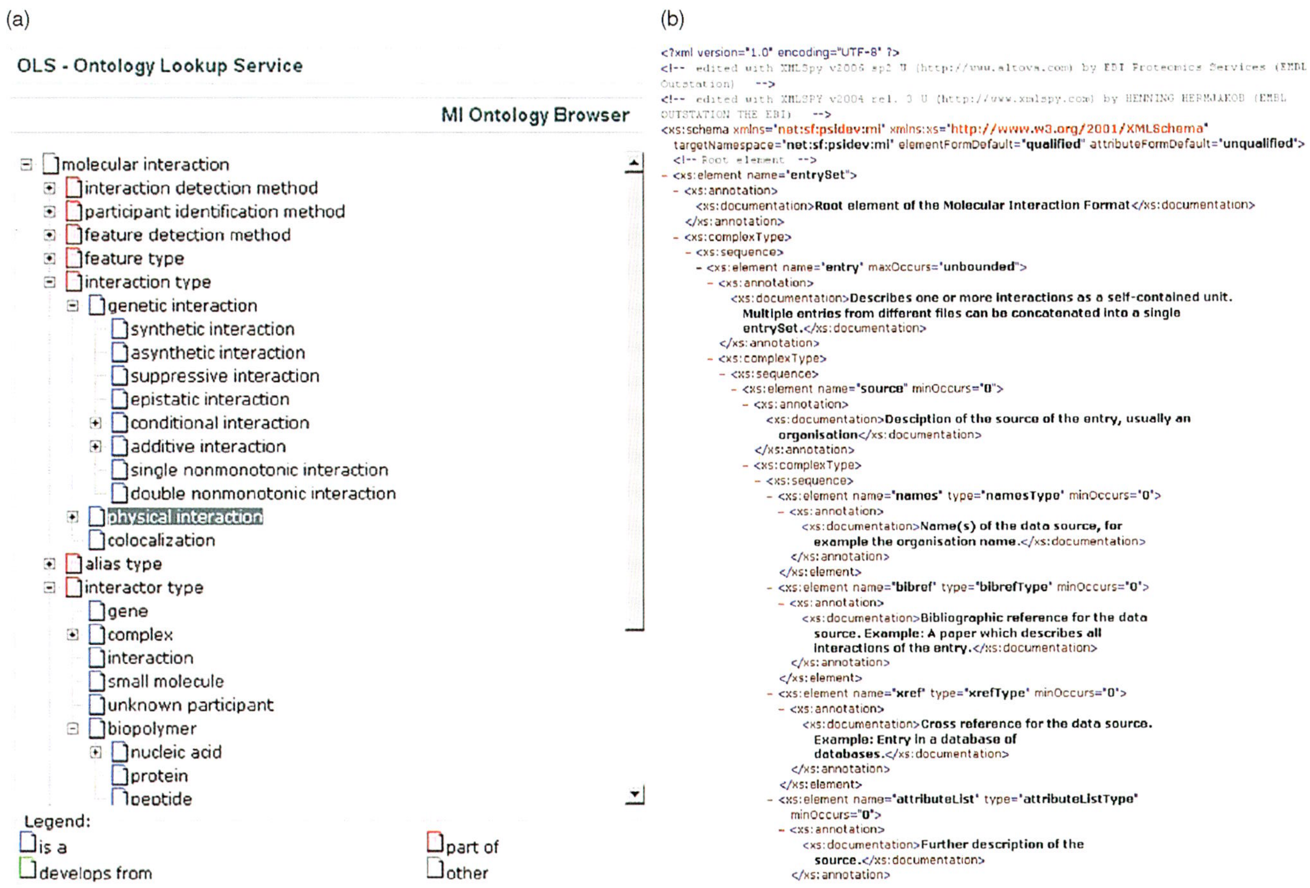

Figure 6.1 *(a) PSI (Proteomics Standards Initiative) – (MI) Molecular Interaction Ontology on OLS (Ontology Lookup Service) (http://www.ebi.ac.uk/ontology-lookup/browse.do?ontName=MI). (b) PSI-MI 2.5 XML schema (http://psidev.sourceforge.net/mi/rel25/src/MIF25.xsd)*

Table 6.1 *Molecular interaction databases with the data formats supported and the statistics (Listed by issue year)*[a]

Name	Issue year	Web site	Data format	Statistics (as of March 2008)	
				Interactor	Interaction
DIP	2000	http://dip.doe-mbi.ucla.edu/	XIN, Tab, PSI-MI 2.5	19 647	56 384
STRING	2000	http://string.embl.de/	Tab	1.5×10^6	Unknown
PIMRider	2000	http://pim.hybrigenics.com/pimriderext/common/	(Commercial)	2404	3182
BOND (BIND)	2001	http://bond.unleashedinformatics.com/	(Commercial)	>60 000 (37 291)	Unknown (63 365)
I2D (OPHID)	2001	http://128.100.65.8/ophidv2.201/index.jsp	Tab, PSI-MI 2.5	Unknown (28 442)	420 727 (161 579)
MINT	2001	http://mint.bio.uniroma2.it/mint/Welcome.do	Tab, PSI-MI 1.0, PSI-MI 2.5	28 186	103 808
HPRD	2003	http://www.hprd.org/	Tab, PSI-MI 2.5	25 611	38 167
BioGrid (GRID)	2003	http://www.thebiogrid.org/	Tab, PSI-MI 1.0, PSI-MI 2.5	529 018	203 767
IntAct	2004	http://www.ebi.ac.uk/intact/site/index.jsf	Tab, PSI-MI 1.0, PSI-MI 2.5	62 672	162 553
HPID	2004	http://wilab.inha.ac.kr/hpid/	Online Entry	6636	725 160
AANT	2004	http://aant.icmb.utexas.edu/	Online Entry	Unknown	24 331
MIPS	2005	http://mips.gsf.de/proj/ppi/	PSI-MI 2.5	>900	>1800
HAPPI	2006	http://bio.informatics.iupui.edu/HAPPI/	Tab, GML, PSI-MI 2.5	70 829	1 209 463
IntNetDB	2006	http://hanlab.genetics.ac.cn/IntNetDB.htm	Tab	9901	180 010
MiMI	2007	http://mimi.ncibi.org/MiMI/home.jsp	GML	119 880	330 152
UniHi	2007	http://theoderich.fb3.mdc-berlin.de:8080/unihi/home	Online Entry	55 942	241 900

[a]XIN: Extensible Interaction Network (http://dip.doe-mbi.ucla.edu/XML/xin.xsd); Tab: Tab-delimited plain text file format (Simple TXT file); PSI-MI 1.0/2.5: Proteomics Standards Initiative (PSI) Molecular Interaction (MI) XML Format Version 1.0 (http://psidev.sourceforge.net/mi/xml/doc/user/) or 2.5 (http://www.psidev.info/index.php?q=node/60); GML: Graph Markup Language Format (http://www.infosun.fim.uni-passau.de/Graphlet/GML/).

including GML (Graph Markup Language), XIN (Extensible Interaction Network) and Tab-delimited plain text file formats. Systems supporting domain-specific molecular interaction standard formats have the highest likelihood of adoption by future user community, as users usually prefer systems that can address specific biological questions to generic systems directly borrowed from other domains.

6.1.1.3 Databases

The experimental collection and study of protein interactions have gained momentum in recent years. As of 2006, high-throughput protein interaction mapping projects alone have generated 6000 interactions for *Saccharomyces cerevisiae* [9, 10], 1465 interactions for *Helicobacter pylori* [11], 20 405 for *Drosophila melanogaster* [12], 5500 interactions for *Caenorhabditis elegans* [13] and approximately 18 000 interactions for *Homo sapiens* [14]. In these projects, high-throughput experimental techniques, for example high-throughput yeast two-hybrid (Y2H) screenings [15], protein arrays and mass spectrometry (MS), have been developed to measure physical bindings between proteins in parallel. By combining data from existing experimental sources of more than 100 organisms, with curated interaction data from PubMed literature, the Database of Interacting Proteins (DIP) records more than 44 000 protein interactions [16]. The Biomolecular Interaction Network Database (BIND) records an even broader range of protein complex and pathway information, to reach 84 000 interactions. There are also other database development efforts similar to DIP and BIND, such as GRID, MIPS, IntAct and MINT; or those which adopt a computational inference approach, such as OPHID [17]. Most recently, highly integrated database software platforms such as UniHI have been developed to facilitate single gateway access to organism-specific protein interaction data retrieval and network analysis [18]. (For an updated comprehensive overview of these database resources, see Table 6.1. Most molecular interaction databases can also be accessed at http://www.pathguide.org.) Additional detailed information concerning these databases can be found in a recent review by Han *et al.* [19]. Also worthy of mention here is that, although many molecular interaction databases have been built, several comparisons have revealed a limited overlap between different databases, which implies that these databases still suffer from considerable poor data coverage and detection biases [20].

6.1.2 Mathematical Data Abstraction

6.1.2.1 Graph Abstraction

Collections of molecular interactions linking biological entities can be abstracted mathematically as a *graph* $G(V, E)$ [21], where V is a set of *nodes* (or *vertices*, or *points*) and E is a set of *edges* (or *links*, or *lines*). A graph abstraction is especially useful for the analysis of molecular interaction networks, which may take many different forms including protein–protein interaction networks, gene–gene coexpression networks, genetic interaction networks, molecular coannotation networks, literature co-occurrence networks and molecular entity association networks. Regardless of the form of the molecular interaction networks, graph abstraction simplifies the representation of network of interactions by reducing all data related to molecular interactions into four basic types: nodes, edges, node properties and edge properties. Depending on the particular characteristics of the underlying molecular interaction networks, different forms of graph may be used. For example, directed acyclic graphs (DAGs) – directed graphs without looping – may be used to represent gene regulatory networks, whereas Petri Nets – graphs with two different classes of nodes (one representing original biomolecular entities and the other representing transitional states of biomolecular entity complex in reaction) – may be used to represent metabolic control networks. The biggest advantage of using graphs to represent biomolecular networks is that many well-developed theorems and algorithms in *graph theory* can be readily applied to the analysis of molecular interaction networks, once the graph abstraction is made. Other basic concepts on graph representation of molecular interaction network are shown in Table 6.2.

6.1.2.2 Graph Implementation: Adjacency List

An adjacency list is an implementation of graph abstraction, and represents each set of edges connected to a node i in a graph as an unordered list of adjacent nodes $J = \{j | j \ in \ Adj(i)$, where $Adj(i)$ are the set of adjacent neighbors of $i\}$ of node i [22]. An adjacency list can be used to represent a directed graph of n nodes with an array of n lists of nodes. An undirected graph may be represented by having node j in the list for node i and node i in the list for node j [23]. A weighted graph may also be represented with a list of node/weight pairs. The biggest advantage of representing a network as an adjacency list is that *combinatorics* can be used for problem solving. For example, combinatorics can be used for analyzing the structure of regulatory interaction networks or gene regulation pathways [24, 25]; two such examples are shown in Figure 6.2a,b.

6.1.2.3 Graph Implementation: Matrix

A matrix is another implementation of graph abstraction, and represents a network as a two-dimensional (2D) adjacency matrix, in which each dimension of the matrix represents a vector of nodes and the cells in the matrix contains binary values for a Boolean network (1 for the presence of an edge between two nodes represented in each dimension of the matrix, and 0 for absence of an edge), or numeric values for a probabilistic network (numeric values representing edge weights). One can derive an adjacency matrix from adjacency lists and an adjacency list from adjacency matrix. Two examples of graph matrix implementations are shown in Figure 6.2c,d. In practice, an adjacency matrix has been used successfully to predict protein functions globally from protein–protein interaction networks [26], to correlate network data with gene expression data [27], and to perform an evolutionary analysis of functional modules in the yeast interaction networks with near-optimal efficiency [28]. The biggest advantage of representing a molecular interaction network as an adjacency matrix is that many theorem and techniques in *matrix theory* can be used for problem solving. For example, the *graph spectrum* of a network corresponds to a set of eigenvalues λ_i $(i = 1, 2, \ldots, N)$ of its adjacency matrix [29]. While the adjacency matrix of a graph depends on the node labeling, its spectrum is a graph invariant. The *spectral density* of the network can also be defined as:

$$\rho(\lambda) = \frac{1}{N} \sum_i \delta(\lambda - \lambda_i) \tag{6.1}$$

where $\delta(\,)$ is the Dirac delta function and ρ approaches a continuous function as $N \to \infty$. These features may also be useful when applied to the description of molecular interaction networks.

Table 6.2 *Basic concepts on graph representation of molecular interaction network*

Name	Description
Graph	Generally, network can be regarded as a graph $G(V, E)$. Here, V is a set of nodes (or vertices, or points) and E is a set of edges (or links, or lines).
Node	Node can denote DNA, RNA, gene, protein, peptide, small molecular or protein complex, and so on. In some molecular interaction networks, node can link itself.
Edge	Edge can denote reaction, interaction, coexpression, coannotation, association or literature co-occurrence, and so on.
Labeled	Nodes in a graph can have different labels, which may represent different subcellular localization, molecular function, biological process or other annotations.
Directed/ undirected	Edge can be undirected or directed. If all the edges of a graph are undirected, this graph is called an undirected graph. If at least one edge of a graph is directed, this graph is called a directed graph. In signaling pathways, directed edge with blunt end denotes inhibition, directed edge with pointed denotes activation and dotted directed edge with pointed end denotes causation. Directed edge also can be called arc.
Weighted/ unweighted	Edge can be unweighted or weighted. In some cases, the weight on an edge can denote the confidence score of this interaction.
Order	The number of nodes of a graph G is its order, written as $\|G\|$ and its number of edges is denoted by $\|\|G\|\|$.
Degree	The degree $d(v)$ of a node v can be defined as the number of its nearest neighbors $N(v)$, which have at least one edge linked directly to this node. If there are two or more edges between this node and one of its neighbors, it only counts one when calculating the degree according to this definition.
Average degree	Average degree is the average value of degrees of all the nodes in a graph, which is the most important concept used to compute topological properties of molecular interaction network.
Degree distribution	Degree distribution is the distribution about the occurrence number of different node degrees in a graph, usually plotted in log–log coordinates to study complex molecular interaction network.
Connectivity	If no two nodes of a graph G are separated by fewer than k other nodes, G is called k-connected. The greatest integer k such that G is k-connected is the connectivity $\kappa(G)$ of G.
Subgraph	Subgraph is a subset of certain graph. A subgraph can also be regarded as a node, in which the whole graph is called a supergraph. Subgraph can be used to describe several different concepts in molecular interaction network. In molecular biology, network module, protein complex and pathway can be all regarded as subgraphs.
Path	A path is a nonempty graph $P(V, E)$, where $V = \{v_0, v_1, \ldots, v_k\}$, $E = \{v_0v_1, v_1v_2, \ldots, v_{k-1}v_k\}$ and the nodes v_i are all distinct.
Circle	A circle is a graph $C(V, E)$, where $V = \{v_0, v_1, \ldots, v_k\}$, $E = \{v_0v_1, v_1v_2, \ldots, v_{k-1}v_k, v_kv_0\}$, $k \geq 3$ and the nodes v_i are all distinct.
Length	The length of a path is the number of edges in this path.
Shortest path	The shortest path has smallest length in all possible paths in a graph. Shortest path and shortest path length are the bases to define many other useful graph properties.

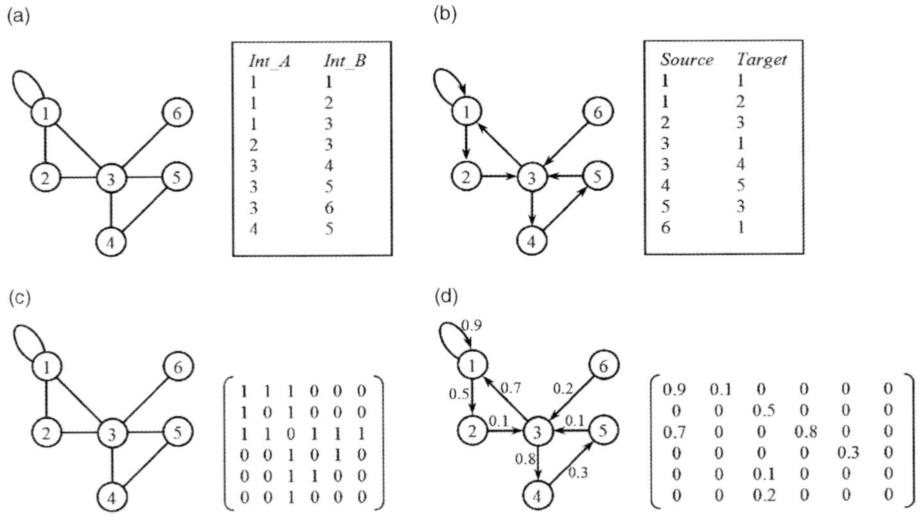

Figure 6.2 *Examples of a mathematical representation of network. (a) The adjacency list of an unweighted undirected graph; (b) The adjacency list of an unweighted directed graph; (c) The adjacency matrix of an unweighted undirected graph; (d) The adjacency matrix of a weighted directed graph*

6.1.3 Visual Representation

6.1.3.1 Radial Layout

A radial layout is a simple way of visualizing a molecular interaction network, in which all the molecules are drawn as equal size and placed at predetermined positions either along circles (2D) or on spherical surfaces (3D). Several examples of different variants of the radial layout of real networks, using proteins and protein–protein interactions derived from Alzheimer's disease as described in Ref. [30], are shown in Figure 6.3. Radial layouts allow one to highlight the highly connected parts of the network and show how they relate to the remainder of the network [31].

6.1.3.2 Hierarchy Layout

A hierarchy model is a good way of showing the hierarchical information which sometimes is hidden inside the molecular interaction network. Usually, nodes at the same hierarchy are shown on the same horizontal lines; a series of horizontal lines can be shown to indicate the existence of multiple hierarchies, so that edges are directed from nodes on lower horizontal lines to nodes on higher horizontal lines. A snapshot of the hierarchy layout of real networks, using proteins and protein–protein interactions derived from Alzheimer's disease as described in Ref. [30], is shown in Figure 6.4. Hierarchical clusters of the nodes or edges can be very useful for obtaining simplified views of large, complex networks. Schwikowski *et al.* (2000) [32] showed that some levels of visual constraints, when applied

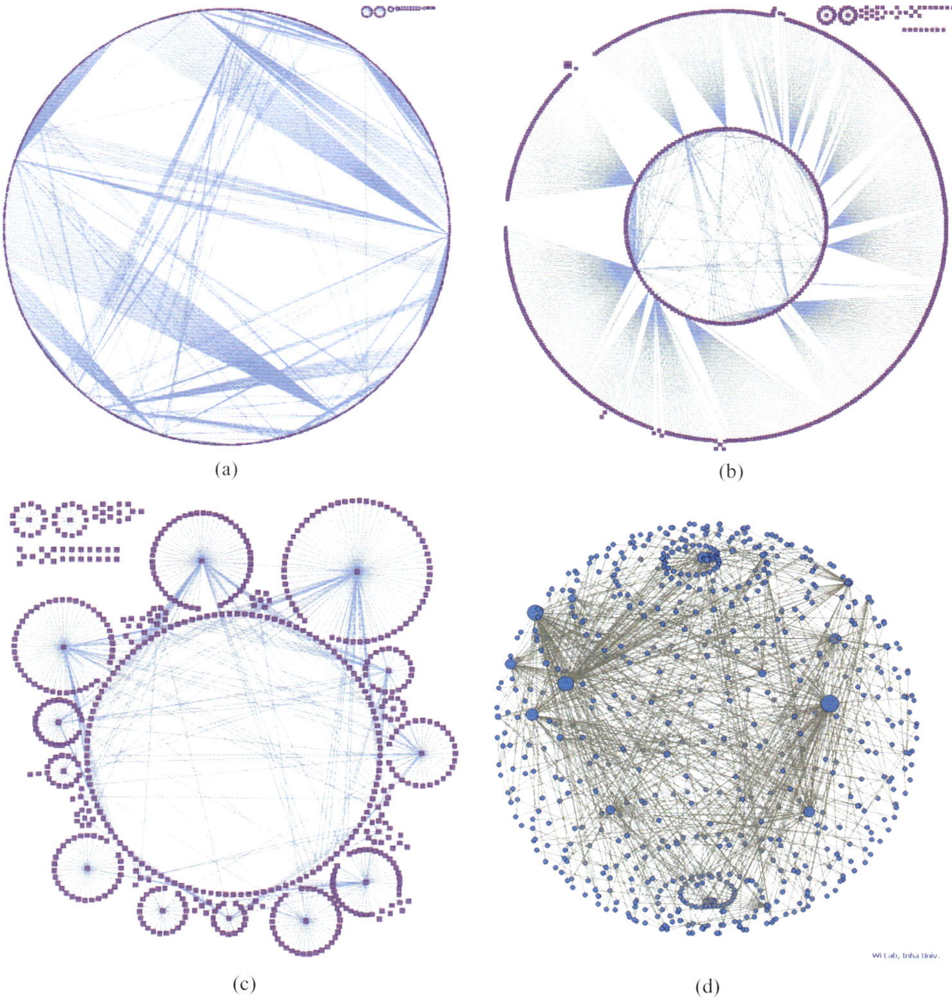

Figure 6.3 *Examples of the radial layout of an Alzheimer's disease-related protein interaction network. (a) Single circle circular layout generated by ProteoLens; (b) BBC compact circular layout generated by ProteoLens; (c) BBC isolated circular layout generated by ProteoLens; (d) Sphere layout generated by Interviewer (3D Engine).*

to a yeast protein–protein interaction network, can be quite effective in revealing network structures that previously were nonobvious [31]. Recent studies that have taken advantage of both radial and hierarchy layouts in 'hierarchical edge bundling' visualizations have been shown to further reduce visual clutter and to reveal implicit adjacency edges between parent nodes, which are the result of explicit adjacency edges between their respective child nodes [33].

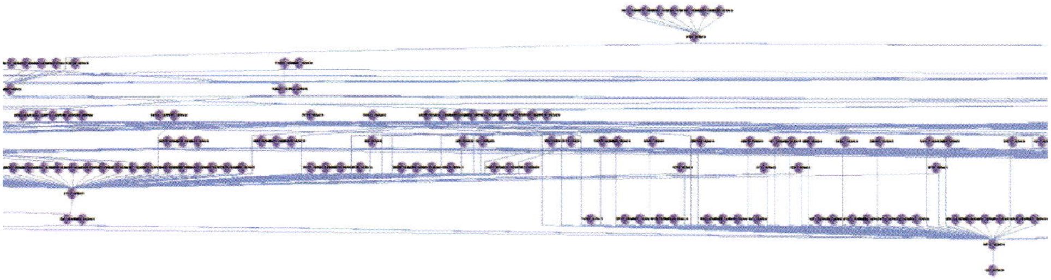

Figure 6.4 *Example of the hierarchy layout of an Alzheimer's disease-related protein interaction network (generated with the ProteoLens software tool). Only a small part of the layout is shown here (the original is too large)*

6.1.3.3 Force-Directed Layout

Force-directed layout is an effective and popular model to produce relatively good network drawings that can highlight the 'centrality' of the network. Also known as 'spring embeddings', it builds a spring force model between each pair of nodes to pull linked nodes together and push unlinked nodes apart, iteratively, until all the forces reach a mechanical equilibrium, although this may take a very long time to achieve [31]. Examples for the force-directed layout of an Alzheimer's disease-related protein interaction network is shown in Figure 6.5. Most biological network visualization tools implement a variant of the initial force-directed layout algorithms as described by Frick, Sander and Wang [34], and use either animation or resource-constrained incremental calculations to strike a balance between an optimal equilibrium and a timely layout of the network.

6.1.3.4 Visualization Software Tools

Many visualization software tools are available that can support the user visual exploration of biological networks. Examples include Cytoscape, Interviewer, Osprey, Pajek, InteractionNetwork, Patika, VisANT and, most recently, ProteoLens [35]. A detailed review that compares miscellaneous features of well-publicized tools was prepared by Suderman *et al.*, in 2007 [31]. Whilst these tools complement each other in their respective performance, user query capability, declarative query capability, flexibility and ease of integration of network biology data management, the trend of future tools is to enable not only 'dynamic' or 'integrative' aspects of the visual networks but also 'data-driven' and knowledge discovery-oriented tasks. In order to accomplish this goal, all visualization software tools require significant further development.

6.2 Network Properties

The common concepts for the study of molecular interaction networks will be introduced in this section. These concepts include network topological properties, network functional properties and network dynamical properties. In this context, the basic concepts used to

Figure 6.5 *Examples of the force-directed layout of an Alzheimer's disease-related protein interaction network. The nodes colored pink are the key proteins (seed proteins)*

characterize topological features of molecular interaction networks, which include graph diameter, clustering coefficient, small-world property, scale-free property, centrality and modularity, will be described. The basic concepts used to characterize network functions, which include molecular function annotation, subcellular localization annotation, biological process annotation, lethality/essentiality, date/party hub, protein complex and biomolecular pathways, will then be explored. Finally, a few network dynamical properties such as entropy, fractal, robustness and complexity will be introduced to provide the reader with a 'snapshot' of active research in this topic.

6.2.1 Topological Properties

6.2.1.1 Graph Diameter

Graph diameter (GD) is a basic distance-related measurement, which is often used to measure the size of a molecular interaction network represented with the graph abstraction [36, 37]. Distance here is referred to the shortest path length between two nodes in a connected graph. By definition, it is defined as the maximum distance among all the

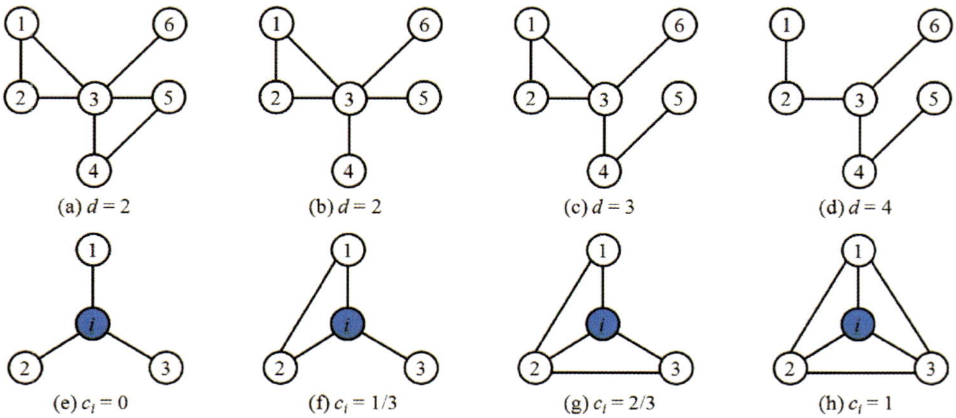

Figure 6.6 *Examples of calculating the graph diameter d of network and clustering coefficient c_i of node i*

shortest paths between all pairs of nodes in a connected graph [38], calculated with the formula below, with examples further illustrated in Figure 6.6a–d:

$$\text{GD} = \frac{\max[d(v_i, v_j)]}{N} \tag{6.2}$$

where $d(v_i, v_j)$ is the length of the shortest path between nodes i and j, and N is the number of all the nodes in a graph. The GD is calculated for all pairs (v_i, v_j), and reflects the longest path. Other size measurements have been proposed and used in molecular interaction network, including:

- The *Wiener Index* [39], which is the sum of distances between all pairs of nodes in a connected graph.
- The *average Graph Radius* [36], which is the average of distances between all pairs of nodes in a connected graph.
- The *Index of Aggregation* [30], which is defined as the ratio of the total number of nodes in the largest connected subgraph existing in a graph to the total number of nodes in the graph.
- *Graph Node Eccentricity* which, in contrast to GD, measures the greatest distance between a node and all other nodes in a connected graph. Graph radius can be defined as the minimum eccentricity of any node in the graph. Graph diameter can also be defined as the maximum eccentricity of any node in the graph [36].

6.2.1.2 Node Clustering Coefficient

The node clustering coefficient (CC) is a local density measurement of the presence of loops in a network [40]. The CC of node i is calculated using Equation 6.3, with examples illustrated in Figure 6.6e–h:

$$c_i = \frac{2l_i}{k_i(k_i - 1)} \tag{6.3}$$

Here, l_i denotes the number of edges between neighbors of node i, and k_i is the number of neighbors of node i. The concept of *clustering* (also known as *transitivity*) is widely used in nonbiological domains such as the acquaintance network, in which two individuals with a common friend are believed to have a high likelihood of knowing each other. The CC of nodes in a network, therefore, can reveal the presence or absence of a heightened number of triangles in the network [41]. In order to measure the topological property of an entire network, the average CC of the network may be calculated to gauge the tendency of proteins in a network to form clusters or groups [42]. Other variants of the concepts also exist, for example a *cyclic coefficient* that measures the cyclicity of a network [29]. Also worthy of mention here is a less well-noted concept, the *rich-club coefficient* [29], which measures the level of tendency of network hubs (highly connected nodes in a network; this will be discussed in detail later) to be connected with one another. This phenomenon, which is known as the *rich-club* [43], can also be traced back to social network studies, which reveal that influential researchers in certain scientific areas tend to form collaborative groups and to publish papers together.

6.2.1.3 Small-World

A *small-world* network, as described by Watts and Strogatz [44], refers to a network in which most nodes can be reached from other nodes by a small number of hops or steps. A small-world network is a type of *complex network* [29] with a smaller average shortest path but with a significantly greater average CC than random networks. The small-world property is commonly referred to as '*six degrees of separation*' in social networks [41]. Biomolecular interaction networks exhibit the small-world property, which may reflect an evolutionary advantage that this type of network is more robust to random attacks than are other types of network [45].

6.2.1.4 Scale-Free

The *scale-free* property implies that the development of biological networks is likely governed by robust self-organizing phenomena [46]. It is one of the key characteristics of protein–protein interaction networks, in which node degree distributions obey a *power-law* form shown in Equation 6.4:

$$P(k) = k^{-\gamma} \tag{6.4}$$

where $P(k)$ is the node degree distribution of a network and γ is the slope of the distribution under log-log scale plot. Different types of network in the real world may have different slopes, which could be used as network classifier [45]. Power-law distribution has a particular role in complex systems because of their connections to fractals and phase transitions [47]. The scale-free property, which initially was described thoroughly by Barabasi and Albert [45], has become one of the key properties of many types of *complex network* [29]. In these networks, there are a small yet significant number of 'highly connected hubs' (high degree), while there are a large number of sparingly connected proteins. Scale-free networks are characterized by self-similarity – taking a constant γ and, therefore, the same functional form at all network scales. Another important characteristic of scale-free networks is the CC distribution, which decreases as the node degree increases; this distribution also follows a power law [48]. The properties of scale-free networks suggest that the low-degree nodes form dense subnetworks, while subnetworks are connected to each other through hubs. Although the emergence of a power-law degree distribution in complex networks is interesting, the degree exponent γ is not universal and should not be used solely for the basis of classifying scale-free networks.

6.2.1.5 *Centrality*

In molecular interaction networks, the greater the number of shortest paths in which a molecule or interaction participates, the greater the importance of this molecule or interaction for the network. The property regarding the importance of a molecule or an interaction can be quantified as *betweenness centrality* [29], which is defined as:

$$\mathrm{BC}_u = \sum_{i,j} \frac{\sigma(i, u, j)}{\sigma(i, j)} \tag{6.5}$$

where $\sigma(i, u, j)$ is the number of shortest paths between nodes i and j that pass through node or edge u, $\sigma(i, j)$ is the total number of shortest paths between i and j and the sum is over all pairs i, j of distinct nodes. The betweenness centrality of a scale-free network also follows a power-law distribution, which has a more robust exponent that has been used for analyzing protein interaction and metabolic networks [49]. Other centrality measurements of a network, for example *degree centrality*, *closeness centrality*, *stress centrality*, *eigenvector centrality*, *subgraph centrality* and *graph centrality*, are beyond the scope of this chapter, although details are available in a review [50]. These topological properties have also been applied successfully to the computational analysis of protein interaction networks in tumors [37].

6.2.1.6 *Modularity*

Modular network structures, which are generally referred to as *network modularity/community* or *modularity/community*, have been found in many types of bimolecular interaction network, as well as social and computer networks [51]. Most real-world networks contain parts in which the nodes (units) are more highly connected to each other than to the rest of the network. Therefore, an intuitive test of network modularity is to compare the edge density inside the subnetwork group with the edge density outside the subnetwork group, or, whether the sum of all node degrees inside the subgraph is larger than that outside. The sets of such nodes are usually called *clusters*, *communities* or *modules*. The presence of modules in networks implies the hierarchical nature of complex systems [52]. Although many networks are found to divide naturally into communities or modules, their active detection remains an outstanding research issue in the study of networked systems [53]. Module identification in large networks is particularly useful because nodes belonging to the same module are more likely to share properties and dynamics. In addition, the number and characteristics of existing modules provide subsidies for identifying the category of a network, as well as understanding its dynamical evolution and organization of the entire network [29]. Another fundamentally related problem involves how to divide a network into its constituent modules. In real networks, the number of existing modules is usually unknown and therefore a measurement of the quality of a particular division of networks is especially important [54]. Another approach to estimating modularity is the use of *information entropy* (also called a *Network Information Bottleneck*). This approach can achieve better performance than the algorithm based on betweenness centrality [55].

6.2.2 Functional Properties

6.2.2.1 Molecular Function

Today, the functions of many genes remain uncharacterized. Although time-consuming experimental or homology-based sequence analysis techniques to characterize gene functions are still the primary techniques of choice, the increasing availability of molecular interaction network data has made it possible to predict gene functions through *guilt-by-association* [56]. By using molecular interaction networks, it is possible to understand the function of biomolecules in their functional contexts. Hence, an expanded view rather than the classic focused view of molecular functions, such as transcription factors, may be assessed both precisely and holistically. This expanded view of function may be achieved through different types of biomolecular interaction network data, including protein–protein interaction networks, gene coexpression networks, gene regulation networks and microRNA-mRNA regulation networks. When the functional links between pairs of biomolecules have been established, it is possible to begin understanding the biological connections including metabolic reactions, protein complexes or signaling cascade events between the anonymous molecule and the known molecule, or by traversing the links until significant clusters of molecules with common known functions are discovered. Despite being an emerging computational technique, the use of networks of functional linkages is expected to provide a new perspective for protein functions, and ultimately widen our understanding of the functioning of cells [57].

6.2.2.2 Subcellular Localization

The analysis of high-resolution, high-coverage molecular localization data set in the context of transcriptional, genetic and protein–protein interaction data, may help to reveal the patterns of transcriptional coregulation and provide a comprehensive view of interactions within and between organelles in eukaryotic cells [58]. For example, localization data from the green fluorescent protein (GFP) library can confirm and extend predictions based on trends within a single dataset, if proteins grouped together in a given dataset have a common localization. Interacting proteins are known to be more likely to have the same subcellular localization than proteins that do not interact. A recent study examined the human interactome for the enrichment or depletion of interactions in which both partners were localized to the same subcellular compartment. The study results showed a statistically significant enrichment of interactions for most subcellular compartments studied [59].

6.2.2.3 Biological Process

The combination of different states of biomolecules and their interactions ensures that the representation and subsequent analysis of biological processes is a clear challenge. A set of notations for a process diagram has been proposed to enhance the formality and richness of the information represented. A biological process diagram is a full state-transition-based diagram that can be translated into machine-readable forms in a straightforward way [60]. Drawing diagrams with nodes and connecting arrows is a common practice for representing interacting biomolecules and, although such diagrams are useful, the information that they contain is often imprecise, as the syntax and semantics of the symbols used are often too limited to describe the real-world biological processes unambiguously. In the real-world scenario, arrows would adopt multiple different meanings, making any correct interpretation of the diagram guesswork. For example, in a signal transduction network a directed arrow could be interpreted in four different ways: activation, translocation, dissociation of protein complex, and residue modification. Moreover, such problems would become magnified as the number of genes, proteins and their interactions was scaled up. Therefore, the most sophisticated diagrams, such as Petri Net and standard machine-readable codes such as Systems Biology Mark-up Language (SBML) (http://www.sbml.org), may represent the key to taking advantage of biological process information for subsequent computational analysis [60].

6.2.2.4 *Lethality/Essentiality*

Lethality and centrality in a scale-free biomolecular interaction network have recently been studied in a systematic manner [46]. This study revealed that highly connected proteins in the cells (e.g. yeast cells) are likely to play pivotal roles in the cell's survival. Therefore, highly connected proteins would be particularly resistant to random node removal yet be extremely sensitive to targeted manipulation, such as gene mutation, and even causing lethal phenotypes upon targeted removal. Ongoing evolutionary comparisons of large-scale biomolecular interaction networks have suggested that future systematic protein–protein interaction studies could uncover similar network topology with evolutionarily preserved essential proteins as network hubs. The correlation between the connectivity and essentiality of a given protein confirms that the robustness of a cell could also be derived from interaction organization and network topology, although individual biochemical function and genetic redundancy are still very important. A comprehensive understanding of cell dynamics and robustness would benefit from an integrated approach combining the individual and contextual properties of all constituents in complex cellular networks [46].

6.2.2.5 *Party Hub and Date Hub*

In scale-free protein interaction networks, most proteins interact with few partners, whereas a small but significant proportion of proteins – the 'hubs' – interact with many partners [46]. Two types of network hub have been described in a protein interaction network: the *party hub*, which refers to a highly connected protein that interacts with most of its partners simultaneously; and the *date hub*, which refers to a highly connected protein that binds its different partners at different times or locations [61]. The classification of network hubs into party hubs and date hubs is useful when the network topology and biological conditions for the static network structures has been determined. Additional studies of network connectivity and genetic interactions described *in vivo* support a model of organized modularity in which the date hubs organize the proteome, connecting biological processes – or modules – to each other, whereas party hubs function inside modules [61]. In contrast to previous studies which focused solely on the partners of a hub or the individual proteins around the hub, a recent investigation [62] used the network motifs concept of a hub or interactions among individual proteins, including the hub and its neighbors.

Depending on the relationship between a hub's network motifs and protein complexes, two new types of hub – *motif party hubs* and *motif date hubs* – were defined by modeling based on the original party hub and date hub concepts. The network motifs of these two types of hub display significantly different features in subcellular localizations, coexpression in microarray data, controlling topological structure of network and organizing modularity. Such different features merit ongoing research.

6.2.2.6 *Protein Complex*

Many cellular processes involving proteins are carried out by *protein complexes*. The identification and analysis of their components in protein interaction networks will provide an insight into how the ensemble of expressed proteins is organized into functional units in concert with each other [63]. The identification of protein–protein interactions often provides clues as to which sets of proteins may be involved in forming protein complexes, although such clues are often incomplete and noisy [64]. The challenges and opportunities in deciphering protein complexes lie in the development of high-throughput computational and experimental validation techniques.

6.2.2.7 *Biomolecular Pathways*

A biomolecular pathway (hereafter abbreviated as 'pathway') is a series of biochemical reactions that are linked by sharing the product of one reaction in either a reactant or an enzyme of a subsequent reaction. There are three major classes of pathway [65]:

- *Metabolic pathways* usually consist of a series of chemical reactions that provide basic biochemical functions to maintain metabolite/protein synthesis and energy metabolisms in cells.
- *Signal transduction pathways* act to send signals between cellular locations such as cell membrane to cytoplasm and from the cytoplasm to the nucleus.
- *Gene regulatory pathways* are responsible for converting genetic information into proteins (gene products) and controlling when and how genetic information is released in response to intracellular signals.

Each pathway's connections can be characterized as the collection of component molecules (DNA, genes, proteins, snRNAs, metabolites and drug compounds) and component molecule reaction/interactions. The study of biomolecular pathways is essential to both network biology and systems biology. Biomolecular pathways are normally defined according to the experimental evidence of signal transduction paths of a given biological process (e.g. insulin binding to receptors are discovered and collected). As new evidence linking biomolecules together, these pathways grow increasingly complex and are often interconnected [19]. Yet, understanding what these pathways are – and how they relate to each other – represents a major step forward for network biology to serve future systems biology. This in turn holds great future promise for the development of *in silico* models and engineering solutions for biomedical applications.

6.2.3 Dynamic Properties

6.2.3.1 Entropy

As a key concept in thermodynamics, statistical mechanics and information theory, *entropy* (also known as *information entropy* or *Shannon entropy*) is used to describe the amount of 'disorder' and information present in a dynamic system, or how much randomness is present in a signal or a process. The concept has been shown to be useful for the study of complex networks [29], including molecular interaction networks [61]. There are different definitions for the concept, including *degree distribution entropy*, *search information*, *target entropy* and *road entropy* [29]. The *entropy of the degree distribution* provides an average measurement of the heterogeneity of the network, which can be defined as:

$$H = -\sum_{k} P(k) \log P(k) \tag{6.6}$$

The maximum value of entropy is obtained for a uniform degree distribution; for this reason, complex networks – such as scale-free networks and small-world networks – may occasionally also be known as *nonuniform networks* [66].

6.2.3.2 Fractal

Evidence exists that there is a close relationship between scale-free property and *fractal* features in molecular interaction networks [67]. Fractals, which are also known as multiscale self-similarity or self-repeating patterns, refer to objects or quantities that display self-similarity in all scales. For complex small-world networks, the concept of self-similarity under a length-scale transformation is not expected, mainly because the small-world property implies that the average shortest path length of a network increases logarithmically with the number of nodes [29]. However, Song *et al.* [68] analyzed complex networks by using fractal methodologies, and verified that real complex networks may consist of self-repeating patterns on all length scales. Intuitively, this can be interpreted by the power law properties of node degree distribution and the CC distribution of a complex scale-free network. Quantitatively, this can be measured by *fractal dimensionality* [69], which can be obtained from a 'box counting method' used in fractal theory. If the network is covered with N_B boxes, and all nodes in each box are connected by a minimum distance smaller than l_B, the relationship between them is shown in Equation 6.7:

$$N_B \propto l_B^{-d_B} \quad \text{or} \quad d_B = \lim_{l_B \to 0} \frac{\ln N_B}{\ln l_B} = \lim_{l_B \to 0} \log_{l_B}^{N_B} \tag{6.7}$$

where d_B is known as the *fractal box dimension* of the network. An example of fractal network generated by self-repeating process is shown in Figure 6.7. The basic unit (motif) of this fractal network is from Figure 6.6a.

6.2.3.3 Robustness

Molecular interaction networks with scale-free or small-world properties tend to be robust against external perturbations and evolutionary innovations, which makes the network *robust* [70, 71]. The robustness of a molecular interaction networks can be related to

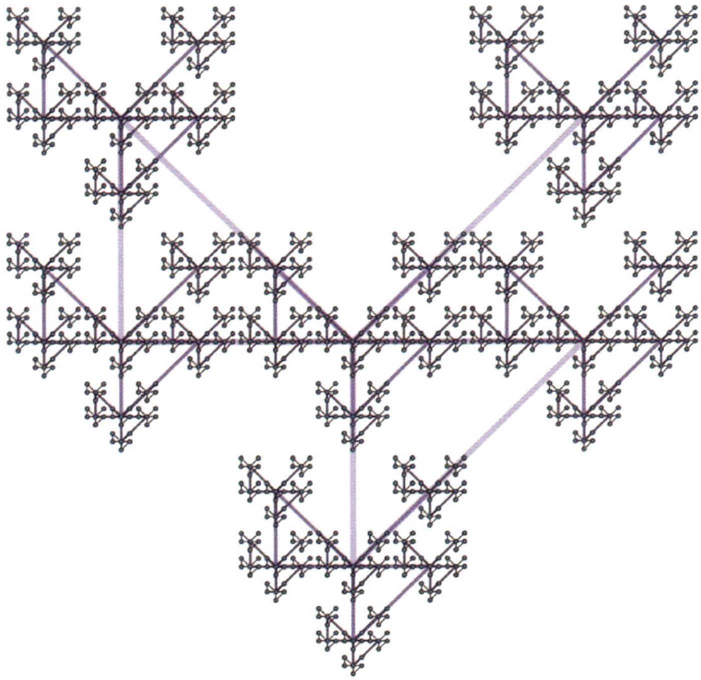

Figure 6.7 An example of a fractal network (fractal box dimension $d_B = \log(6)/\log(3) \approx 1.63$) generated by self-repeating process

network entropy – that is, their resilience to attacks [72] – and the contribution of nodes to the network entropy has been correlated with lethality in protein interactions networks [73]. The scale-free nature of networks also has important consequences for network robustness. For example, in a scale-free network, if the network nodes fail randomly then the network should fall apart only after a significant fraction (rather than a finite small fraction) of the nodes is eliminated. This robustness is accompanied by a relative vulnerability to systematic attacks through network hubs removal, unless the network hubs are replicated. For the same reason, a scale-free network is more vulnerable to virus attacks, as electronic or biological viruses spread more quickly through hubs. In real-world biomolecular interaction networks, the replication of genes/proteins encoding network hubs, and the local organization of groups of nodes into tightly connected modules, may also help to improve network robustness.

6.2.3.4 Complexity

Although the complexity of biological networks has been widely studied [36, 74, 75], the theory of complex network only addresses the emergence and structural evolution of the skeleton of a complex system, and is not a proxy for a theory of complexity. The overall behavior of a complex system is nonetheless rooted in the nature of the dynamic processes that take place [76]. The collective behavior of most processes occurring in scale-free networks is drastically different from their behavior in random or structured networks, and this provokes new thoughts on network-based dynamical processes [76]. For example, why do complex systems show patterns of organization without any central rules, and how is the emergence generated only through local rules? In order to address these questions, complex systems properties such as emergence and its behavior are currently under study [77].

6.3 Network Modules

In this section, both topological and functional network module identification techniques will be described. A biomolecular network module is a subnetwork consisting of biomolecular nodes that are highly interconnected, yet sparsely connected to the remainder of the network. Computational methods to identify/predict network modules in two types of network modules, topological modules and functional modules will be presented here. For network topological module identification, graph-based partitioning and clustering methods will be discussed, while for network functional module identification methods, to perform clustering of genes/proteins in the network based on coexpression and/or coannotation patterns to predict protein complexes and functional modules will be described.

6.3.1 Topological Module Identification

Computational methods for identifying network modules (or communities) based on topological features of the network can be categorized as two major types:

- *Graph partitioning methods*: these have been pursued primarily in computer science, with conventional applications in parallel computing, data mining and integrated circuit design. These methods aim to divide a graph into two or more large pieces, while minimizing the size of the 'interface' between them is a fundamental combinatorial problem, which is normally NP-complete.
- *Graph clustering methods*: these were initially developed by sociologists (they are also called *community detection*) but have recently been adopted by biologists, physicists and applied mathematicians to solve problems in social and biological networks [53]. Graph clustering is a commonly used computational technique to identify modules in large-scale molecular interaction networks. Similar to traditional clustering techniques, the graph clustering technique aims to group nodes into modules that are more densely connected to each other within the module than to other nodes outside the module [66]. It addresses the computational problem of how to identify the best graph node distance measure so as to group nodes into modules algorithmically. Unlike graph partitioning, in graph clustering the number and size of the groups are determined by the network topology itself, and users usually do not need to assume a good division of the network beforehand [53].

Some common computational techniques to find network modules are explored next.

6.3.1.1 Graph Cuts

An intuitive approach for graph partitioning is to look for the best *graph cuts* in the input graph. The concept of a graph cut, which can be naturally defined for directed and undirected graphs as well as weighted or unweighted graphs, refers to a partition of graph nodes into two groups by 'cutting through' edges that connect the two groups of nodes. The minimum cut (i.e. to find the smallest number of edges for such grouping) in a given graph can be found efficiently with a maximum flow algorithm [78]. *Spectral graph partitioning* is another method with conceptual simplicity and excellent overall performance [79].

6.3.1.2 *Spectral Graph Partitioning*

Spectral graph partitioning algorithm is based on spectral graph theory [80]. It tries to divide the graph using the eigenvector associated with the second smallest eigenvalue of the Laplacian matrix of a graph; hence, it is computationally demanding. A typical computational task in graph partitioning is to use a 'divide-and-conquer' strategy to optimize the parallel computational execution and to minimize interprocessor communication. Due to computational resource constraints, the number and size of partitions are usually set in advance; therefore, the goal is usually to find the optimal division of the network given a specific partitioning parameters, regardless of whether a good solution even exists [53]. Many domain-neutral software tools can be used for this purpose, including CHACO, GOBLIN, JOSTLE, LINK, METIS, PARTY and SCOTCH.

6.3.1.3 *Hierarchical Clustering*

In hierarchical clusters, the top level clusters have a hierarchical structure, each of which can consist of subclusters with additional hierarchical structures. This representation is useful in situations where the graph structure itself is hierarchical, and a single cluster can naturally be composed further to obtain a more fine-grained clustering or, alternatively, merged with another cluster to obtain a coarser division into clusters. The root cluster contains at most all of the data, and each of the leaf clusters contains at least one data element [66]. The principles used in spectral graph partitioning can also be used in graph clustering; this is known as *spectral graph clustering*, and is also very time-consuming.

6.3.1.4 *Spectral Graph Clustering*

Spectral graph clustering is typically based on computing the eigenvectors corresponding to the second-smallest eigenvalue of the normalized Laplacian, or certain eigenvectors of matrices representing the graph structure. Possible matrices include modifications of the adjacency matrix such as the transition matrix of a blind random walk on the graph. The component values of the resulting eigenvector are used as node-similarity values to determine the clustering [66, 80]. Quality measures (if feasible, the visual representation of a network) will help to determine whether there are significant clusters present in the graph, and whether a given clustering reveals them, or not. In fact, the spring-force or other energy models for network visualization can naturally achieve the goal of graph clustering, especially when the scale of a network is not too large [66].

6.3.1.5 *Electrical Circuits-Based Clustering*

Electrical circuits also provide a reasonable intuition for graph clustering. First, consider the graph as a circuit that has a unit resistor on each edge, which is called a *resistor network*. Then, calculate the potentials at all of the nodes (i.e. the voltages for all the edges), and cluster the nodes based on the potential differences [54]. However, in order to have a current (this can be seen also as a flow-based method in this sense) in the circuit, a battery must be introduced. The problem is not only the placement of the battery but also how to choose the source and the sink of the current.

6.3.1.6 Markov Clustering

Based on *flow simulation*, Dongen presented an interesting graph clustering algorithm named the Markov clustering algorithm (MCL) [81], which can be also seen as *random walk* in a graph. The components of the eigenvector corresponding to the second eigenvalue of the transition matrix of a random walk on a graph serve as 'proximity' measures for how long it takes for the walk to reach each node [66].

6.3.1.7 Agent-Based Graph Clustering

Agent-based graph clustering is very similar to the method based on flow simulation, and also can be regarded as a random walk model. However, in agent-based approaches, agents walking in a graph or network could change their populations, leave some tracks, communicate, or learn from each other. As an example, ant colony optimization (ACO), which also is known as the *ant colony algorithm*, is a dynamic stochastic searching algorithm for finding optimal paths, that is based on the behavior of ants searching for food. In an ACO-based clustering algorithm, ants roam all possible network paths iteratively. Yet, by designing various strategies of ants for each step taken to walk in a network, the iteration process can be manipulated to obtain the density distribution of ants crowding on each node. According to this density distribution, the adjacency matrix of the network with ranked nodes is shown as a map in order to reveal the system-level features of the network.

6.3.1.8 Relationship and Efficiency

Other interesting graph clustering algorithms have been found in a useful survey [66]. In general, many graph partitioning/clustering algorithms seem to be related. Spectral graph partitioning is one of the cut-based methods. Spectral graph partitioning and spectral graph clustering are both using the eigenvalue and eigenvector of the Laplacian matrix of a graph. Spectral graph clustering is related to random walk, which can model the behavior of both circuit networks and betweenness-like computations [53]. Circuits-based clustering is a special form of flow simulation, which is the basis of Markov clustering. Random walk can be also seen as a discrete flow simulation, while agent-based method can be regarded as some kind of half-intelligent half-random walk. Although further study is required, many graph-clustering algorithms are available for the general-purpose domain, including GraphClust, Graclus, CCVisu and MCL.

One straightforward way to demonstrate the efficiency of a graph-clustering result is to rerank the nodes according to the clustering result, and show the ordered adjacency matrix of the network. The results from two-dimensional (2D) hierarchical clustering and ACO-based clustering (with ant population increasing in each step) of an Alzheimer's disease-related protein interaction network are shown in Figure 6.8a and b, respectively. By comparing the two results, it can be seen that an agent-based approach can reveal clearer patterns in this type of disease-specific protein interaction network, although hierarchical clustering is very efficient for analyzing gene expression profiles.

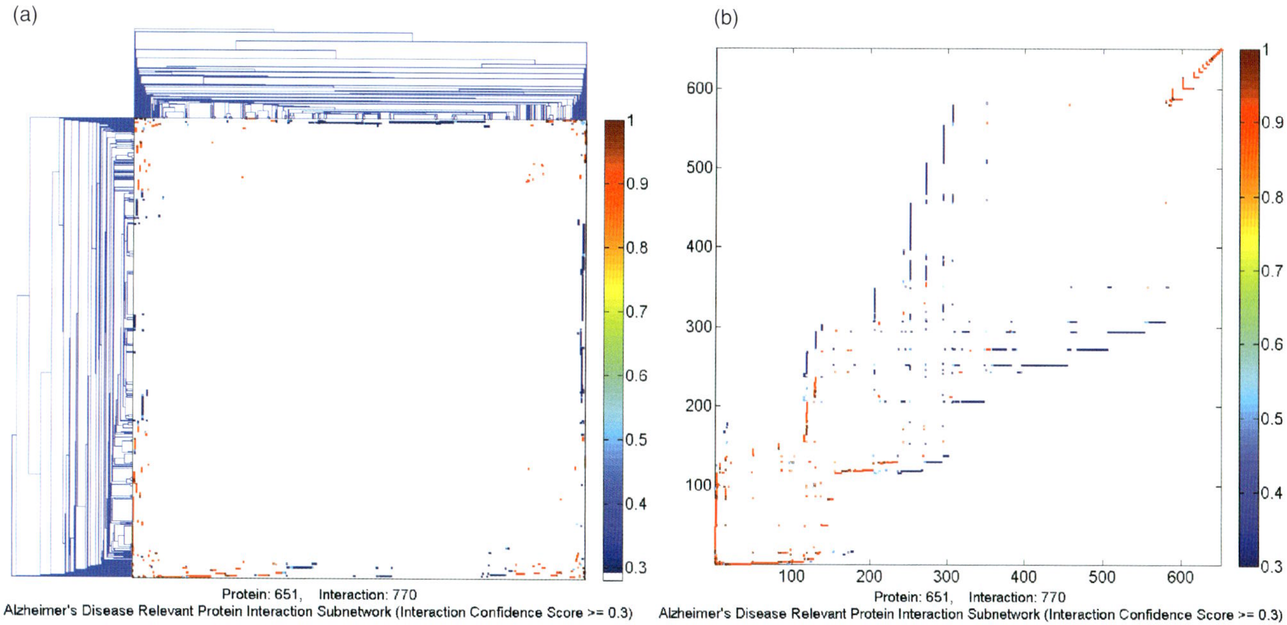

Figure 6.8 *Clustering results of the Alzheimer's disease-related protein interaction network. (a) Ordered adjacency matrix by hierarchical clustering; (b) Ordered adjacency matrix by ACO-based clustering*

6.3.2 Functional Module Prediction

Genes/gene products are organized on their functional interaction networks, which include metabolic pathways, signaling pathways and myriad biomolecular regulatory networks. Integrating the information from different types of network may lead new functional insights of biomolecular interactions network and their functional modules [82]. Towards this end, topological modules seem inefficient, as they need to be mapped to biological knowledge beyond network topological features, due to the noise and incompleteness inherent in network data. Based on coexpression and/or coannotation patterns, protein complexes (splicing machinery, transcription factors, etc.) and dynamic functional units (signaling cascades, cell-cycle regulation, etc.) can be well predicted by applying various graph clustering methods into molecular interaction networks [83]. Two examples of this technique are described in the following section.

6.3.2.1 Clustering Based on Coexpression

It remains a challenging task to interpret expression data in the context of known biomolecular interactions. Systematic general-purpose approaches that integrate different genetic information with expression profiles are required in the post-genomic era to understand the functional context of genes/proteins, to predict functional modules and to expand biological pathways [82]. The correlation of protein complexes significantly overlapped with interaction data appears to be a logical consequence of the necessity for cells to coexpress tightly interacting and functionally dependent proteins. Recent studies have shown how to combine protein interaction networks and gene expression data, including transitive coexpression data, to reveal hypothetical functional modules from independent experiments [82]. The method starts from different groups of proteins with known interacting partners, and examines whether they are also significantly related in terms of other types of experimental data. If correlations are found in a dense coexpression subnetwork, then the genes/proteins in such a subnetwork become candidate components of the functional modules. By calculating distribution of the correlation strength of all groups of gene expression profiles (nodes of the coexpression network), any module of a given size can be evaluated [82].

6.3.2.2 Clustering Based on Coannotation

By integrating information from sequence analysis and gene ontology (GO) analysis into a Bayesian inference framework, the results of a recent study have shown that functional modules can be successfully predicted [84]. This study presented a computational method for the prediction of functional modules encoded in microbial genomes. The researchers developed a formal statistical measure and used it to quantify the degree of consistency between predicted modules and known modules. They evaluated the functional relationship between two genes from three different perspectives: phylogenetic profile analysis; gene neighborhood analysis; and GO analysis. Next, they integrated three different sources of information using a Bayesian inference method, and applied the integrated information into measuring the strength of biomolecular functional relationship. Finally, predicted functional modules can be selected out by setting a certain functional relationship threshold. When the method was applied to the genome of *Escherichia coli* K12, the results showed that: (i) the predicted modules were consistent with known pathways; (ii) the neighborhood profiles or GO annotation significantly outperformed phylogenetic profiles in determining functional modules; (iii) by combining GO annotation, phylogenetic and neighborhood profile methods using Bayesian inference achieved higher degrees of consistency than single methods for known functional module predictions; (iv) potentially new interesting gene functional relationships that deserved further experimental investigations were discovered; and (v) different threshold values could be used to predict functional modules at different resolution levels. Methods in this direction are expected to play significant roles in the accurate prediction of functional annotations.

6.4 Discussion

With the advancement of both high-throughput experimental data capturing methods for identifying biomolecular interactions and computational methods for mining hidden relationships from biological literature and genomic/proteomic experimental data, there are accelerated opportunities for understanding molecular function in the complex biological context. At the same time, there have been significant challenges in using new network biology methods, including knowledge representation, concepts and analytical techniques, to unravel the complexity of biomolecular interaction networks. The end goal is for researchers to develop network biology models by moving from coarse and static protein interaction network models to refined and dynamic gene regulatory network models. Towards this goal, the concept of many methods used in network biology analysis has been introduced.

There are many potential applications of biomolecular interaction networks in translational systems biology, such as *network biomarkers* for disease molecular diagnosis and *network pharmacology* for therapeutic drug developments. By using network biology data and methods, several studies have already shown that it is possible to discover (or even to 'rediscover') candidate disease-related genes/proteins from networks implicated in a given complex condition such as the Alzheimer's disease [30, 85]. Several recent studies in network biology have also shown that network and pathway modeling might be the 'enabling technology' for identifying highly specific biomarkers in breast cancer [86, 87]. Another application is to introduce network concepts into computational pharmacology studies, including drug target identification and drug discovery [88], which attempt to gain an understanding of drug actions through various biochemical networks. With the continued discovery of new topological/functional network properties, and the development of computational tools for network biology data representation, integration, analysis and visualization, network biology will surely lead the way for ongoing systems biology studies and future-generation personalized medicine applications.

Abbreviation List

ACO: Ant Colony Optimization
CC: Clustering Coefficient
ChIP-Chip: Chromatin Immunoprecipitation on Chip
DAG: Directed Acyclic Graphs
DNA: Deoxyribonucleic Acid
GD: Graph Diameter
GFP: Green Fluorescent Protein
GML: Graph Markup Language

GO: Gene Ontology
HUPO: Human Proteome Organization
MCL: Markov Clustering
MI: Molecular Interaction
MIMIx: Minimum Information required for reporting a Molecular Interaction experiment
miRNA: micro Ribonucleic Acid
mRNA: messenger Ribonucleic Acid
MS: Mass Spectrometry
OLS: Ontology Lookup Service
PSI: Proteomics Standards Initiatives
RNA: Ribonucleic Acid
SBML: Systems Biology Mark-up Language
snRNA: small nuclear Ribonucleic Acid
XIN: eXtensible Interaction Network
XML: eXtensible Markup Language
Y2H: Yeast 2-Hybrid

References

1. Barabasi, A.L. and Oltvai, Z.N. (2004) Network biology: understanding the cell's functional organization. *Nature Reviews Genetics*, **5**(2), 101–13.
2. Palsson, B. (2002) In silico biology through "Omics"? *Nature Biotechnology*, **20**(7), 649–50.
3. Kiemer, L. and Cesareni, G. (2007) Comparative interactomics: comparing apples and pears? *Trends in Biotechnology*, **25**(10), 448–54.
4. Hermjakob, H., Montecchi-Palazzi, L., Bader, G. *et al.* (2004) The HUPO PSI's molecular interaction format – community standard for the representation of protein interaction data. *Nature Biotechnology*, **22**(2), 177–83.
5. Orchard, S., Salwinski, L., Kerrien, S. *et al.* (2007) The minimum information required for reporting a molecular interaction experiment (MIMIx). *Nature Biotechnology*, **25**, 894–8.
6. Kitano, H. (2002) Computational systems biology. *Nature*, **420**(6912), 206–10.
7. Ideker, T. (2004) Systems biology 101: What you need to know. *Nature Biotechnology*, **22**(4), 473–5.
8. Cote, R.G., Jones, P., Apweiler, R. and Hermjakob, H. (2006) The Ontology Lookup Service, a lightweight cross-platform tool for controlled vocabulary queries. *BMC Bioinformatics*, **7**(1), 97.
9. Uetz, P., Giot, L., Cagney, G. *et al.* (2000) A comprehensive analysis of protein-protein interactions in *Saccharomyces cerevisiae*. *Nature*, **403**(6770), 623–7.
10. Ito, T., Chiba, T., Ozawa, R. *et al.* (2001) A comprehensive two-hybrid analysis to explore the yeast protein interactome. *Proceedings of the National Academy of Sciences of the United States of America*, **98**(8), 4569–74.
11. Rain, J.C., Selig, L., De Reuse, H. *et al.* (2001) The protein-protein interaction map of *Helicobacter pylori*. *Nature*, **409**(6817), 211–15.
12. Giot, L., Bader, J.S., Brouwer, C. *et al.* (2003) A protein interaction map of *Drosophila melanogaster*. *Science*, **302**(5651), 1727–36.
13. Li, S., Armstrong, C.M., Bertin, N. *et al.* (2004) A map of the interactome network of the metazoan *C. elegans*. *Science*, **303**(5657), 540–3.
14. Myriad Genetics Pronet (2003) Web Site, http://www.myriad-pronet.com/. (last accessed January 2003); Available from: http://www.myriad-pronet.com/

15. Bartel, P. and Fields, S. (eds) (1997) The yeast two-hybrid system, in *Advances in Molecular Biology*, Oxford University Press.
16. Xenarios, I., Salwinski, L., Duan, X.J. *et al.* (2002) DIP, the Database of Interacting Proteins: a research tool for studying cellular networks of protein interactions. *Nucleic Acids Research*, **30**(1), 303–5.
17. Brown, K.R. and Jurisica, I. (2005) Online predicted human interaction database. *Bioinformatics*, **21**(9), 2076–82.
18. Chaurasia, G., Iqbal, Y., Hanig, C. *et al.* (2007) UniHI: an entry gate to the human protein interactome. *Nucleic Acids Research*, **35** (Database issue), D590.
19. Han, J.D.J. (2008) Understanding biological functions through molecular networks. *Cell Research*, **18**, 224–37.
20. Futschik, M.E., Chaurasia, G. and Herzel, H. (2007) Comparison of human protein-protein interaction maps. *Bioinformatics*, **23**(5), 605–11.
21. Diestel, R. (2005) *Graph Theory (3rd Version)*, Springer, New York.
22. Cormen, T.H., Leiserson, C.E., Rivest, R.L. and Stein, C. (2001) *Introduction to Algorithms*, 2nd edn, MIT Press and McGraw-Hill, pp. 527–29 of section 22.1: Representations of graphs. ISBN 0-262-03293-7.
23. Bockholt, B. and Black, P.E. (2004) Adjacency-list representation, in *Dictionary of Algorithms and Data Structures* [online] (ed. P.E. Black), U.S. National Institute of Standards and Technology, 17 December 2004. (accessed TODAY) Available from: http://www.nist.gov/dads/HTML/adjacencyListRep.html.
24. Wagner, A. (2001) How to reconstruct a large genetic network from *n* gene perturbations in fewer than *n* (2) easy steps. *Bioinformatics*, **17**(12), 1183–97.
25. Krishnamurthy, L., Nadeau, J., Ozsoyoglu, G. *et al.* (2003) Pathways database system: an integrated system for biological pathways. *Bioinformatics*, **19**(8), 930–7.
26. Vazquez, A., Flammini, A., Maritan, A. and Vespignani, A. (2003) Global protein function prediction from protein-protein interaction networks. *Nature Biotechnology*, **21**(6), 697–700.
27. Lappe, M. and Holm, L. (2004) Unraveling protein interaction networks with near-optimal efficiency. *Nature Biotechnology*, **22**(1), 98–103.
28. Wuchty, S., Oltvai, Z.N. and Barabasi, A.L. (2003) Evolutionary conservation of motif constituents in the yeast protein interaction network. *Nature Genetics*, **35**(2), 176–9.
29. Costa, L.D.F., Rodrigues, F.A., Travieso, G. and Villas Boas, P.R. (2007) Characterization of complex networks: A survey of measurements. *Advances in Physics*, **56**(1–2), 167–242.
30. Chen, J.Y., Shen, C. and Sivachenko, A.Y. (2006) Mining Alzheimer disease relevant proteins from integrated protein interactome data. Biocomputing 2007-Proceedings of the Pacific Symposium, Vol. **11**, pp. 367–78.
31. Suderman, M. and Hallett, M. (2007) Tools for visually exploring biological networks. *Bioinformatics*, **23**(20), 2651.
32. Schwikowski, B., Uetz, P. and Fields, S. (2000) A network of protein- protein interactions in yeast. *Nature Biotechnology*, **18**, 1257–61.
33. Holten, D. (2006) Hierarchical Edge Bundles: visualization of adjacency relations in hierarchical data. *IEEE Transactions on Visualization and Computer Graphics*, **12**(5), 741–8.
34. Frick, A., Sander, G. and Wang, K. (1999) Simulating graphs as physical systems: a spring-embedder system for force-directed layout. *Dr. Dobb's Journal*, **24**(8), 1–7 (http://www.ddj.com/architect/184411016)
35. Huan, T., Sivachenko, A., Harrison, S. and Chen, J.Y. (2008) ProteoLens: a visual analytic tool for multi-scale database-driven biological network data mining. *BMC Bioinformatics*, **9**, S5.
36. Bonchev, D. (2004) Complexity analysis of yeast proteome network. *Chemistry and Biodiversity*, **1**(2), 312–26.
37. Platzer, A., Perco, P., Lukas, A. and Mayer, B. (2007) Characterization of protein interaction networks in tumors. *BMC Bioinformatics*, **8**(1), 224.

38. Chin, C.S. and Samanta, M.P. (2003) Global snapshot of a protein interaction network – percolation based approach. *Bioinformatics*, **19**(18), 2413–19.
39. Mohar, B. and Pisanski, T. (1988) How to compute the Wiener index of a graph. *Journal of Mathematical Chemistry*, **2**(3), 267–77.
40. Newman, M.E.J. (2003) The structure and function of complex networks. *SIAM Review*, **45**(2), 167–256.
41. Albert, R. and Barabasi, A.L. (2002) Statistical mechanics of complex networks. *Reviews of Modern Physics*, **74**(1), 47–97.
42. Stelzl, U., Worm, U., Lalowski, M. *et al.* (2005) A human protein-protein interaction network: A resource for annotating the proteome. *Cell*, **122**(6), 957–68.
43. Colizza, V., Flammini, A., Serrano, M.A. and Vespignani, A. (2006) Detecting rich-club ordering in complex networks. *Nature Physics*, **2**(2), 110–15.
44. Watts, D.J. and Strogatz, S.H. (1998) Collective dynamics of 'small-world' networks. *Nature*, **393**(6684), 409–10.
45. Barabasi, A.L. and Albert, R. (1999) Emergence of scaling in random networks. *Science*, **286**(5439), 509.
46. Jeong, H., Mason, S.P., Barabasi, A.L. and Oltvai, Z.N. (2001) Lethality and centrality in protein networks. *Nature*, **411**(6833), 41–2.
47. Strogatz, S.H. (2001) Exploring complex networks. *Nature*, **410**, 268–76.
48. Szabo, G., Alava, M. and Kertesz, J. (2003) Structural transitions in scale-free networks. *Physical Review E*, **67**(5), 56102.
49. Goh, K.I., Oh, E., Jeong, H. *et al.* (2002) Classification of scale-free networks. *Proceedings of the National Academy of Sciences of the USA*, **99**(20), 12583–8.
50. Koschützki, D., Lehmann, K.A., Peeters, L. *et al.* (2005) Centrality indices. *Lecture Notes in Computer Science*, **3418**, 16–61.
51. Schlosser, G. and Wagner, G.P. (2004) *Modularity in Development and Evolution*, University of Chicago Press.
52. Palla, G., Derenyi, I., Farkas, I. and Vicsek, T. (2005) Uncovering the overlapping community structure of complex networks in nature and society. *Nature*, **435**(7043), 814–18.
53. Newman, M.E.J. (2006) Modularity and community structure in networks. *Proceedings of the National Academy of Sciences of the United States of America*, **103**(23), 8577–82.
54. Newman, M.E.J. and Girvan, M. (2004) Finding and evaluating community structure in networks. *Physical Review E*, **69**(2), 26113.
55. Ziv, E., Middendorf, M. and Wiggins, C.H. (2005) Information-theoretic approach to network modularity. *Physical Review E*, **71**(4), 46117.
56. Hu, P., Bader, G., Wigle, D.A. and Emili, A. (2007) Computational prediction of cancer-gene function. *Nature Reviews Cancer*, **7**, 23–34.
57. Eisenberg, D., Marcotte, E.M., Xenarios, I. and Yeates, T.O. (2000) Progress protein function in the postgenomic era. *Nature*, **405**, 823–6.
58. Huh, W.K., Falvo, J.V., Gerke, L.C. *et al.* (2003) Global analysis of protein localization in budding yeast. *Nature*, **425**(6959), 686–91.
59. Gandhi, T.K.B., Zhong, J., Mathivanan, S. *et al.* (2006) Analysis of the human protein interactome and comparison with yeast, worm and fly interaction datasets. *Nature Genetics*, **38**, 285–93.
60. Kitano, H., Funahashi, A., Matsuoka, Y. and Oda, K. (2005) Using process diagrams for the graphical representation of biological networks. *Nature Biotechnology*, **23**, 961–6.
61. Han, J.D.J., Bertin, N., Hao, T. *et al.* (2004) Evidence for dynamically organized modularity in the yeast protein-protein interaction network. *Nature*, **430**(6995), 88–93.
62. Jin, G., Zhang, S., Zhang, X.S. and Chen, L. (2007) Hubs with network motifs organize modularity dynamically in the protein-protein interaction network of yeast. *PLoS ONE*, **2**(11), e-1207.
63. Gavin, A.C., Boesche, M., Krause, R. *et al.* (2002) Functional organization of the yeast proteome by systematic analysis of protein complexes. *Nature*, **415**(6868), 141–7.

64. Krogan, N.J., Cagney, G., Yu, H. *et al.* (2006) Global landscape of protein complexes in the yeast Saccharomyces cerevisiae. *Nature*, **440**, 637–43.
65. Cary, M.P., Bader, G.D. and Sander, C. (2005) Pathway information for systems biology. *FEBS Letters*, **579**(8), 1815–20.
66. Schaeffer, S.E. (2007) Graph clustering. *Computer Science Review*, **1**(1), 27–64.
67. Song, C., Havlin, S. and Makse, H.A. (2006) Origins of fractality in the growth of complex networks. *Nature Physics*, **2**(4), 275–81.
68. Song, C., Havlin, S. and Makse, H.A. (2005) Self-similarity of complex networks. *Nature*, **433**(7024), 392–5.
69. Goh, K.I., Salvi, G., Kahng, B. and Kim, D. (2006) Skeleton and fractal scaling in complex networks. *Physical Review Letters*, **96**(1), 18701.
70. Kitano, H. (2004) Biological robustness. *Nature Reviews Genetics*, **5**(11), 826–37.
71. Ciliberti, S., Martin, O.C. and Wagner, A. (2007) Innovation and robustness in complex regulatory gene networks. *Proceedings of the National Academy of Sciences of the United States of America*, **104**(34), 13591.
72. Wang, B., Tang, H., Guo, C. and Xiu, Z. (2006) Entropy optimization of scale-free networks' robustness to random failures. *Physica A: Statistical Mechanics and its Applications*, **363**(2), 591–6.
73. Demetrius, L. and Manke, T. (2005) Robustness and network evolution-An entropic principle. *Physica A: Statistical Mechanics and its Applications*, **346**(3-4), 682–96.
74. Claussen, J.C. (2007) Offdiagonal complexity: A computationally quick complexity measure for graphs and networks. *Physica A: Statistical Mechanics and its Applications*, **375**(1), 365–73.
75. Neutel, A.M., Heesterbeek, J.A., van de Koppel, J. *et al.* (2007) Reconciling complexity with stability in naturally assembling food webs. *Nature*, **449**(7162), 599–602.
76. Barabasi, A.L. (2005) Taming complexity. *Nature Physics*, **1**(2), 68–70.
77. Manrubia, S.C., Mikhailov, A.S. and Zanette, D. (2004) Emergence of dynamical order: synchronization phenomena in complex systems. *World Scientific*, Singapore.
78. Feige, U., Peleg, D. and Kortsarz, G. (2001) The dense k-subgraph problem. *Algorithmica*, **29**(3), 410–21.
79. Elsner, U. (1997) *Graph partitioning: A survey. Technical Report, Preprint SFB 393/97-27, Technische Universitat Chemnitz, Chemnitz, Germany.*
80. Chung, F.R.K. (1997) *Spectral Graph Theory*, American Mathematical Society.
81. van Dongen, S.M. (2000) Graph clustering by flow simulation, Ph.D. Thesis, Universiteit Utrecht, Utrecht, The Netherlands.
82. Tornow, S., Mewes, H.W. and Journals, O. (2003) Functional modules by relating protein interaction networks and gene expression. *Nucleic Acids Research*, **31**(21), 6283–9.
83. Spirin, V. and Mirny, L.A. (2003) Protein complexes and functional modules in molecular networks. *Proceedings of the National Academy of Sciences of the United States of America*, **100**(21), 12123.
84. Wu, H., Su, Z., Mao, F. *et al.* (2005) Prediction of functional modules based on comparative genome analysis and Gene Ontology application. *Nucleic Acids Research*, **33**(9), 2822–37.
85. Morrison, J.L., Breitling, R., Higham, D.J. and Gilbert, D.R. (2005) GeneRank: Using search engine technology for the analysis of microarray experiments. *BMC Bioinformatics*, **6**(1), 233.
86. Pujana, M.A., Han, J.D.J., Starita, L.M. *et al.* (2007) Network modeling links breast cancer susceptibility and centrosome dysfunction. *Nature Genetics*, **39**, 1338–9.
87. Chuang, H.Y., Lee, E., Liu, Y.T. *et al.* (2007) Network-based classification of breast cancer metastasis. *Molecular Systems Biology*, **3**, 140–9.
88. Hopkins, A.L. (2007) Network pharmacology. *Nature Biotechnology*, **25**, 1110–11.

Section 3
Design via Automation

7
DNA Synthesis

Jingdong Tian
Department of Biomedical Engineering & Institute for Genome Sciences and Policy, Duke University, USA

7.1 Design Methods

7.1.1 Chemical Synthesis of Oligodeoxynucleotides

7.1.1.1 The Standard Phosphoramidite Chemistry

Chemical oligodeoxynucleotide synthesis is a cyclical process that elongates a chain of nucleotides from the 3′-end to the 5′-end. Currently, the phosphoramidite four-step process, which was developed during the early 1980s, is the method of choice and is used by all commercial DNA synthesizers [1–3]. This process couples an acid-activated deoxynucleoside phosphoramidite to a deoxynucleoside on a solid support.

A phosphoramidite is a nucleotide monomer that is fully protected at all reactive positions on the ribose sugar, the phosphate group and the base. These reactive groups interfere with the phosphate trimester reactions used to couple the nucleotide monomers, and should be carefully blocked. The 5′ hydroxyl (5-OH) group of the ribose sugar is protected with a dimethoxytrityl (DMT) ether moiety, which is removed by the action of a mild acid at the start of each coupling cycle. The phosphate oxygen is usually protected by diisopropylamine (iPr2N) and β-cyanoethoxy groups; all reactive sites on the bases are also protected. The common protecting groups for the exocyclic amine are *N*-benzoyl on deoxyadenosine (dA) and deoxycytidine (dC), and *N*-isobutyryl on deoxyguanosine (dG). The deoxythymidine (dT) base does not need any protecting group. Other base-protecting groups include *N*-acetyl dC ('Fast C') or *N*-dmf dG ('Fast G'). Upon completion of the synthesis cycles, the remaining protecting groups are easily removed to yield almost lesion-free natural nucleic acids, with high efficiency.

Automation in Proteomics and Genomics: An Engineering Case-Based Approach Edited by Gil Alterovitz, Roseann Benson and Marco Ramoni

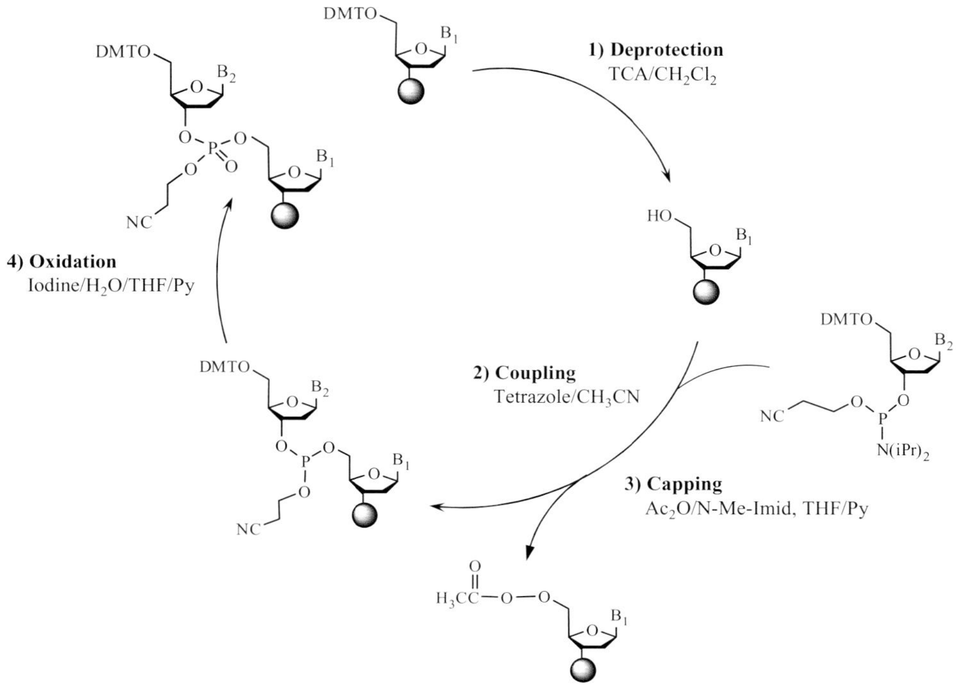

Figure 7.1 *Schematic illustration of steps for oligonucleotide chemical synthesis using standard phosphoramidite chemistry. Shown here is the cycle of coupling the second nucleotide to the first nucleotide attached to the bead (shaded sphere)*

In the first synthesis cycle, the nucleotide chain grows from an initial protected nucleoside tethered to a solid support via its terminal 3′ hydroxyl. The commonly used supports include the controlled pore glass (CPG) or polystyrene beads. Chemicals and solvents are pumped onto and through the support to induce the stepwise addition of nucleotide monomers to the growing oligonucleotide chain. The addition of each nucleotide monomer to the elongating oligonucleotide chain is carried out in four steps: deprotection; activation/coupling; capping; and oxidation (Figure 7.1). When the synthesis is complete, cleavage and deprotection results in the product being cleaved from the solid support, while the remaining protection groups are removed to reveal the normal synthetic nucleic acid product:

- *Step 1* – Deprotection: The first step in the synthesis cycle is removal of the acid-labile 5′-*O*-DMT group from the first deoxynucleoside linked to the solid support or the 5′-end deoxynucleoside on the growing oligodeoxynucleotide chain. This is achieved by using a large excess of a weak acid, such as trichloroacetic acid (TCA) or dichloroacetic acid (DCA), in an organic solvent. The resultant 5′-OH group becomes the only reactive nucleophile capable of participating in the subsequent coupling step. The deprotection step is kept short in order to prevent any possible acid-catalyzed depurination of DNA.

A subsequent rinse with acetonitrile removes the acid from the support and prevents any premature detritylation of the incoming phosphoramidite monomer.

- *Step 2* – Coupling: In the coupling step, the 5′-OH group generated from the deprotection step reacts with an activated monomer created by simultaneously adding the desired phosphoramidite and an appropriate activator, the weakly acidic tetrazole ($pK_a = 4.8$). Because the activated phosphoramidite is very reactive, the coupling reaction is usually complete within 30 s. An excess of tetrazole over phosphoramidite ensures complete activation, while an excess of phosphoramidite over free 5′-OH of the growing chain promotes efficient coupling (>99%).
- *Step 3* – Capping: Even with high coupling efficiencies, a small proportion of the 5′-OH groups fails to couple to the incoming activated phosphoramidite. These remaining reactive hydroxyl groups on the 5′ end of the growing oligonucleotide chains must be rendered inactive to minimize deletion products. This is accomplished by adding acetic anhydride and *N*-methylimidazole dissolved in pyridine and tetrahydrofuran (THF) to create an acylating agent that 'caps' the free, unextended 5′-OH groups. After an acetonitrile wash, the 5′-acetyl ester cap remains unreactive in all subsequent cycles and is removed during the final ammonia deprotection step.
- *Step 4* – Oxidation: After coupling and capping, the unstable phosphite triester internucleotide linkages are oxidized to a more stable phosphotriester. This step is carried out using 0.02 *M* iodine dissolved in water/pyridine/THF. Water in the oxidizer is thoroughly removed with acetonitrile washes following the reaction.

This completes one cycle of monomer addition, whereupon the next cycle starts over with removal of the 5′-DMT from the newly added nucleotide. The released trityl cation chromophore can be quantitated to determine the coupling efficiency.

7.1.1.2 Cleavage/Deprotection

The four-step cycle is repeated for the addition of each nucleotide in the sequence and, when the synthesis is complete, the oligonucleotide chain is cleaved from the solid support and deprotected using concentrated ammonium hydroxide. All of the protection groups on the bases and the phosphate backbone are removed with this treatment. Nonetheless, if the final trityl group is left on ('trityl-on') at the end of the synthesis, it can be used for purification purposes to enrich for the full-length products.

7.1.1.3 Purification

At this point the native oligonucleotide can be further purified by a variety of strategies. While the method of choice will depend on the purity required, time considerations and availability of resources, the following isolation methods may be considered, or even combined:

- *Methods for desalting*: Contaminating chemicals can be quickly removed by direct precipitation with ethanol or sizing columns. However, these methods do not separate abortive synthesis products from their full-length counterparts. Purified oligonucleotides can be used for routine molecular biology tasks, such as sequencing, cloning and PCR. Lingering ammonium ions after precipitation may inhibit certain enzymatic reactions, such as phosphorylation by T4 polynucleotide kinase. A more thorough purification procedure may be required for such applications.

- *Methods for isolating full-length products*: If the final trityl group is left on following the final coupling reaction, the hydrophobically tagged full-length 'trityl-on' oligonucleotide may be separated from failure sequences using a reversed-phase cartridge. Failure sequences, which lack trityl groups, do not bind to the hydrophobic matrix efficiently. Denaturing polyacrylamide gel electrophoresis (PAGE) or high-performance liquid chromatography (HPLC) can be used to separate oligonucleotides with single-residue resolution, and is the method of choice for purifying full-length oligonucleotides. HPLC can also be used to purify full-length 'trityl-on' oligonucleotides by charge differences through ion-exchange or hydrophobicity.

7.1.1.4 Alternative Synthesis Chemistry

Most recently, an alternative solid-phase phosphoramidite-based oligodeoxynucleotide synthesis method has been developed that involves only two steps [4]. This approach utilizes a peroxy anion as nucleophile during each synthetic cycle such that a 5′-carbonate is removed and the internucleotide phosphite triester is oxidized, simultaneously. The cyclical removal of the 5′-protecting group with a peroxy anion under mildly basic conditions is essentially nonreversible and quantitative. This procedure can therefore completely eliminate depurination and reduce mutation frequencies in synthetic DNA. As the two-step procedure also simplifies oligodeoxynucleotide synthesis by eliminating several reagents, this should allow for a simpler – and potentially more robust – automation. It should also result in dramatic cost savings for the large-scale synthesis of oligodeoxynucleotides.

7.1.1.5 Automation of Oligonucleotide Synthesis

A solid-phase synthesis makes automation possible because it eliminates the need to purify synthetic intermediates or unreacted reagents. Rather, the reagents are simply rinsed from the column at the end of each step. Based on the four-step synthesis procedure, fully automated DNA synthesizers have been developed with throughputs ranging from one to 1536 sequences [5–8]. The first such machines were built and sold by Applied Biosystems, but were capable of synthesizing only two to four independent sequences at a time, using relatively large reaction volumes (ca. 1 ml). A parallel synthesis machine capable of synthesizing oligonucleotides in a 96-well plate format was reported in 1995 at the Stanford Genome Research Center [7]. In order to synthesize a large number of oligonucleotides in a multiplexed fashion using the phosphoramidite synthesis chemistry with this design, the reagent bottles were connected by Teflon tubing to multiple solenoid valves that were individually controlled by computer to deliver the reagents into wells of a reaction plate. Individual valve control was essential in order to prepare oligonucleotides of different lengths and sequences. Such a design allowed the highly parallel synthesis of oligonucleotides by multiplexing the reagent delivery, and without any sacrifice of product quality. Based on similar designs, additional parallelization further raised the throughput of synthesis of these instruments to 192 and 384 (and more) independent reaction wells.

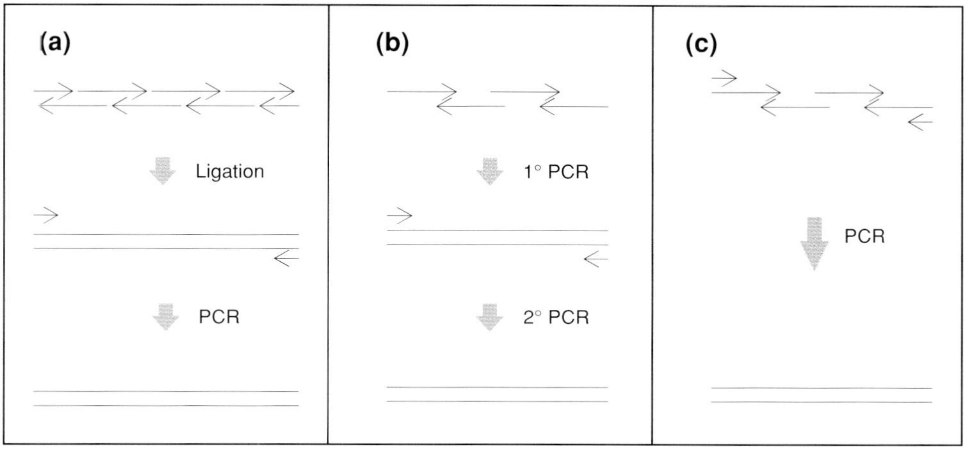

Figure 7.2 *Commonly used methods for gene synthesis. (a) The ligation-based assembly usually involves two steps, namely ligation and PCR amplification; (b) PCR-driven assembly can be carried out in two PCR reactions or (c) in a one-tube, single-step reaction. In (b), the gene-end primer pair is added after the first PCR assembly reaction*

7.1.2 Gene Synthesis

Chemical DNA synthesis is typically used for the synthesis of oligonucleotide sequences shorter than ~150 bases. However, for the synthesis of longer, gene-sized DNA, a number of enzymatic methods have been used over the past decades. Among these are included two general techniques that are most often used today for the convenient synthesis of individual genes, starting from short synthetic oligonucleotides (Figure 7.2).

7.1.2.1 Ligation-Based Assembly

The joining of oligonucleotides with DNA ligase to form longer genes was used in the earlier examples of gene synthesis [9–13], and also in certain commercial solid-phase gene synthesis set-ups. With the discovery of thermostable DNA ligases and development of the ligase chain reaction (LCR) [14–16], thermoligase or LCR-based gene assembly methods have become very convenient [17–19]. One advantage of using a thermoligase over T4 DNA ligase is that less oligonucleotide secondary structure will form at elevated ligation temperatures.

In this approach (Figure 7.2a), carefully designed overlapping oligonucleotides that completely cover both strands of the gene sequence are chemically synthesized and phosphorylated at the 5′-ends. The oligonucleotides are then mixed together in buffer with a thermoligase and heat-denatured; the mixture is then cooled slowly to a temperature suitable for proper annealing and ligation. The denaturation and annealing/ligation steps can be repeated for a number of cycles. In order to produce enough quantities of the full-length gene product, the ligation reaction is usually coupled with a polymerase chain reaction (PCR), using a pair of specific gene-end primers to amplify the full-length gene sequence.

7.1.2.2 PCR-Driven Assembly

Without ligation, a procedure similar to PCR alone can also assemble overlapping oligonucleotides into full-length gene constructs, either in two steps or in a single step [20–23]. These methods have been named variously 'assembly PCR', 'overlapping PCR', 'polymerase chain assembly (PCA)', and so on.

In the two-step procedure (Figure 7.2b), overlapping oligonucleotides that together cover the whole construct are mixed together in equal, low concentrations with a PCR mixture, including buffer, dNTPs and a polymerase. The thermal cycling steps are then carried out similar as for a normal PCR. During the first PCR reaction, overlapping oligonucleotides will anneal and extend, using each other as a template, to form increasingly longer DNA fragments until eventually they reach full length. A second PCR is then necessary to amplify the full-length construct; this second PCR will use a pair of end primers and a small fraction of the first PCR mixture as template.

In the one-step procedure (Figure 7.2c), the pair of end primers is added from the start, and at a higher concentration than the remainder of the oligonucleotides. Hence, extra cycles may be needed to assemble and amplify the full-length construct.

Compared to the LCR-based approach, a difference may exist in the design of oligonucleotides. In the PCR-driven assembly, gaps are allowed between adjacent oligonucleotides that belong to the same sense or antisense strand. This gives the PCR-driven assembly a slight advantage in terms of the amount of chemical DNA synthesis required over the ligation-based assembly, where no gap is allowed. The speed and convenience of the single-step PCR-driven assembly is another attractive feature. Unfortunately, however, not all sequences can be assembled by PCR, and for some difficult constructs which involve repetitive sequences or excessive DNA secondary structures, ligation may be the only option.

7.1.3 Error Removal

Both chemical oligonucleotide synthesis and enzymatic gene assembly reactions will introduce errors to the final synthetic gene product. Thus, a variety of error-removal strategies must be in place to eliminate errors during the different stages of the gene synthesis process.

7.1.3.1 Error Removal from Synthetic Oligonucleotides

Because the chemical reactions and washing steps are rarely 100% efficient, the coupling efficiency for each monomer is typically 98.5–99.5% during chemical nucleic acid synthesis. Deletions and insertions are the most frequent error types in oligonucleotide synthesis. Typically, the deletion rate (which is due largely to incomplete capping) will be up to 0.5% per position, while the insertion rate (which is due largely to DMT cleavage by tetrazole) is approximately 0.4% per position. As a consequence, for an oligonucleotide which is 100 bases in length, only about 30–40% of the sequences will be correct.

Besides perfecting the DNA synthesis chemistry to improve oligonucleotide quality, an effective method to reduce deletion/insertion is that of size selection, including HPLC and PAGE purification (as discussed previously). Although approximately 90% of the impurities can be removed by using these methods, they are not effective against other types of mutation that do not involve size change. Recently, a hybridization-based oligo error reduction approach using DNA microarray was reported (this will be discussed later) [24].

7.1.3.2 *Error Removal from Synthetic Genes*

The errors that remain in the synthetic oligonucleotides will be carried over and subsequently accumulate in longer synthetic DNA constructs. PCRs are also error-prone and may introduce additional errors. The identification of error-free sequences by cloning and sequencing is both time-consuming and costly, although in some cases expression or functional screens can be implemented to eliminate errors that will cause frame shift or function loss [19, 23, 25]. However, such targets are limited to protein-coding sequences and functional DNA elements. A more general approach is to use DNA mismatch binding or cleaving proteins, such as the MutS or MutHLS complex [26,27]; these proteins are able to bind selectively to mismatches generated by hybridization between correct and incorrect sequences. The resultant binding complex, which contains incorrect sequences, can then be removed from the pool by using gel-shift assays or affinity columns.

7.1.4 Gene Design Tools

On the assumption that the goal of these investigations is to design and build a biological system with predictable behavior(s), the complexity of biological systems requires the use of design tools at several different levels.

7.1.4.1 *The Organism or Whole-System Level*

All of the components must function synchronously within the genetic and biochemical context of the organismal chassis, without causing any unpredicted behaviors. Besides more intelligent mathematical frameworks and tools for accurately predicting the behavior of genome or gene circuits, better information is needed on genome function and regulation.

7.1.4.2 *The Genetic Circuit Level*

Better physical and mathematical models and tools need to be developed to design genetic circuits with desired and predictable behaviors. Although some encouraging developments have been made, significant limitations still exist, and these must be overcome in order for the field to move forward [28]. Some major limitations include: (i) an incomplete list of all the functional or structural components in most biological systems; (ii) an incomplete understanding of the functions or physical characteristics of most genes, genetic elements or proteins, and how these elements interact; and (iii) how to define and simulate complex interactions and crosstalks in a noisy, crowded and compartmented environment such as the cell.

7.1.4.3 *The Component Level*

The creation of better or novel genetically encoded components is an important goal for the field of synthetic biology. Improving the function of existing parts or components is achievable using molecular evolution approaches, with or without the aid of rational design.

However, rational *de novo* design in order to create novel genetic elements, proteins or enzymes is still a developing art. Today, most current designs are effected based on existing sequence and structure information. One foundation for *de novo* design is an understanding of the sequence–structure–function relationship of biological molecules, including mostly proteins and RNAs, although this at present is far from being either complete or accurate.

7.1.4.4 The DNA Sequence Level

Unlike standardized mechanical or electrical components, which can be conveniently assembled into different devices, sequence-specific genetic elements or genes usually do not behave exactly as predicted in different hosts or genetic environments. Part of the reason for this is the codon bias of different organisms or systems in which the genes are used. If a designed gene does not express at all, or does not express at the desired level, then the whole system may fail if there are no built-in redundancy or compensation mechanisms. Our current understanding is still too patchy to provide any accurate predication of the expression level of a gene in a specific host or system. In addition, a lack of comprehensive understanding makes the development of reliable and effective 'codon optimization' algorithms difficult, if not impossible.

Once codon usage is selected, then a number of computer algorithms exist to design oligonucleotide sequences for gene construction [24, 29, 30]. The main functions of these programs include: (i) the integration of design features and the combination of building blocks with protein-coding sequences, such as regulatory DNA elements, cloning strategies and affinity tags; and (ii) to design oligonucleotide sequence sets for gene construction.

7.2 Applications

In the past, synthetic oligonucleotides have played a critical role in modern biotechnology by enabling PCR, mutagenesis and cloning. Synthetic oligonucleotides have also been used as building blocks for the construction of genes and longer DNA constructs. In principle, the ability to design and write DNA blueprints freely from scratch will provide the opportunity to create novel biological systems, and in so doing will revolutionize biomedical research.

7.2.1 DNA Microarray Synthesis

DNA microarray is a powerful technology in the study of genomics. Depending on the type of probe printed on the surface, such arrays can be divided into two general categories, namely cDNA microarrays [31, 32] and oligonucleotide microarrays [33–35]. Oligonucleotide microarrays, being flexible in sequence design and more specific in hybridization, are popular for gene expression profiling, mutation detection, genotyping, sequencing and a variety of other applications [36]. Regular phosphoramidite chemistry, with minor modifications, is used for the automated *in situ* synthesis of oligonucleotide microarrays [33, 37–40].

Unlike oligo synthesis on a DNA synthesizer – where the reactions occur in separate compartments – microarray synthesis takes place on the surface of a silicon chip or glass slide, with the oligo growth being confined to specific spots or regions. This is achieved through a variety of different mechanisms, including photolithography with physical masks, digital photolithography, and electrode array or inkjet printing. It is these mechanisms which control whether or not a phosphoramidite monomer will be coupled to growing oligonucleotide chains on a particular spot during each synthesis cycle, during which the oligonucleotide chains are anchored to the surface via a special noncleavable chemical linker.

7.2.1.1 *Photolithography with Physical Masks*

The earliest systems created to synthesize oligonucleotide microarrays included photolithography using physical masks and photolabile nucleoside monomers [33, 41, 42]. Here, the mask is used to generate a light pattern that dictates which areas on the array are to be activated for chemical coupling. Consequently, a stack of masks needs to be prefabricated according to the oligo sequences to be synthesized on the chip. Light exposure in specified areas removes photolabile protecting groups from the growing chains and, after deprotection, a selected phosphoramidite monomer is added onto the entire surface, although the coupling reactions only occur in areas exposed to light. The cycle is then repeated until the entire synthesis is complete.

When optimized, Affymetrix, Inc. applied this technology to the large-scale fabrication of high-density GeneChip probe arrays for nucleic acid sequence analysis [36, 42–44].

7.2.1.2 *Digital Photolithography*

The cost of using large numbers of prefabricated photomasks is high, and is probably justifiable only for large-scale, high-volume gene chip fabrication. A relatively low-cost and flexible alternative is digital photolithography, using Texas Instrument's digital micromirror device (DMD), which is based on digital light processing (DLP) technology. DMD, which is normally used in commercial projectors, is a reflective display device which consists of an electromechanically controlled array of micromirrors. The resolutions may be as high as 307 200 pixels (VGA) or 2 073 600 pixels (SVGA), with a $16 \times 16\,\mu m$ area per pixel. One major advantage of using a DMD over a physical mask is that it is programmable, with high-resolution, precisely controllable light patterns being generated in an automated manner.

In an early version of the maskless array synthesizer (MAS), a DMD consisting of a 600×800 array of 16 μm-wide micromirrors was used [45,46]. The mirrors were individually controlled by computer signal, and could be used to generate any given pattern up to 480 000 pixels simultaneously. In theory, the device allowed the synthesis of almost half a million different oligonucleotide sequences on the substrate.

Different photochemistry can be used with digital photolithography for DNA microarray synthesis. Some designs have directly used a photolabile protecting group (PLPG), such as (*R*,*S*)-1-(3,4-(methylenedioxy)-6-nitrophenyl)ethyl chloroformate (MeNPOC) or 2-(2-nitrophenyl) propoxycarbonyl (NPPOC), to protect the hydroxyl groups on the linker or on the phosphoramidite monomers [45, 46]. Other designs have used photogenerated acid (PGA) in solution to perform the deprotection step in conventional nucleotide phosphoramidite chemistry [47]. At millimolar concentrations, the PGA solution can effectively remove the DMT group to free the 5′-OH group of nucleosides or nucleotides.

7.2.1.3 Electrode Array

Instead of using photochemical methods to generate acid for the deprotection step, localized electrochemical reactions can also be used for DNA microarray synthesis. Acid is produced only at specified sites by electrochemical oxidation, using an array of individually addressable microelectrodes. The electrolyte used in one design was 25 m*M* hydroquinone and 25 m*M* benzoquinone with 25 m*M* tetrabutylammonium hexafluorophosphate in anhydrous acetonitrile [48]. When current is applied to the microelectrodes, the electrolyte is oxidized at the anodes; this causes acid to be released, which in turn diffuses to the substrate in which the oligonucleotide was synthesized. The acid is confined in the region by adjacent cathodes that consume it by reduction. Alternatively, the synthesis can be performed on a porous polymeric layer, which slows down the diffusion of the acid generated from a local electrode and increases the amount of oligonucleotides synthesized per unit area. The deprotection step can be completed in seconds, such that any side reactions between the chemical and synthesized oligonucleotides are minimal.

7.2.1.4 Inkjet Printing

Inkjet printer heads can be used to deliver small drops of reagents to a chemically modified slide surface, where they react to synthesize DNA. Piezoelectric jetting, high-quality motion controllers and standard phosphoramidite oligonucleotide synthesis chemistry together allow the synthesis of arrays of oligonucleotide sequences at specific, closely spaced features on suitable solid substrates. This technology is primarily commercialized by Agilent.

7.2.2 Multiplex Gene Synthesis from DNA Microchips

Today, at a current average price of US$ 1–2 per base pair, gene synthesis is still very expensive and, perhaps more importantly, the process is also very difficult to automate. In an initial attempt to reduce the cost and increase the throughput of synthesizing oligonucleotide building blocks, special customized oligonucleotide arrays were adapted as an economic source for large numbers of different oligonucleotide sequences [24, 46, 49]. Synthesis on DNA chips not only offers advantages in throughput, cost and speed but also dramatically reduces the consumption of toxic organic solvents and reagents. The introduction of microfluidic plumbing, which can be fabricated directly on top of, or adjacent to, the synthesis reaction champers, will further reduce human handling and lead to savings in reagent costs.

The strategy of using a chip-synthesized oligonucleotide pool for gene synthesis is illustrated in Figure 7.3. In order to harvest oligonucleotides made on DNA microchips, cleavable linkers were used in DNA microarray synthesis to anchor oligonucleotides on the surface [24, 46, 49]. Following treatment with ammonium hydroxide or enzyme to release the oligonucleotides from the chip, they were collected and purified. Current DNA microarrays are capable of synthesizing 10^3 to 10^6 different oligo sequences although at very low yields (i.e. $\sim 10^6$ molecules for each sequence). When using a microliter scale gene assembly reaction, an oligo preamplification step is usually required in order to achieve optimal oligo concentrations for the reaction. In this case, sequence features necessary for postsynthesis enzymatic amplification must be designed into the oligo sequences to be synthesized

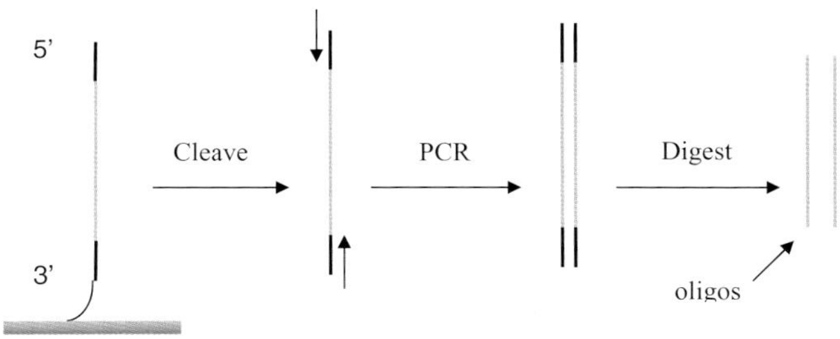

Figure 7.3 *Gene assembly from a DNA microchip. Gene-construction oligonucleotides flanked by universal primer sequences are synthesized* in situ *on the microchip surface. After cleavage, the oligonucleotides are amplified by PCR. The PCR primer sequences are removed by digestion with type IIs restriction enzymes. Clean oligonucleotides are used for subsequent gene construction steps*

on the chip, an example being short universal PCR primers which flank the sequences. Importantly, these primer sequences would need to be removed after amplification; one way of achieving this would be to use a type IIs restriction enzyme digestion [24].

Subsequently, a chip hybridization-based method was designed as a preliminary means of removing errors from chip-synthesized and PCR-amplified oligonucleotides [24]. For this, two pools of error-correction oligonucleotides were synthesized from two DNA chips, with each pool consisting of short oligonucleotides complementary to approximately one half of the 'gene-construction' oligonucleotides released from the first chip. After hybridization and appropriate washes, the mismatched sequences were selectively removed from the pool and the correct sequences preferentially enriched. In this way, a multiplexed gene assembly reaction could be used to assemble multiple genes from the same large pool of oligonucleotides [24]. Moreover, these gene fragments could be further assembled into increasingly longer sequences, either *in vitro* or *in vivo*.

A typical DNA chip with between 10^3 and 10^6 different oligonucleotides is capable of constructing megabases of DNA sequence, equivalent to or surpassing the lengths of microbial genomes. However, special measures must be taken to fully utilize this capacity. In addition to multiplexing, a combination of spatial separation, selective releasing or amplification, microfluidic or bioinformatic designs can all be explored.

7.2.3 Applications in Bioengineering

Conventional DNA synthesis, when combined with recombinant DNA techniques, has been used in a wide range of applications, including biomolecular engineering, DNA nanotechnology and computing, gene circuit construction, metabolic engineering and genome synthesis [50–54]. Due to the high cost and low throughput nature of conventional gene synthesis, most applications mainly require oligonucleotide primers to be synthesized. Yet, even in a complex gene circuit, pathway or metabolic engineering, the *de novo* synthesis accounted for only a small fraction of the total sequence constructed. In fact, in such cases most of the protein-coding genes were PCR-copied from natural sources.

Although, today, the *de novo* synthesis of error-free long DNA sequences remains a major challenge, a number of successful attempts have been made in this respect. In

2002, the chemical synthesis of a functional poliovirus genome was demonstrated using conventional DNA synthesis and gene assembly technology [55], while in 2003 Smith and colleagues reported the synthesis of a functional bacteriophage genome which took two weeks to complete [55]. Today, attempts at the *de novo* synthesis of bacterial genomes are underway and, with a further drop in price and increase in throughput, the time will surely come when *de novo* DNA writing will become a routine and standard method in molecular biology and bioengineering. Given time, *de novo* DNA writing should offer the freedom of obtaining any DNA molecule in convenient manner and, in so doing, will transform biomedical research in the near future.

Abbreviation List

DMT:	dimethoxytrityl
iPr2N:	diisopropylamine
CPG:	controlled pore glass
DMT:	5′-*O*-dimethoxytrityl
TCA:	trichloroacetic acid
DCA:	dichloroacetic acid
THF:	tetrahydrofuran
PAGE:	polyacrylamide gel electrophoresis
LCR:	ligase chain reaction
PCA:	polymerase chain assembly
DMD:	digital micromirror device
DLP:	digital light processing
MAS:	maskless array synthesizer
PLPG:	photolabile protecting group
MeNPOC:	(*R*,*S*)-1-(3,4-(methylenedioxy)-6-nitrophenyl)ethyl chloroformate
NPPOC:	2-(2-nitrophenyl) propoxycarbonyl
PGA:	photogenerated acid

References

1. Caruthers, M.H., Beaucage, S.L., Becker, C. *et al.* (1983) Deoxyoligonucleotide synthesis via the phosphoramidite method. *Gene Amplification and Analysis*, **3**, 1–26.
2. Caruthers, M.H., Barone, A.D., Beaucage, S.L. *et al.* (1987) Chemical synthesis of deoxyoligonucleotides by the phosphoramidite method. *Methods in Enzymology*, **154**, 287–313.
3. Caruthers, M.H. (1985) Gene synthesis machines: DNA chemistry and its uses. *Science*, **230**, 281–5.
4. Sierzchala, A.B., Dellinger, D.J., Betley, J.R. *et al.* (2003) Solid-phase oligodeoxynucleotide synthesis: a two-step cycle using peroxy anion deprotection. *Journal of the American Chemical Society*, **125**, 13427–41.
5. Horvath, S.J., Firca, J.R., Hunkapiller, T. *et al.* (1987) An automated DNA synthesizer employing deoxynucleoside 3′-phosphoramidites. *Methods in Enzymology*, **154**, 314–26.
6. Sindelar, L.E. and Jaklevic, J.M. (1995) High-Throughput DNA-Synthesis in a Multichannel Format. *Nucleic Acids Research*, **23**, 982–7.

7. Lashkari, D.A., Hunickesmith, S.P., Norgren, R.M. *et al.* (1995) An automated multiplex oligonucleotide synthesizer – development of high-throughput, low-cost DNA-synthesis. *Proceedings of the National Academy of Sciences of the United States of America*, **92**, 7912–15.
8. Cheng, J.Y., Chen, H.H., Kao, Y.S. *et al.* (2002) High throughput parallel synthesis of oligonucleotides with 1536 channel synthesizer. *Nucleic Acids Research*, **30**, e93.
9. Agarwal, K.L., Buchi, H., Caruthers, M.H. *et al.* (1970) Total synthesis of the gene for an alanine transfer ribonucleic acid from yeast. *Nature*, **227**, 27–34.
10. Heyneker, H.L., Shine, J., Goodman, H.M. *et al.* (1976) Synthetic lac operator DNA is functional in vivo. *Nature*, **263**, 748–52.
11. Itakura, K., Hirose, T., Crea, R. *et al.* (1977) Expression in *Escherichia coli* of a chemically synthesized gene for the hormone somatostatin. *Science (New York)*, **198**, 1056–63.
12. Goeddel, D.V., Kleid, D.G., Bolivar, F. *et al.* (1979) Expression in *Escherichia coli* of chemically synthesized genes for human insulin. *Proceedings of the National Academy of Sciences of the United States of America*, **76**, 106–10.
13. Khorana, H.G. (1979) Total synthesis of a gene. *Science*, **203**, 614–25.
14. Barany, F. (1991) Genetic disease detection and DNA amplification using cloned thermostable ligase. *Proceedings of the National Academy of Sciences of the United States of America*, **88**, 189–93.
15. Wiedmann, M., Wilson, W.J., Czajka, J. *et al.* (1994) Ligase chain reaction (LCR)–overview and applications. *PCR Methods and Applications*, **3**, S51–64.
16. Barany, F. (1991) The ligase chain reaction in a PCR world. *PCR Methods and Applications*, **1**, 5–16.
17. Jayaraman, K., Shah, J. and Fyles, J. (1989) PCR mediated gene synthesis. *Nucleic Acids Research*, **17**, 4403.
18. Au, L.C., Yang, F.Y., Yang, W.J. *et al.* (1998) Gene synthesis by a LCR-based approach: high-level production of leptin-L54 using synthetic gene in *Escherichia coli*. *Biochemical and Biophysical Research Communications*, **248**, 200–3.
19. Smith, H.O., Hutchison, C.A. 3rd, Pfannkoch, C. and Venter, J.C. (2003) Generating a synthetic genome by whole genome assembly: phiX174 bacteriophage from synthetic oligonucleotides. *Proceedings of the National Academy of Sciences of the United States of America*, **100**, 15440–5.
20. Ho, S.N., Hunt, H.D., Horton, R.M. *et al.* (1989) Site-directed mutagenesis by overlap extension using the polymerase chain reaction. *Gene*, **77**, 51–9.
21. Vallejo, A.N., Pogulis, R.J. and Pease, L.R. (1994) In vitro synthesis of novel genes: mutagenesis and recombination by PCR. *PCR Methods and Applications*, **4**, S123–30.
22. Chen, G., Choi, I., Ramachandran, B. and Gouaux, J.E. (1994) Total gene synthesis: novel single-step and convergent strategies applied to the construction of a 779 base pair bacteriorhodopsin gene. *Journal of the American Chemical Society*, **116**, 8799–800.
23. Stemmer, W.P., Crameri, A., Ha, K.D. *et al.* (1995) Single-step assembly of a gene and entire plasmid from large numbers of oligodeoxyribonucleotides. *Gene*, **164**, 49–53.
24. Tian, J., Gong, H., Sheng, N. *et al.* (2004) Accurate multiplex gene synthesis from programmable DNA microchips. *Nature*, **432**, 1050–4.
25. Cox, J.C., Lape, J., Sayed, M.A. and Hellinga, H.W. (2007) Protein fabrication automation. *Protein Science*, **16**, 379–90.
26. Modrich, P. (1991) Mechanisms and biological effects of mismatch repair. *Annual Review of Genetics*, **25**, 229–53.
27. Carr, P.A., Park, J.S., Lee, Y.J. *et al.* (2004) Protein-mediated error correction for de novo DNA synthesis. *Nucleic Acids Research*, **32**, e162.
28. Elowitz, M.B. and Leibler, S. (2000) A synthetic oscillatory network of transcriptional regulators. *Nature*, **403**, 335–8.

29. Hoover, D.M. and Lubkowski, J. (2002) DNAWorks: an automated method for designing oligonucleotides for PCR-based gene synthesis. *Nucleic Acids Research*, **30**, e43.
30. Rouillard, J.M., Lee, W., Truan, G. *et al.* (2004) Gene2Oligo: oligonucleotide design for in vitro gene synthesis. *Nucleic Acids Research*, **32**, W176–80.
31. Schena, M. (1996) Genome analysis with gene expression microarrays. *BioEssays*, **18**, 427–31.
32. Schena, M., Shalon, D., Heller, R. *et al.* (1996) Parallel human genome analysis: microarray-based expression monitoring of 1000 genes. *Proceedings of the National Academy of Sciences of the United States of America*, **93**, 10614–19.
33. Southern, E.M., Maskos, U. and Elder, J.K. (1992) Analyzing and comparing nucleic acid sequences by hybridization to arrays of oligonucleotides: evaluation using experimental models. *Genomics*, **13**, 1008–17.
34. Maskos, U. and Southern, E.M. (1992) Parallel analysis of oligodeoxyribonucleotide (oligonucleotide) interactions. I. Analysis of factors influencing oligonucleotide duplex formation. *Nucleic Acids Research*, **20**, 1675–8.
35. Case-Green, S.C., Mir, K.U., Pritchard, C.E. and Southern, E.M. (1998) Analysing genetic information with DNA arrays. *Current Opinion in Chemistry and Biology*, **2**, 404–10.
36. Lipshutz, R.J., Fodor, S.P., Gingeras, T.R. and Lockhart, D.J. (1999) High density synthetic oligonucleotide arrays. *Nature Genetics*, **21**, 20–4.
37. Maskos, U. and Southern, E.M. (1992) Oligonucleotide hybridizations on glass supports: a novel linker for oligonucleotide synthesis and hybridization properties of oligonucleotides synthesised in situ. *Nucleic Acids Research*, **20**, 1679–84.
38. Blanchard, A.P. and Hood, L. (1996) Sequence to array: probing the genome's secrets. *Nature Biotechnology*, **14**, 1649.
39. Hughes, T.R., Mao, M., Jones, A.R. *et al.* (2001) Expression profiling using microarrays fabricated by an ink-jet oligonucleotide synthesizer. *Nature Biotechnology*, **19**, 342–7.
40. Gao, X., Gulari, E. and Zhou, X. (2004) In situ synthesis of oligonucleotide microarrays. *Biopolymers*, **73**, 579–96.
41. Fodor, S.P.A., Read, J.L., Pirrung, M.C. *et al.* (1991) Light-directed, spatially addressable parallel chemical synthesis. *Science*, **251**, 767–73.
42. Barone, A.D., Beecher, J.E., Bury, P.A. *et al.* (2001) Photolithographic synthesis of high-density oligonucleotide probe arrays. *Nucleosides Nucleotides and Nucleic Acids*, **20**, 525–31.
43. Fodor, S.P., Rava, R.P., Huang, X.C. *et al.* (1993) Multiplexed biochemical assays with biological chips. *Nature*, **364**, 555–6.
44. Pease, A.C., Solas, D., Sullivan, E.J. *et al.* (1994) Light-generated oligonucleotide arrays for rapid DNA sequence analysis. *Proceedings of the National Academy of Sciences of the United States of America*, **91**, 5022–6.
45. Singh-Gasson, S., Green, R.D., Yue, Y. *et al.* (1999) Maskless fabrication of light-directed oligonucleotide microarrays using a digital micromirror array. *Nature Biotechnology*, **17**, 974–8.
46. Richmond, K.E., Li, M.H., Rodesch, M.J. *et al.* (2004) Amplification and assembly of chip-eluted DNA (AACED): a method for high-throughput gene synthesis. *Nucleic Acids Research*, **32**, 5011–18.
47. Gao, X.L., LeProust, E., Zhang, H. *et al.* (2001) A flexible light-directed DNA chip synthesis gated by deprotection using solution photogenerated acids. *Nucleic Acids Research*, **29**, 4744–50.
48. Egeland, R.D. and Southern, E.M. (2005) Electrochemically directed synthesis of oligonucleotides for DNA microarray fabrication. *Nucleic Acids Research*, **33**, e125.
49. Zhou, X., Cai, S., Hong, A. *et al.* (2004) Microfluidic PicoArray synthesis of oligodeoxynucleotides and simultaneous assembling of multiple DNA sequences. *Nucleic Acids Research*, **32**, 5409–17.
50. Martin, V.J., Pitera, D.J., Withers, S.T. *et al.* (2003) Engineering a mevalonate pathway in *Escherichia coli* for production of terpenoids. *Nature Biotechnology*, **21**, 796–802.

51. Fishman, A., Tao, Y., Rui, L. and Wood, T.K. (2005) Controlling the regiospecific oxidation of aromatics via active site engineering of toluene para-monooxygenase of *Ralstonia pickettii* PKO1. *Journal of Biological Chemistry*, **280**, 506–14.
52. Dueber, J.E., Yeh, B.J., Chak, K. and Lim, W.A. (2003) Reprogramming control of an allosteric signaling switch through modular recombination. *Science*, **301**, 1904–8.
53. Park, S.H., Zarrinpar, A. and Lim, W.A. (2003) Rewiring MAP kinase pathways using alternative scaffold assembly mechanisms. *Science*, **299**, 1061–4.
54. McDaniel, R. and Weiss, R. (2005) Advances in synthetic biology: on the path from prototypes to applications. *Current Opinion in Biotechnology*, **16**, 476–83.
55. Cello, J., Paul, A.V. and Wimmer, E. (2002) Chemical synthesis of poliovirus cDNA: generation of infectious virus in the absence of natural template. *Science*, **297**, 1016–18.

of the structural and functional properties of the building blocks, as well as the final form that would be prevalent in the presence of environmental factors, is very important. Therefore, general experimental experiences combined with computational design methodologies represent an approach that can greatly speed up the development of functional RNA-based nanoparticles.

The goal of RNA rational nanodesign can be approached from at least two different directions. The first direction assumes that one has in mind particular topological and functional properties of the target structure, and then seeks to find the appropriate building blocks (and associated sequences) that can assemble into the target structure(s). The second approach, which involves more computation, involves a combinatoric search based on a library of building blocks to pregenerate numerous structures; these can be deposited in a database, later allowing a database search for structures that adhere to the required properties. The advantage of the first approach is that one might be able to design a very specific RNA particle based on a preconceived notion of what it should contain and look like. The alternate approach leaves the door open for new structures that were never conceived of before, and ultimately might act as useful functional particles or serve as building components for more complex structures.

In order to rationally design RNA architectures, multiple considerations derived from the understanding of how natural RNA molecules fold and assemble need to be taken in account. The hierarchical folding process of natural RNA molecules is what makes RNA an especially suitable polymer for inverse folding design. Therefore, the principles and concepts that presently emerge from RNA architectonics [2] are at the root of RNA nanodesign.

In this chapter, the general approach that has been defined over the past eight years to rationally generate self-assembling RNA architectures is outlined. A general overview of the methodology behind RNA nanodesign and a review of the necessary criteria that make this approach feasible at an experimental level will be presented. Following this, some of the state-of-the-art bioinformatics tools that presently facilitate the automation of RNA rational design at a computational level will be discussed. After providing a framework that allows the implementation of this technology at an experimental level, specific examples of rationally designed RNA nanoparticles that have been successfully generated and tested in the laboratory will be explored.

8.2 Rational Design of RNA Nanoparticles

8.2.1 Inverse RNA Folding Design Method: RNA Architectonics

The design of new RNA structures via the architectonics methodology is achieved through inverse folding, a multistep process in its simplest form (see Figures 8.1 and 8.2):

1. Take a 3-D shape from an atomic resolution structure, treating it as a rigid building block.
2. Attach helical stems to all branch points of the molecular building block.
3. Attach connection points to the ends of the stems to allow quaternary assembly. For example, one half of a kissing loop motif when connected via inserted helices to other

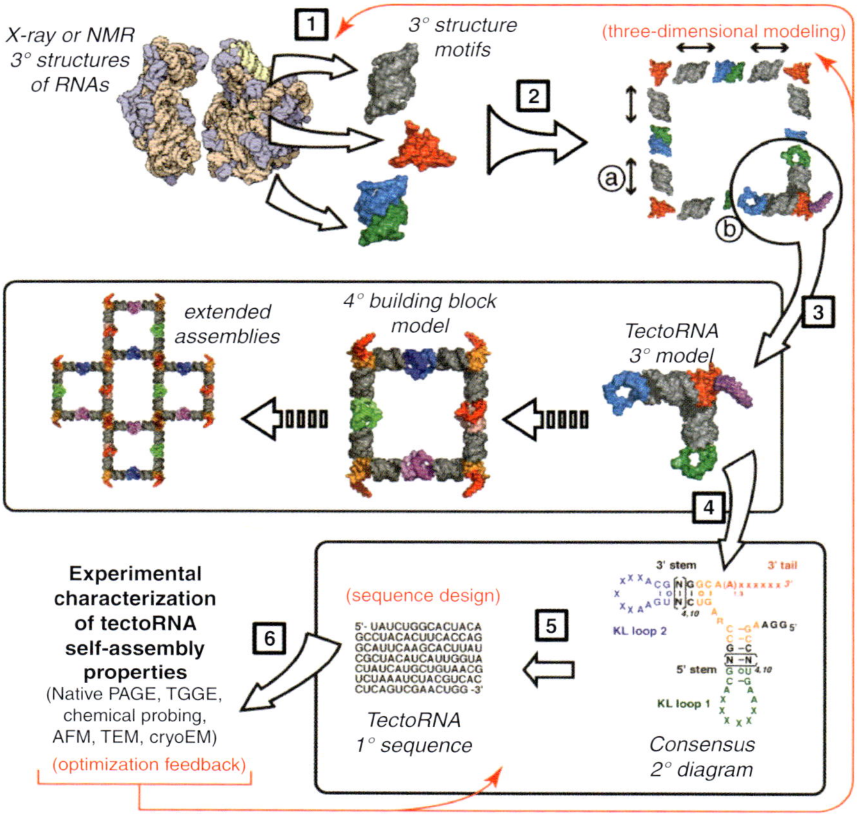

Figure 8.1 The general concept of RNA nanodesign (the RNA architectonics methodology [2]). The process of engineering artificial tectoRNA architectures is a multistep procedure. First, RNA fragments are extracted from the PDB database [1]. The structural fragments are then reassembled into artificial RNA molecules by computer 3° modeling [2]. This process involves optimizing the length of helices that connect motifs [2a] and assigning connection points for quaternary assembly [2b]. The final tectoRNA model [3] can then be used for the hierarchical assembly of multimeric RNA structures. The computer-generated tertiary models are then used as scaffold to define consensus 2° diagrams [4], which are then used as blueprints for designing RNA 1° sequences [5]. During the sequence design process the 1° sequences are optimized to maximize their thermodynamic stability. The RNA sequences are synthesized by either chemical or enzymatic methods and characterized for their expected folding and self-assembly properties [6]. The rational design of tectoRNAs can be further optimized by returning to the sequence design or 3° modeling steps based on experimental results (red arrows)

motifs serves as point for self-assembly into larger structures by docking with its partner (the other half of the kissing loop) that is contained in another component, thus forming a very stable, noncovalently linked interaction (see Refs [26, 27] for examples).

4. Generate a secondary structure diagram describing the 3-D model, where specific nucleotides are fixed and other nucleotides are allowed to be variable.

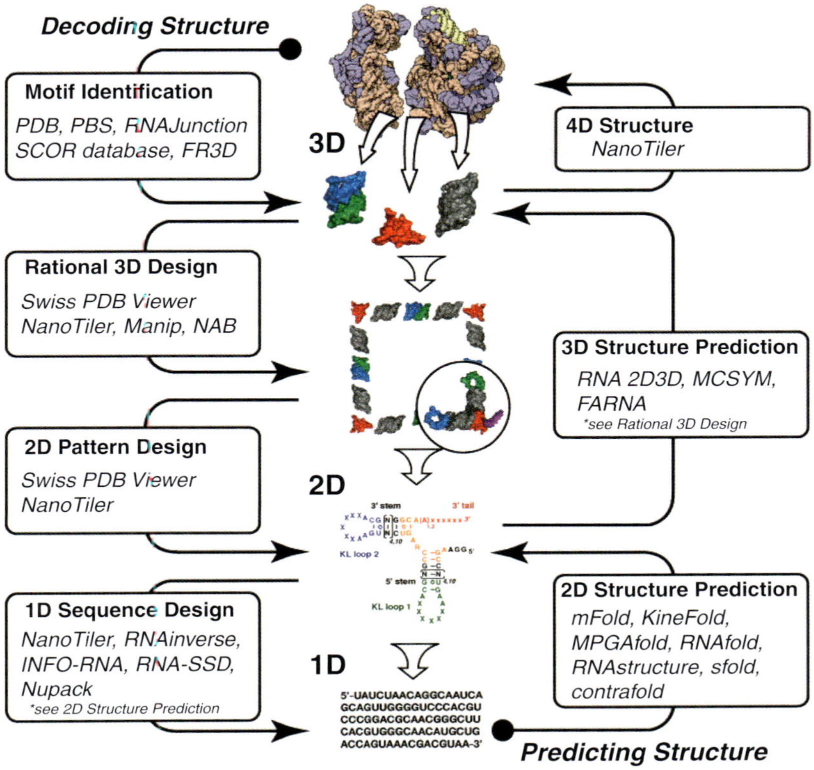

Figure 8.2 *The various steps of RNA nanodesign that are presently facilitated by computer programs. The majority of programs mentioned in the figure boxes are discussed in Ref. [28], except NanoTiler (see text and Ref. [33]), FARNA [34], FR3D [35] and Nupack (http://www.nupack.org; J.N. Zadeh, R.M. Dirks and N.A. Pierce, unpublished results)*

5. Design sequences for the variable residues that stabilize the desired secondary structure and incorporate the chosen nucleotide sequences into the 3-D structure.
6. Relax the structure using molecular mechanics, and test for 3-D structural stability using molecular dynamics.
7. Further optimize the primary sequence through experimental testing.

Many of the steps of inverse folding (Figure 8.1), which originally were performed by hand, can now be automated through an assortment of highly useful bioinformatics computer programs (Figure 8.2) [28]. Computational methods derived from experimental data gathered on RNA and DNA base pair (bp) energies have been developed over the years which can predict with reasonable accuracy, especially for shorter sequences (<100 nt), the secondary structure of an RNA [6, 28–31] and with much more difficulty the 3-D atomic structure of an RNA (see Section 8.3) [28, 32].

8.2.2 Necessary Criteria for Rational Design

The ability to rationally design RNA-based nanoparticles is grounded in part by the notion that some sequences of RNA have the ability to fold autonomously into a precise 3-D structure outside of their natural context in Nature. These autonomous folding RNA sequences are called motifs, and are often highly prevalent in the database of solved RNA crystal and NMR structures. However, not all solved RNA structures are autonomous folding domains; it is still not yet possible to tell if a sequence of RNA appearing in a high-resolution X-ray structure will fold into the same shape in a different molecular context simply by examining its structure. Thus, it is necessary to consider several criteria for ranking RNA motifs during the initial stages of RNA design:

- Can a given structure of RNA be considered a motif?
- Is the motif recurrent within multiple RNA structures?
- Is the motif able to fold outside of its natural context?
- What is the stability of the motif outside of its natural context?
- What is the stability of the motif when associated with other motifs or helical connectors?
- What is the relative flexibility of a motif within the desired design framework?
- How do environmental factors such as temperature and ion concentration affect stability?

An important concern regarding the use of 3-D RNA motifs to form cyclic nanoparticles is the number of helices that must be connected using a given motif. This is important, since with an increasing number of 3-D constraints it becomes progressively more unlikely that one can find an appropriate motif in a motif database, especially if motifs are seen as rigid building blocks. As will be shown later, this is less problematic for motifs that connect two helices (kissing loops, internal loops and bulges). Connecting three-way or higher-order junctions can become difficult to the extent that the motif-approach might have to be augmented or generalized, for example, through the use of synthetic motifs computationally generated upon request or a facility to account for the flexibility inherent in many motifs. As a rule of thumb, it can be said that the smaller the number of helices emanating from the building blocks, the easier the design task. It is, however, important to bear in mind that, rather than being extremely rigid, RNA structural motifs have an inherent flexibility that allows room for adjustment (see Section 8.3.4.2).

8.2.3 Towards a Better Understanding of RNA Parts

A RNA motif is a structure of RNA that has a similar 3-D shape according to X-ray crystal structure and nuclear magnetic resonance (NMR) data in ideally more than one example. The motif corresponds to a specific set of nucleotides that define a sequence signature, or the minimal information necessary for folding a nucleic acid sequence into a specific 3-D structure.

A useful and simple criterion for choosing a RNA motif to test experimentally is its recurrence within natural RNA molecules that have been solved either by X-ray crystallography or NMR. *Recurrence* is defined as a motif found in multiple locations within a single structure, or in different molecules, that seems to be naturally selected for a specific biophysical property, such as structural rigidity or flexibility. A higher recurrence of a motif, especially within different structures, is evidence that a structure is robust in its surrounding environment and will not likely deviate from its native fold in a new rationally designed environment. This is the case for numerous RNA interactions that have already been classified as motifs in the literature (11 nt motif, sarcin loop, T-loop, etc.) [36, 37]. However, some motifs may be stabilized outside of their normal environment by the addition of helical segments that emanate from their junction stubs.

The other criteria listed cannot easily be determined from the available atomic resolution structures of RNA, as these structures cannot provide information about how an RNA sequence will behave when removed from its native environment. Therefore, once a motif is identified, experimental data must be acquired to address the additional criteria for rational design [2]. Additionally, while experimental techniques may allow us to explore a handful of new RNA motifs systematically, this is both very time-consuming and expensive. Despite the prevalence of identified RNA motifs to work with, there are certainly numerous RNA motifs that remain unidentified or untested. For this purpose, computerized algorithms can aid us in identifying potential RNA motifs, saving both time and energy for the RNA designer.

8.2.4 RNA Motif Detection and Search

Several tools are presently available to facilitate the identification of prevalent structural motifs in atomic RNA structures. Backbone-based motifs are very useful for classifying structural fragments that consist of only one RNA strand. Several methods have been described for finding RNA structural motifs that are defined through their backbone conformation [38–42].

Other programs such as MC-Annotate and MC-Search, as developed by Major and coworkers, can also be used to analyze and search RNA 3-D structures for structural motifs. These programs are based on a 4×4 homogeneous transformation matrix (HTM) that can store the relative position and orientation of any two base pairs [43, 44]. By using the HTM formalism, it is possible to apply subgraph isomorphism algorithms to detect 3-D motifs. Harrison and coworkers [45] have used a similar approach to detect small RNA motifs and non-Watson–Crick base pairs. JunctionScanner – a program that uses HTMs to describe helix orientation – is used to identify the motifs that are present in the RNAJunction database (see below) [46].

The alignment of RNA tertiary structures (ARTS) can be used to pair RNA or DNA 3-D structures [47], allowing the potential identification of new structural motifs. More recently, a suite of Matlab programs, called FR3D, was developed to search for structural motifs within X-ray structures, given a query motif that consists of a set of nucleotides in addition to information about base pairing or base stacking [35].

8.2.5 RNA Parts Databases

Several databases are dedicated to providing and annotating RNA tertiary structures. For example, SCOR is a database that contains RNA structures classified by either function, 3-D motif or tertiary interaction type [48]. Among other things, it contains internal loops, hairpin loops, kissing loops, pseudoknots and several other types of motifs. Alternatively, the nucleic acid database (NDB) contains annotated and categorized RNA and DNA structures from X-ray crystallography and NMR experiments. NDB also contains several RNA junctions [49].

The ability to build RNA-based nanoparticles can be greatly facilitated, however, by the existence of databases which focus more on the topological characteristics of RNA building blocks, such as RNAJunction [46]. The RNAJunction database was built by scanning the PDB database for various RNA building blocks (junctions, kissing loops, internal loops and bulges). The extracted building block junctions are classified according to the number of helices that emanate from the junction. Thus, two-way junctions are derived from internal and bulge loop structures. The latter was further characterized in Ref. [50]. Higher-order emanations give rise to three-way, four-way, and up to nine-way junctions, while kissing loop structures are also found and included in the database. A very important characteristic of these junctions is the angles that the emanating branches form with one another. Whilst the dynamics of junctions or the range of angles that they can accommodate are still not known, this forms a starting point for investigation. Building blocks containing appropriate angles can be searched for in the database and then used to form the desired shapes. Currently, the RNAJunction database contains over 13 000 entries, thus permitting the potential building of a very large number of structures over a wide variety of shape characteristics. This large number of structural elements is also due to redundancy between deposited PDB structures. The database contains a scheme to reduce the redundancy to some extent by providing clusters of structural elements. All elements of a cluster containing more than one element are required to consist of the same sequences and to be no more then 3 Å DRMSD different from at least one cluster member. This simple definition reduces the number of 13 108 structural elements to 2672 structure clusters.

Clearly, more investigations must be completed in order to further reduce the redundancy of the RNAJunction database. The simple rule of requiring all structural elements of a cluster to consist of the exact same sequences is arguably too strict, and might not reflect the structure and sequence variability of an RNA motif. Categorizing motifs according to their phylogenetically recurrent sequence patterns could facilitate this endeavor. Moreover, some of the entries may not be stable if isolated from their environment. A computational or experimental assessment of the stability and flexibility of a large set of RNA structural elements would be highly beneficial for the rational design of RNA nanoscale structures.

8.3 Computational Approaches for Automation

Numerous computational tools have been developed to aid in the creation of RNA 3-D structure models, and several of these are reviewed in Ref. [28]. The program ERNA-3D allows the real-time interactive manipulation of protein and RNA structures, whereby an RNA 3-D structural model can be generated using a secondary structure representation [51]. The Nucleic Acid Builder (NAB) is a programming environment for generating nucleic acid structures [52], while the make-na web server uses the NAB functionality and provides for user-friendly generation of simple tertiary structures (helices) from a primary structure (sequence). The program MANIP can rapidly assemble motifs into a complex 3-D structure [53], while the software S2S can display, manipulate and interconnect RNA 3-D structures, multiple sequence alignments and secondary structures [54]. More recently, the Shapiro laboratory developed RNA2D3D [55] (see below), an interactive program for generating, visualizing, editing and comparing RNA 3-D structures (including RNA nanodesigns) generated from secondary structure representations. MC-SYM is a program that uses a constraint satisfaction algorithm to generate sets of RNA 3-D structures that are compatible with constraints defined through the RNA sequence and secondary structure [56]. FARNA is a *de novo* RNA structure prediction method derived from the successful Rosetta protein structure prediction program. It is based on the rapid assembly of RNA structural fragments, combined with a relatively simple scoring function [34].

8.3.1 Computational Design Methodology

The design of RNA-based nanoparticles requires more than a programmatic means for the user to place helices and structural elements in a 3-D workspace. Many steps of the design process still require a high amount of human judgment. In order to expedite the design of RNA-based nanoparticles, a software system called NanoTiler was developed [33]. This program can be used in a variety of ways, including the two alternative approaches of design mentioned, and it is intended to be very versatile in order to accommodate the particular engineering needs of the user. NanoTiler assumes that each building block motif is a rigid body, which greatly speeds up the processes involved as each atom of the building block does not have to be considered.

Given that one has a rough specification of a target shape and the building blocks to be used in its design, NanoTiler allows the user to specify which helical 'stubs' from the RNAJunction database motif to connect to linker helices, as well as to specify the size of the used helices. The program also allows the user to specify base pair constraints within the design. NanoTiler then attempts to connect the helices to the stubs using a simulated annealing algorithm [57] to optimize the attachment point connections while maintaining proper steric properties. The simulated annealing algorithm optimizes the distances between the connecting elements, and then allows uniform bending in the helices to optimize steric interactions.

NanoTiler also allows RNA nanoparticles to be designed through the automatic combinatoric exploration of shapes generated from RNA motifs derived from the RNAJunction database and variable-sized connecting helices. In this methodology, a list of motifs can be specified, where linking helices of various sizes are attached to specified stubs of the seeded motifs. These connectivity specifications are iterated, and a fitting process is applied to ensure reasonable steric qualities at the points of connection. As a result of the process, various shapes are generated. Some resemble dendrameric structures, while others produce closed rings. Additionally, a closure checking procedure can be activated to detect structures that form rings or cycles within a specified tolerance level. Likewise, a related methodology checks for general collisions, and then terminates the procedure. An example of one of these generated structures is illustrated in Figure 8.3.

The process starts by applying the JunctionScanner program (part of the NanoTiler package) to a set of RNA 3-D coordinate files in order to obtain a set of building blocks (Figure 8.3; Detect Motifs). Alternatively, the NanoTiler program can read building blocks that are downloaded from the web site of the RNAJunction database [46]. By using a graphical user interface or a scripting language it is possible to allow the program to perform a combinatorial search, for example, for ring structures (Figure 8.3; Ring Search). Ring structures found in this manner often contain a gap or a small collision at one point in the ring, but this can be improved by optimizing the fit of the helices using constraint

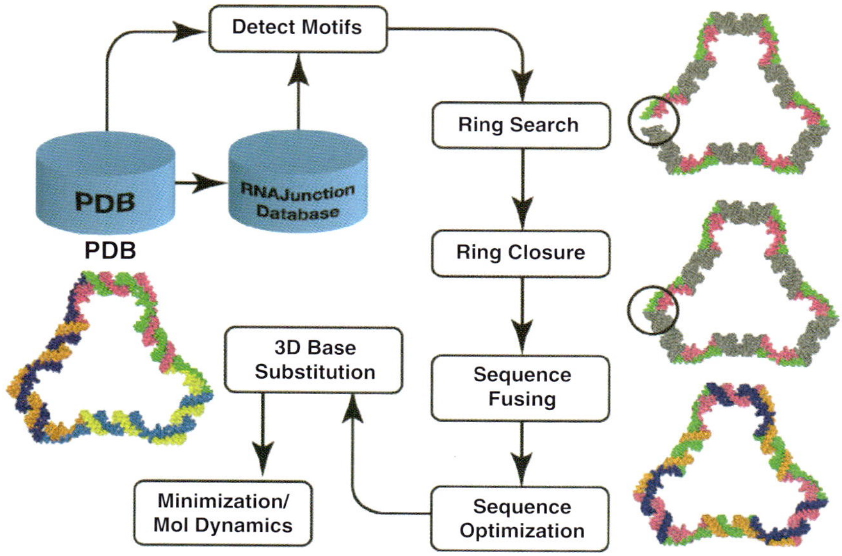

Figure 8.3 *A real example of the workflow showing how RNA ring structures can be designed in an automated fashion, starting from building blocks that are extracted from PDB coordinate structures [33]. The steps are:* ***Detect Motifs*** *– molecular fragments representing junctions, kissing loops, internal loops or bulges are extracted from RNA coordinate structures either manually or with programs such as JunctionScanner;* ***Ring Search*** *– using molecular modeling or a combinatorial search procedure to find a combination of motifs that together with connecting helices can be assembled into approximately closing ring structures;* ***Ring Closure*** *– modify the computational model to improve ring closure (interactive molecular modeling or constraint satisfaction);* ***Sequence Fusing*** *– use interactive molecular modeling or an algorithm (NanoTiler) to fuse the sequences of the molecular model in order to obtain building blocks that are connected through either 'sticky ends' or kissing loop interactions;* ***Sequence Optimization*** *– apply a sequence optimization algorithm to the set of sequences and target secondary structure;* ***3D Base Substitution*** *– modify the 3-D coordinate model by mutating the bases according to the optimized sequence;* ***Minimization/Mol Dynamics*** *– apply physics-based minimization and simulation methods such as molecular dynamics in order to refine the structure. Copyright (2008), with permission from Elsevier. Ref. [33]*

satisfaction. The user can either specify these constraints individually or allow an algorithm to suggest constraints that, in a subsequent step, can be used to improve the ring closure by a constraint satisfaction approach (Figure 8.3; Ring Closure).

The sequences of the structure with improved ring closure still correspond to the initial small fragments or the linker helices. The sequences can be fused or split individually using the NanoTiler program, or an algorithm can be used to suggest a set of fused sequences (result shown in Figure 8.3; Sequence Fusing).

Sequence optimization takes the initial sequences of the structure and optimizes them as described in Section 8.3.2. The resulting changed nucleotides lead to an automatically modified 3-D model (Figure 8.3; Sequence Optimization and Base Substitutions). This model, when generated in various steps, should be of sufficient quality to be subjected to energy minimization and molecular dynamics with packages such as AMBER [58] (Figure 8.3; Minimization/Mol Dynamics). Between the described steps it is possible for the user to inspect the status of the current molecular model.

Whilst the structures generated by this process might lead to interesting nanoparticles, it is clear that a careful evaluation of their assembly properties will need to be experimentally assessed. For instance, the energies of the loop-closing connectors will need to be considered in order to predict accurately which ensemble of closed rings will form in the test tube.

8.3.2 From 3-D to 2-D Structure Design: Computational Sequence Optimization

One crucial component in the design of RNA nanostructures is the determination of the sequences that will be used to fold into the designated structures. Correctly specified sequences are important because a delicate balance must be struck between local intrastrand interactions and the formation of hybrid cross-strand interactions that could impede the formation of a correct structure. Several methods have been developed for the task of RNA sequence design. The program RNAinverse, which is included in the Vienna RNA package [59], uses a local search strategy to either minimize the distance between the minimum free energy structure of the current sequence and the target structure or, in a different mode, maximize the probability of folding of the current sequence into the target structure. The program INFO-RNA uses two stages: in the first stage, an initial solution in terms of the designed sequence is determined with the help of a dynamic programming algorithm. In the second stage, a local search is applied using an objective function that takes the structure distance between the current and target structure into account [60]. The RNA Secondary Structure Designer (RNA-SSD) attempts to minimize a structure distance based on a recursive stochastic local search [61].

The NanoTiler software also has a facility to optimize an RNA sequence to fold preferentially into a given structure. Unlike the above-mentioned sequence design programs, it is able to optimize a set of different RNA sequences simultaneously. The initial sequences that are produced by piecing together the motifs and linker helices are, by the nature by which they are produced, fragmented into their individual components. A first step in the sequence generation process is to fuse together the individual fragmented strands into one continuous fragment that will constitute a complete unit for self assembly. NanoTiler has this facility, as well as the ability to specify where the 5′ and 3′ ends of the fused fragments should be. In order to limit the degree of cross-talk between the fused fragments, an optimization methodology is employed that uses in part the program RNAcofold [62], an algorithm that when given a pair of sequences attempts to determine the degree to which the two strands will fold independently or will interact. Energy measures are used to indicate the strength of these interactions.

Part of the sequence optimization process is to generate random mutations to specified parts of the sequence, and to declare other portions of the sequence off limits to change because those bases may be involved in complex tertiary interactions which, if disrupted, would cause the structural motif to degenerate. One might allow, for example, linker helical regions and kissing loop interactions to vary, as long as base complementary is preserved. Currently, a score is produced that measures the number of correct base-pair interactions that occur versus the number of incorrect base-pair interactions that occur, including cross-strand interactions. Another term of the scoring function favors designed sequences that have a specified G+C content. This prevents the algorithm from generating sequences that have a very low energy but an impractically high G+C content. The sequence optimization algorithm is run multiple times while continually recording the best scored sequences such that, ultimately, the best sequences are produced.

Once the desired sequences have been generated, the mutated bases are substituted back into the original 3-D structure produced by the above-described methods, using another component of the NanoTiler software. Thus, on completion of this procedure the sequences that were designed to fold into the self-contained assembly units with the appropriate motifs as well as the 3-D model of the RNA nanostructure have been produced.

8.3.3 From 2-D to 3-D RNA Structure Design

Another software system, RNA2D3D [55], permits the modeling of RNA 3-D structures given secondary structure descriptions. The system takes a planar secondary structure layout and initially embeds 3-D representations of the bases perpendicular to the plane. A winding procedure is then invoked, which imparts A-form helicity to the base-paired regions. Various molecular editing facilities and molecular mechanics and dynamics can then be used to interactively refine the model. Within RNA2D3D's interactive environment it is also possible to import motifs from the PDB database and create connectivities between kissing loop structures and single-stranded regions. Thus, for example, it was possible to produce tectosquare models including meshes (see Ref. [55]).

Because the building block elements are initially modeled as rigid bodies, issues concerning ring closure are easily visualized. Interactive molecular editing allows for the exploration of various rotations, base-pair manipulations, stacking and unstacking of helices, all of which give the user very important insights into the structural make-up of the modeled RNA nanoparticles.

RNA2D3D can be particularly useful at times when it may be necessary to construct RNA nanostructural motifs *de novo*. Under these circumstances it can be used in conjunction with secondary structure prediction programs to build a 3-D model of the desired motifs. The design of the entire nanostructure can then proceed using a combination of the facilities that are available in RNA2D3D, or those available in NanoTiler.

8.3.4 Further Optimization of Three-Dimensional RNA Designs

8.3.4.1 Structure Energy Minimization

The procedures that are described above ultimately yield fairly good 3-D all-atom models of the desired RNA nanostructure. However, because the components of the structure have been pieced together, portions of the structure at the joining points may be somewhat distorted; the bond lengths and angles may not be correct. Because of the inherent flexibility found in RNA helices, these distortions can – in principle – be relaxed out (Figure 8.4). To accomplish this, the created sequence-optimized 3-D nanostructure is subjected to molecular mechanics minimization using Amber [58], although other molecular dynamics packages may also be used [63–68]. This process is based on principles derived from Newtonian mechanics and the use of partial charges that are associated with each atom that makes up the nucleic acid structure. It allows relaxation of the structure by optimizing

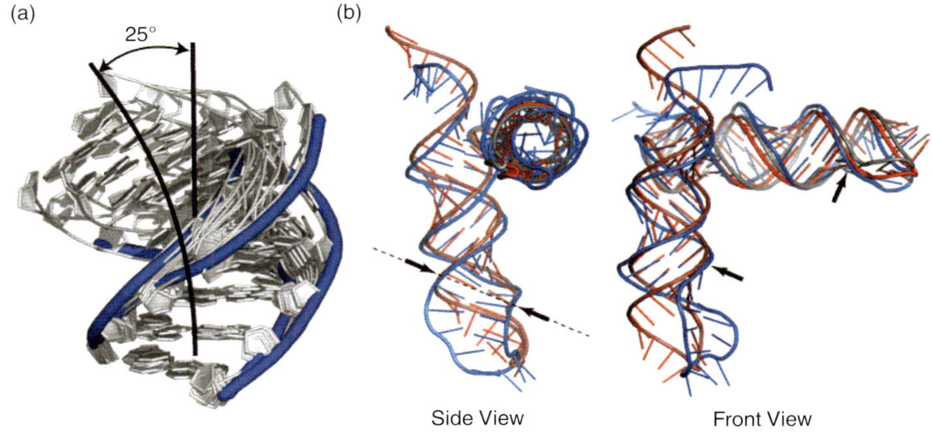

Figure 8.4 *The inherent flexibility of RNA. (a) Inherent flexibility and springiness of regular RNA helices within the natural context of the ribosome. Fifteen helices taken from the 2.4 Å* Haloarcula marismortui *23S rRNA (PDB ID: 1FFK) were superimposed at one WC position. The angle measured between the C1 position of the backbone at the two extremes (shown in blue) was 25° over a 7 bp span of helix; (b) Inherent flexibility of a tectosquare (LT17) building block monomer (A3s) explored with the help of a 30 ns molecular dynamics (MD) simulation. Shown in gray is an idealized monomer created with the help of the program RNA2D3D. An RNA tectosquare built out of such blocks does not close in 3-D modeling. Shown in red is a version of the monomer with a 26° twist added to its 5′ helix. A tectosquare built out of four monomers modified in this fashion does close in 3-D modeling. The modified monomer, therefore, is used as a reference structure. Shown in blue is a structure selected from an MD trajectory of an unmodified monomer (gray), based on its low RMSD value measured relative to the backbone P atoms of the second closing base pairs of both hairpin loops in the reference structure (indicated with black arrows). The flexibility of the monomer (twisting and bending motions of its helices) brings it close to the reference structure at multiple points of the MD trajectory*

the bond lengths and bond angles, in principle moving atoms that are too far apart closer together and moving those atoms that are too close further apart.

8.3.4.2 Molecular Dynamics

In some cases the structure derived from the previous step might be sufficient for use in the next major stage, which is experimental verification (see below). However, as some of the junction motifs that are used in the building process are derived from much larger structures (e.g. the ribosome), there is some question as to the innate stability of the motif out of its original context. In some cases, attaching stable constructs to the stubs of the motif can ensure this stability. These stable constructs might include helical extensions or known stable tetraloops (e.g. the GAAA and UUCG tetraloops). Another approach, which can be applied to the stabilized structures, to the original motif itself or to the entire nanostructure, is to apply molecular dynamics. Again, the Amber package is used to accomplish this. During molecular dynamics, heat is applied to the system with water and ions, thus imparting kinetic energy; atomic molecular motions are then observable and can be analyzed. From this, it is possible to discover points of weakness in the structure or motif if, for example, the structure begins to fall apart and the bonds break. Properties such as the maintenance of angles or planarity can also be observed. However, molecular dynamics, because of the large number of atomic interactions that must be considered (this includes the solvent, solute and ions) can be quite time-consuming and computationally intensive. Structures consisting of hundreds of nucleotides can take several weeks of computing time to observe a molecule's behavior over just 30–50 ns.

Molecular dynamics can also be used to help analyze the inherent flexibility that exists in the structural motifs. This can be important because the requirement that the individual pieces comprising a structure be rigid bodies presents potential problems when trying to determine whether ring structures will close. An example of this arose when building the tectosquares described Section 8.5.3. Ring closure did not occur; however, when molecular dynamics was employed to one right-angle building block (consisting of the ribosomal right-angle motif with a concatenated HIV kissing loop motif), there was found to be enough flexibility to induce closure of the ring (see Figure 8.4b).

8.4 Synthesis and Experimental Characterization of RNA Nanoparticles

8.4.1 RNA Nanoparticles Synthesis

RNA nanoparticles can be produced by either chemical or enzymatic synthesis. The chemical synthesis approach offers the advantage of creating oligonucleotides with a large variety of modified nucleobase analogues that can be incorporated within the sequence with high precision. Presently, DNA oligonucleotides of up to 120 nucleotides can be chemically synthesized using phosphoramidite technology; however, a lower coupling efficiency limits the length of single-stranded RNAs to 45–50 nucleotides [69]. On the other hand, RNA of almost any size and sequence can be obtained by enzymatic synthesis. *In vitro* enzymatic synthesis methods include cloning [70], *in vitro* RNA transcription of plasmid and polymerase chain reaction (PCR) -generated DNA templates using T7 RNA polymerase [71, 72]. DNA templates generated by PCR from synthetic DNA molecules code for the antisense sequence of the desired RNA molecule, and are amplified using a forward primer containing the T7 RNA polymerase promoter in combination with a reverse primer. The forward and reverse primers are designed to hybridize to the template with a T_m $\sim$56 °C, and their sequences should be optimized to eliminate any alternative pairing within themselves [71]. Additional sequence constraints are necessary, especially when the RNA sequences need to be generated by *in vitro* transcription from PCR-generated templates. For instance, most T7 polymerase-generated RNAs will begin with 5′-GGGAAA and end with U-3′. Furthermore, the sequence that is chosen through the optimization procedure needs to be successfully amplified through PCR in its DNA form. Therefore, in order to prevent the formation of stable 2-D structures at the level of the DNA template, the regular helical regions include one G-U wobble base pair for every 5–6 base pairs in the RNA strand (much like in the helical regions of natural stable RNAs, where the maximal length of fully regular Watson–Crick helices is rarely greater than 7 bp). The final RNA products obtained after *in vitro* transcription are expected to fold into the predicted geometric shape and assemble into the desired RNA nanoparticle, according to the design.

8.4.2 Optimization of Folding Protocols

Even if the secondary structure of an RNA nanoparticle is stable and promotes a unique fold, the assembly and folding protocols often must be optimized by empirical trials. One important criterion is to optimize the metal ion concentration. Metal ions essentially screen the negative charges on the phosphate backbone, thus forcing the loosely folded intermediate to go through conformational rearrangements before adopting the final 3-D structure [5, 73]. The presence of divalent metal ions such as magnesium are absolutely required for efficient assembly and to stabilize the tertiary fold of RNA nanoparticles at low monovalent salt concentrations [71, 74, 75]. For this purpose, native polyacrylamide gel electrophoresis (PAGE) must be carried out at various magnesium concentrations to estimate the magnesium dependency of the RNA nanoparticle, which greatly depends on the particular RNA motif it contains.

8.4.3 RNA Programmable Self-Assembly

Programmable self-assembly is defined as the assembly process where molecules can be controlled with high precision to fold and assemble into predefined 3-D architectures [2]. There are two main approaches in the assembly of nucleic acid architectures. The first approach – known as the '*one-pot assembly*' – is a single-step assembly process in which all the units that make up the nanoparticle are mixed together and assembled via a slow annealing process [76, 77]. According to the different energetics of the secondary structure pairings, the most stable substructures fold first. As lower temperatures are reached, larger, complex architectures are formed through weaker interactions. The second approach – 'step-wise assembly' – is a hierarchical self-assembly strategy in which various subunits (tiles) are first separately formed through the formation of long-range RNA interactions such as loop–loop or loop–receptor interactions at low magnesium concentrations, and then mixed together to form the final complex architecture at high magnesium concentrations [26, 78, 79]. These long-range interactions have different thermodynamic strengths and dependencies on divalent ions, thus allowing monitoring of the stepwise assembly of architectures with increasing complexity. Although the stepwise assembly strategy is more time-consuming, it offers the advantages of tuning the assembly protocols by adjusting the magnesium ion concentration and temperature. Thus, it is possible to use a reduced number of loop–loop or tail–tail interactions by uncoupling tile formation from the formation of the supramolecular architecture [26]. For this purpose, the melting temperature of the tiles and the resulting nanoparticle should be kept well separated. Another advantage of stepwise assembly is that it can be used to generate programmable self-assemblies with a finite size in which all the positions of the subunits are precisely known within the context of the nanoparticle [26, 78].

8.4.4 Biochemical Characterizations

The characterization of RNA nanoparticle assemblies can be studied using classical biochemical methods, such as nondenaturing PAGE or agarose gel electrophoresis to verify that nanoparticle subunits can assemble into desired architectures. This step is also essential to optimize folding and assembly protocols.

The thermodynamic stability and robustness of RNA nanoparticles can be assessed using thermal gradient gel electrophoresis (TGGE), a method used to separate different assemblies based on their temperature-dependent conformational changes [80,81]. In brief, during TGGE a linear temperature gradient is applied perpendicular to the electric field. As the temperature is increased, supramolecular architectures first lose their quaternary structures and finally disassemble into monomers. TGGE can be a very useful method for investigating the thermodynamic contribution of a particular RNA motif to the overall geometry and the stability of the resultant supramolecular architecture. This is possible by introducing sequence mutations at key tertiary nucleotide positions within the RNA motif. Mutated RNA assemblies are used as negative controls for comparison with nonmutated assemblies. By performing TGGE with a linear temperature gradient applied parallel to the electric field, it is possible to compare side-by-side the thermal stability of alternative architectures. Thus, this approach represents a powerful means of investigating and characterizing the structural properties of tertiary RNA motifs in their native state [26,82,83].

Probing RNA nanoparticles in solution and the mapping of secondary and tertiary interactions provides valuable insights into the 3-D structure of RNA molecules. In this approach, the RNA molecule is either 5′ or 3′-end labeled prior to RNase hydrolysis or modification. The labeled RNA is then subjected to attack by a chemical or enzymatic probe, thus allowing testing of the reactivity of every nucleotide. Some examples of chemical probes include dimethylsulfate (DMS), which reacts primarily with N7-G, N1-A and N3-C, and diethylpyrocarbonate (DEPC), which reacts primarily with N7-A [84]. Due to their small size, chemical probes are not sensitive to steric hindrance. In contrast, due to their bulky size, the enzymatic probes are sensitive to steric hindrance and thus can be sterically blocked by the particular 3-D structure of the RNA nanoparticle. By comparing the cleavage patterns of a combination of single and double-stranded specific ribonucleases such as RNase T1 and RNase V, it is possible to obtain information on the accessibility of the tertiary structure and the degree of protection towards RNase degradation [84,85]. Alternatively, Pb(II)-induced cleavage experiments can be performed to monitor the folding and assembly of RNA nanoparticles into the expected architecture [71,86]. In this approach, end-labeled RNA subunits are subjected to $Pb(OAc)_2$ cleavage after folding and assembly in the presence of desired $Mg(OAc)_2$ concentration.

8.4.5 Biophysical and Structural Characterizations

The two methods of choice for solving the structure of RNA molecules at atomic resolution are X-ray crystallography and NMR. Whilst these techniques are widely applied to a variety of RNA structures, they are still very time-consuming; moreover, the main limiting factors are a requirement for good diffracting crystals for X-ray crystallography and the need for large quantities of materials, for NMR. Therefore, less-resolving techniques such as atomic force microscopy (AFM) and electron microscopy may be very useful for validating the structure of rationally designed RNA nanoparticles. In recent years, both AFM [26, 87] and transmission electron microscopy (TEM) [72] have been used successfully to study the topology of small RNA nanoparticles that are fairly planar in shape. AFM can be performed either in air or in solution after depositing the nanoparticle onto a mica surface in the presence of magnesium ions. The overall geometric shape and height information can be assessed from AFM characterization; however, the 3-D shape of the RNA nanoparticles may be lost when the molecules are deposited onto the mica surface, or damaged under the force exerted by the AFM tip. Such flattening of 3-D structures makes it difficult to characterize 3-D architectures. In TEM visualization, methods of staining during sample preparation and beam damage represent some limitations that could impede the determination of 3-D structures. At the present time, cryoelectron microscopy (cryo-EM) coupled with single particle reconstruction seems to be one of the most powerful techniques [88] for the structural characterization of polyhedral RNA nanoparticles at low resolution [70]. In cryo-EM, the water is preserved in the specimen, such that the native structure of the macromolecule is maintained, while damage from chemical staining is also avoided. Recently, the cryo-EM technique has been successfully used to determine the 3-D structures of several nucleic acid-based polyhedra [70, 89, 90].

Small-angle X-ray scattering (SAXS) experiments can be performed to obtain additional information about the 3-D shape of RNA nanoparticles that are in the range of 5 to 50 nm in size. For example, diffraction information obtained from nanoparticles can be used to calculate the radius of gyration. Although one major advantage of using SAXS is that the nanoparticles can preserve their native state in solution, large amounts of highly purified RNA nanoparticles are required in order to obtain a good signal.

8.5 Examples of RNA Nanobiology

The rational design of RNA nanoparticles can be envisioned for a multitude of uses, including therapeutic devices, biosensors, substrates for crystallography and nanowires to name a few. Several examples of rationally designed RNA nanoparticles have been recently published Figure 8.5), and some of these are described briefly in the following sections.

8.5.1 RNA Nanoparticles using Loop/Receptor Interfaces

The first RNA nanoparticles generated using RNA architectonics were self-assembled through loop/receptor interfaces to form dimeric nanoparticles [71,82] (Figure 8.5a,b). The assembly of these nanoparticles was mediated by the class GAAA tetraloop/11 nt receptor interaction, which is a highly recurrent motif in large ribozymes. The loop/receptor-based dimer was subsequently characterized by NMR and X-ray crystallography, corroborating at atomic resolution the validity of the initial model obtained by architectonics [93] (Figure 8.6a). The assembling interface of these RNA nanoparticles was further used to build up a more complex system that produced RNA filaments and trimeric particles by combining multiple loop/receptor interactions with a four-way junction motif [72]. Additionally, the loop/receptor dimer was used as a scaffolding for the *in vitro* selection of new RNA tertiary interactions not yet found in Nature [86], as well as for creating new assembling interfaces that would take advantage of the loop C motif [94].

8.5.2 Phi29 Packaging Motor Particles

An RNA nanoparticle designed from a natural RNA motif found in the phi29 packaging motor was used as a therapeutic agent in regulating apoptosis in cell culture and in mice [20, 92]. The motif includes an RNA hairpin bulge–loop interaction and can assemble to form dimer and triangular shapes (Figure 8.5d,e). Each 'corner' of the triangle can be

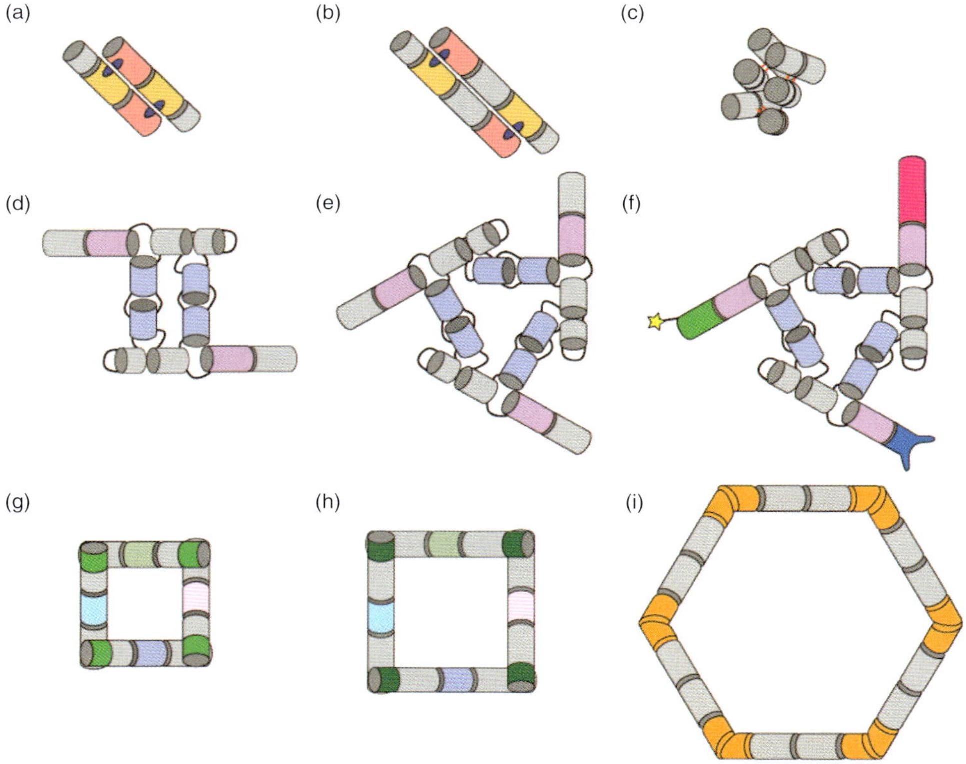

Figure 8.5 *Examples of recently reported RNA nanoparticles. (a) Loop-receptor tectoRNA dimeric particle using the 11 nt/GAAA interaction [71, 82]; (b) Construct in (a) with an additional helical turn added (shown in gray) [71]; (c) H-shaped tectoRNA nanoparticle [72, 82]; (d,e) pRNA dimeric and trimeric particles based on the phi29 packaging RNA [91, 92]; (f) pRNA trimeric particle functionalized on each unit [92] (the blue color shows the aptamer for the CD4 receptor, pink shows the siRNA (BIM) group, green shows the FITC fluorophore (star) attachment); (g,h) Small and large tectosquares based on the 'right-angle' motif (in green) [26]; (i) Computationally designed hexagonal nanoring based on the RNAi/RNAii inverse complex from* Escherichia coli *(orange motif) [27]*

engineered to contain different components with specific targeting or therapeutic properties (Figure 8.5f). For example, one of the nanoparticles had one corner containing a small interfering RNA (siRNA) that targets a specific gene for silencing, thus enabling apoptosis (cell death). A second corner contained an RNA aptamer that was used to target specific cell receptors (e.g. the folate receptor found on cancer cells), while a third corner contained a molecular beacon for visualizing the entry and location of the particle in a cell [92].

8.5.3 RNA Tectosquares

One of the first examples of RNA nanoparticle design incorporating numerous RNA motifs to build up complex assemblies is the tectosquare (Figure 8.5g,h). The tectosquare

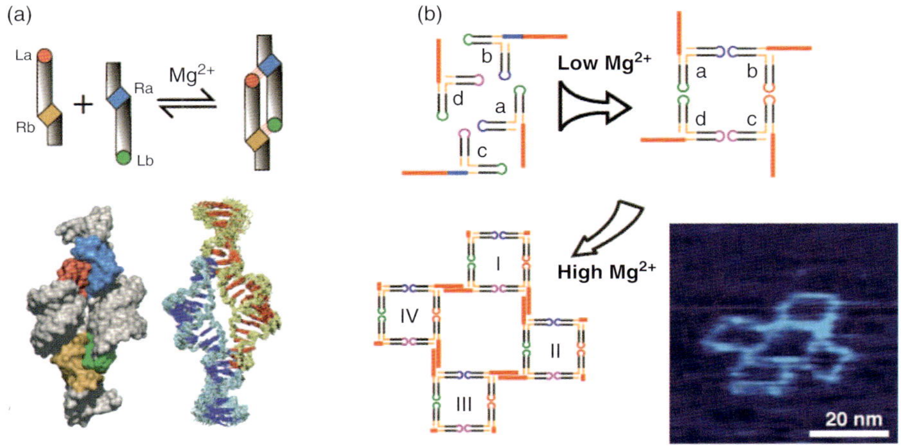

Figure 8.6 *Structural characterization of RNA nanoparticles. (a) Assembly of magnesium dependent loop-receptor dimeric tectoRNA particle. The original tertiary structure model [82] on the left is in remarkable agreement with the recent NMR solution structure [93]; (b) Hierarchical assembly of the first fully programmable RNA nanogrid with 16 distinct, addressable positions. The stepwise assembly process can be controlled by the order of molecule mixing, the temperature and, most importantly, the magnesium concentration [26]. AFM visualization of the nanoparticle assembly represents a means to confirm that RNA molecules assemble into the intended design*

is composed of four building blocks (tectoRNAs) that self-assemble through kissing loop interactions [26,95,96]. The basic assembly unit consists of two helical stems capped with a modified kissing loop derived from the HIV dimerization domain. At the junction between the two helical stems is a 90° bend, comprising a small 11-nucleotide motif taken from the crystallographic structure of the ribosome. The 90° bend motif, in addition to orienting the two stems of the unit, also provides directionality to the 3′ tail that can be used for additional supramolecular assemblies of multiple squares. The first fully programmable assembly of multiple RNA squares was a four-square cross composed of 16 unique tectoRNAs [26] (Figure 8.6b). Further variations on this molecular system permitted the construction of a ladder shape made of RNA, which was further functionalized by binding positively charged modified gold nanoparticles to the center of each tectosquare. The assembly of charged gold nanoparticles to this RNA scaffolding resulted in a self-assembling gold nanowire [96].

8.5.4 Hexagonal Nanoring and Nanotube Design

An example illustrating the rational design process of an RNA hexagonal nanoring (Figures 8.5i and 8.7) and nanotube will now be depicted [27]. Assuming that one wishes to design an RNA nanostructure that is hexagonal (i.e. having six sides and corners) for the potential attachment of functional units (e.g. aptamers, beacons therapeutic agents), the issue becomes one of finding an RNA motif that might satisfy such an initial specification. It transpires that the motif from the colE1 kissing loop structure (PDB entry: 2BJ2), satisfies a requirement that the kissing complex forms an angle of about 120°. Thus, one can use

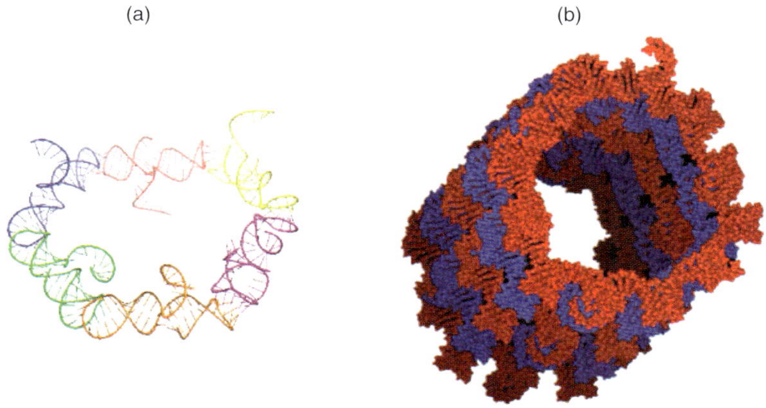

Figure 8.7 Depiction of the hexagonal nanoring and nanotube [27]. (a) The hexagonal ring with each color indicating the six building blocks. Note the single-stranded tails pointing in opposite directions; (b) Hexagonal nanotube built from the hexagonal ring. Each tail forms interlocking double-stranded fragments to permit stacking of the rings. Reprinted from [27] with permission from the American Chemical Society

essentially six copies of this complex to form the corners of the hexagonal ring. Six helical segments can in turn, connect these corners, with each segment having somewhat different sequence compositions to reduce potential cross-talk between the fragments. Molecular dynamics experiments have shown that the 120° angle is reasonably maintained (see Ref. [27]) and in addition, the structure remains relatively planar. It has also been shown [97] that the essential sequence elements of the kissing loop involve not only the loop bases that interact, but also the two base pairs that are part of the helices flanking the kissing loops. Stacking interactions involving these base pairs are important for maintaining the stability of the structure. This exemplifies the importance of maintaining the sequences in the motifs, thus leaving the sequence optimization to those fragments away from the motifs.

There are two ways in which the building blocks can be designed. As there are two halves to the kissing loop, designated as RNAIi and RNAIIi (note the notation 'i' – which actually indicates the inverse sequence which is more stable than the wild-type and was used in the NMR structure), one building block may contain the RNAIi loop motif on both ends while another may contain the RNAIIi loop motif on both ends. Alternatively, a single building block may be used by placing an RNAIi loop on one end and an RNAIIi loop on the other. The initial experimental data suggests that the second alternative works better than the first (I. Severcan, L. Jaeger, Y.G. Yingling and B.A. Shapiro, unpublished results).

Further alterations can be made to the structure that contains dangling single-stranded fragments that emanate from each side of the hexagonal ring. These fragments can be designed so that they are oriented in alternate directions – that is, with some pointing up and others pointing down (Figure 8.7a). Potential therapeutic agents can be attached to these dangling ends, or alternatively elements (such as siRNAs) can be designed into the hexagonal edges.

The dangling single-stranded fragments have yet another potential interesting use. One can conceive of making multiple hexagonal rings each containing single-stranded fragments that point in opposite directions and are complementary to the single strands on other rings.

One can imagine the self-assembly of these individual rings with interlocking single strands to form hexagonal nanotubes (Figure 8.7b). These nanotubes could potentially contain gold particles in their centers forming nanowires, and carry cargo that could be delivered in a controlled way by dissolution of the surrounding nanorings. Alternatively, they could serve as scaffolds for the generation of molecular superstructures by, for example, laying down proteins on the tubes.

The design process described above represents the beginning of a pipeline for the production of RNA-based nanoparticles. Ultimately, the design process as discussed must be used in conjunction with experimental verification, an issue which is best illustrated by the design of hexagonal building blocks for nanorings. The single building block approach, where the building block contained both the RNAIi and RNAIIi motifs, gave better initial experimental results than the double building block approach, where one building block contained the RNAIi loop motif on both ends and the other contained the RNAIIi loop motif on both ends.

8.6 Conclusions and Future Developments

8.6.1 Towards Full Automation of RNA Rational Design

In this chapter, a methodology for the design of RNA nanostructures using 3-D motifs as building blocks was outlined. Many of the steps (detection of 3-D structural elements, sequence optimization, 3-D structure refinement) can be automated to a large extent, while other steps – such as combinatorial searches for closed structures or automated ring-closure – could be automated for the case of RNA ring structures. However, for more complex structures, additional algorithm development will be necessary. One reason – as outlined in Section 8.3.1 – is that for higher-order junctions it is increasingly unlikely that an entry in a structural element database that fits the structural constraints imposed by the designer could be found. One way to improve this situation would be to extend the motif approach to synthetic (*in silico*) structural motifs that fit the target geometry. It might also be possible to modify the 3-D structure of the building blocks and to relax the approximation that they are rigid bodies. Lastly, it should be possible to connect helices with short, noninteracting single-stranded regions in a manner used for DNA nanostructure design [76–78, 98–101].

By using computational methodologies, it is now possible to generate 3-D models of a large number of hypothetical RNA structures. In the future, it will become increasingly important to be able to catalogue these structures, perhaps by their topology. It will also become more important to characterize the designs computationally, which means that, for a set of sequences, one would ideally have a computer program which generated a folding protocol (one-pot or stepwise assembly), and computed estimates for the homogeneity of the target RNA structure, as well as the stability and flexibility of the structure and its components at a defined temperature and salt concentration.

8.6.2 Towards More Complex Nanoparticles for Biomedical and Biological Applications

Today, the field of RNA nanobiology is still in its early stages of research and development. Yet, because of the potential wide range of applications that might result from these

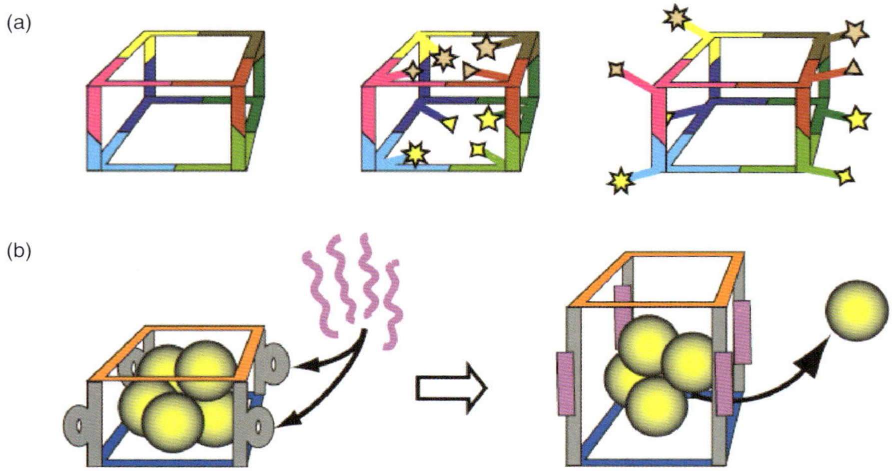

Figure 8.8 Towards multifunctional RNA nanocages. (a) RNA units entering into the composition of a 3-D nanoparticle can potentially be functionalized with various RNA modules (aptamers, ribozymes, siRNAs) that can be oriented either inward or outward; (b) A complementary RNA or DNA strand (in purple) binds to an RNA domain (gray blobs) within the RNA cage, triggering its expansion and letting the cargo free. The concentration of purple strands (e.g. intracellular mRNA triggers) controls the rate of release. The purple DNA or RNA strand could be found only in a specific cell type. A similar approach has been recently used to design expandable DNA-based 3-D nanostructures [101]

investigations, this is clearly an exciting area for pursuit, with many avenues still to be explored [2, 18–22]. These include the expression of self-assembling RNA *in vivo* for the eventual reprogramming of cellular pathways, and the functionalization of RNA-based nanoparticles for potential biomedical and biological applications. As mentioned earlier, aptamers, molecular beacons and siRNAs may be attached to these particles to facilitate cell entry, visualization and therapeutic results. Functionality can be realized by tapping into the many natural capabilities found in RNA. One can imagine RNA-based cages containing multiple therapeutic agents (Figure 8.8) which, if designed properly, could disassociate to release their cargo at predefined times. Functional nanotubes, nanowires and even more complex scaffolds with catalytic and responsive properties might also be realized in the not too distant future. For instance, combining RNA nanodesign with *in vitro* selection and evolution approaches (e.g. [102–106]) might represent a powerful way to create multifunctional catalytic nanofactories from predefined RNA structural scaffolds (e.g. Refs [107–109]). It is, however, clear that more fundamental studies aimed at understanding the principles of nucleic acid architectonics must be conducted in order to achieve better control over the movement, dynamics and responsiveness of these RNA-based nanomachines. Rather than a thermodynamic control of RNA folding and assembly, a critical step for achieving the expression of artificial RNA nanoparticles within a cell would be to control the kinetics of RNA folding under isothermal conditions. Likewise, due to the enzymatic instability of RNA towards ribonucleases, it would be necessary to develop the design and engineering of nanoparticles, taking advantage of RNA analogues such as LNA [110] or 2′OMe RNA [111].

In order to accomplish these goals, the computational and experimental procedures discussed above have still to be more fully automated to enable rapid design and experimentation. To this end, a database of functional building blocks must be developed in addition to the structural motif databases mentioned above: a combination of these two databases should then significantly expedite the design process. Coarse-grained computational techniques must be pursued to improve the speed associated with the 3-D modeling aspects of the described pipeline, as all atom molecular dynamics is very time-consuming. Databases incorporating the functional, structural and toxicological characteristics of the designed particles must also be established, and include information on the heterogeneity of specified RNA nanoparticles as found *in vitro* or *in vivo*. Moreover, databases incorporating information relative to nucleic acid structural analogues would be of prime importance in order to increase the chemical stability of RNA-based nanoparticles towards ribonucleases. The incorporation of human insights from experimentation into heuristic rules would also ultimately aid in the design process, by permitting 'educated guesses' with regards to many of the required properties. An important outcome of these studies would be a much wider comprehension of the functional and folding properties of RNA which, by themselves, are important if the general role that RNA plays in normal cellular function and disease processes is to be understood.

Acknowledgments

L.J. wishes to dedicate this book chapter to Edith Stein and Dietrich Bonhoeffer. These studies were partially funded by a NIH grant (R01 GM079604-01) to L.J. The authors wish to thank the Advanced Biomedical Computing Center (ABCC) at the NCI for their computing support. The studies were funded in part with Federal funds from the National Cancer Institute, National Institutes of Health, under Contract No. NO1-CO-12400. This research was also supported by the Intramural Research Program of the NIH, National Cancer Institute, Center for Cancer Research.

The content of this chapter does not necessarily reflect the views or policies of the Department of Health and Human Services, nor does the mention of tradenames, commercial products or organizations imply endorsement by the U.S. Government.

References

1. Gesteland, R., Cech, T. and Atkins, J. (2005) *The RNA World*, 3rd edn, Cold Spring Harbor Laboratory Press, Cold Spring Harbor, New York.
2. Jaeger, L. and Chworos, A. (2006) The architectonics of programmable RNA and DNA nanostructures. *Current Opinion in Structural Biology*, **16**, 531–43.
3. Westhof, E., Masquida, B. and Jaeger, L. (1996) RNA tectonics: towards RNA design. *Fold and Design*, **1**, R78–88.
4. Tinoco, I. Jr. and Bustamante, C. (1999) How RNA folds. *Journal of Molecular Biology*, **293**, 271–81.

5. Woodson, S.A. (2005) Metal ions and RNA folding: a highly charged topic with a dynamic future. *Current Opinion in Chemical Biology*, **9**, 104–9.
6. Mathews, D.H. (2006) Revolutions in RNA secondary structure prediction. *Journal of Molecular Biology*, **359**, 526–32.
7. Meyer, I.M. and Miklos, I. (2004) Co-transcriptional folding is encoded within RNA genes. *BMC Molecular Biology*, **5**, 10.
8. Mir, M.A., Brown, B., Hjelle, B. *et al.* (2006) Hantavirus N protein exhibits genus-specific recognition of the viral RNA panhandle. *Journal of Virology*, **80**, 11283–92.
9. Wong, T.N., Sosnick, T.R. and Pan, T. (2007) Folding of noncoding RNAs during transcription facilitated by pausing-induced nonnative structures. *Proceedings of the National Academy of Sciences of the United States of America*, **104**, 17995–18000.
10. Pan, T. and Sosnick, T. (2006) RNA folding during transcription. *Annual Review of Biophysics and Biomolecular Structure*, **35**, 161–75.
11. Maity, T.S. and Weeks, K.M. (2007) A threefold RNA-protein interface in the signal recognition particle gates native complex assembly. *Journal of Molecular Biology*, **369**, 512–24.
12. Schroeder, R., Grossberger, R., Pichler, A. and Waldsich, C. (2002) RNA folding in vivo. *Current Opinion in Structural Biology*, **12**, 296–300.
13. Jaeger, L. (1997) The new world of ribozymes. *Current Opinion in Structural Biology*, **7**, 324–35.
14. Chen, X., Li, N. and Ellington, A.D. (2007) Ribozyme catalysis of metabolism in the RNA world. *Chemical Biodiversity*, **4**, 633–55.
15. Linnstaedt, S.D., Kasprzak, W.K., Shapiro, B.A. and Casey, J.L. (2006) The role of a metastable RNA secondary structure in hepatitis delta virus genotype III RNA editing. *RNA*, **12**, 1521–33.
16. Lee, J.F., Stovall, G.M. and Ellington, A.D. (2006) Aptamer therapeutics advance. *Current Opinion in Chemical Biology*, **10**, 282–9.
17. Winkler, W.C. and Breaker, R.R. (2005) Regulation of bacterial gene expression by riboswitches. *Annual Review of Microbiology*, **59**, 487–517.
18. Isaacs, F.J., Dwyer, D.J. and Collins, J.J. (2006) RNA synthetic biology. *Nature Biotechnology*, **24**, 545–54.
19. Davidson, E.A. and Ellington, A.D. (2007) Synthetic RNA circuits. *Nature Chemical Biology*, **3**, 23–8.
20. Guo, S., Tschammer, N., Mohammed, S. and Guo, P. (2005) Specific delivery of therapeutic RNAs to cancer cells via the dimerization mechanism of phi29 motor pRNA. *Human Gene Therapy*, **16**, 1097–109.
21. Saito, H. and Inoue, T. (2007) RNA and RNP as new molecular parts in synthetic biology. *Journal of Biotechnology*, **132**, 1–7.
22. Keasling, J.D. (2008) Synthetic biology for synthetic chemistry. *ACS Chemical Biology*, **3**, 64–76.
23. Chworos, A. and Jaeger, L. (2007) Nucleic acid foldamers: design, engineering and selection of programmable bio-materials with recognition, catalytic and self-assembly properties, in *Foldamers: Structure, Properties, and Applications* (ed. S.H.I. Hecht), Wiley-VCH Verlag GmbH, pp. 291–330.
24. Shapiro, B.A., Bindewald, E., Kasprzak, W. and Yingling, Y.G. (2008) Protocols for the in silico design of RNA nanostructures, in *Nanostructure Design* (ed. R.N.A.E. Gazit), Humana Press, Inc., Totowa, N.J.
25. Madaio, M.P., Hodder, S., Schwartz, R.S. and Stollar, B.D. (1984) Responsiveness of autoimmune and normal mice to nucleic acid antigens. *Journal of Immunology*, **132**, 872–6.
26. Chworos, A., Severcan, I., Koyfman, A.Y. *et al.* (2004) Building programmable jigsaw puzzles with RNA. *Science*, **306**, 2068–72.

27. Yingling, Y.G. and Shapiro, B.A. (2007) Computational design of an RNA hexagonal nanoring and an RNA nanotube. *Nano Letters*, **7**, 2328–34.
28. Shapiro, B.A., Yingling, Y.G., Kasprzak, W. and Bindewald, E. (2007) Bridging the gap in RNA structure prediction. *Current Opinion in Structural Biology*, **17**, 157–65.
29. Andronescu, M., Condon, A., Hoos, H.H. *et al.* (2007) Efficient parameter estimation for RNA secondary structure prediction. *Bioinformatics*, **23**, i19–28.
30. Gardner, P.P. and Giegerich, R. (2004) A comprehensive comparison of comparative RNA structure prediction approaches. *BMC Bioinformatics*, **5**, 140.
31. Mathews, D.H. and Turner, D.H. (2006) Prediction of RNA secondary structure by free energy minimization. *Current Opinion in Structural Biology*, **16**, 270–8.
32. St-Onge, K., Thibault, P., Hamel, S. and Major, F. (2007) Modeling RNA tertiary structure motifs by graph-grammars. *Nucleic Acids Research*, **35**, 1726–36.
33. Bindewald, E., Grunewald, C., Boyle, B. *et al.* (2008) Computational strategies for the automated design of RNA nanoscale structures from building blocks using NanoTiler. *Journal of Molecular Graphics and Modelling*, **27**(3), 299–308.
34. Das, R. and Baker, D. (2007) Automated de novo prediction of native-like RNA tertiary structures. *Proceedings of the National Academy of Sciences of the United States of America*, **104**, 14664–69.
35. Sarver, M., Zirbel, C.L., Stombaugh, J. *et al.* (2008) FR3D: finding local and composite recurrent structural motifs in RNA 3D structures. *Journal of Mathematical Biology*, **56**, 215–52.
36. Hendrix, D.K., Brenner, S.E. and Holbrook, S.R. (2005) RNA structural motifs: building blocks of a modular biomolecule. *Quarterly Review of Biophysics*, **38**, 221–43.
37. Leontis, N.B., Lescoute, A. and Westhof, E. (2006) The building blocks and motifs of RNA architecture. *Current Opinion in Structural Biology*, **16**, 279–87.
38. Duarte, C.M., Wadley, L.M. and Pyle, A.M. (2003) RNA structure comparison, motif search and discovery using a reduced representation of RNA conformational space. *Nucleic Acids Research*, **31**, 4755–61.
39. Hershkovitz, E., Tannenbaum, E., Howerton, S.B. *et al.* (2003) Automated identification of RNA conformational motifs: theory and application to the HM LSU 23S rRNA. *Nucleic Acids Research*, **31**, 6249–57.
40. Murray, L.J., Arendall, W.B. 3rd, Richardson, D.C. and Richardson, J.S. (2003) RNA backbone is rotameric. *Proceedings of the National Academy of Sciences of the United States of America*, **100**, 13904–9.
41. Murray, L.J., Richardson, J.S., Arendall, W.B. and Richardson, D.C. (2005) RNA backbone rotamers–finding your way in seven dimensions. *Biochemical Society Transactions*, **33**, 485–7.
42. Schneider, B., Moravek, Z. and Berman, H.M. (2004) RNA conformational classes. *Nucleic Acids Research*, **32**, 1666–77.
43. Olivier, C., Poirier, G., Gendron, P. *et al.* (2005) Identification of a conserved RNA motif essential for She2p recognition and mRNA localization to the yeast bud. *Molecular and Cellular Biology*, **25**, 4752–66.
44. Gendron, P., Lemieux, S. and Major, F. (2001) Quantitative analysis of nucleic acid three-dimensional structures. *Journal of Molecular Biology*, **308**, 919–36.
45. Harrison, A.M., South, D.R., Willett, P. and Artymiuk, P.J. (2003) Representation, searching and discovery of patterns of bases in complex RNA structures. *Journal of Computer Aided Molecular Design*, **17**, 537–49.
46. Bindewald, E., Hayes, R., Yingling, Y.G. *et al.* (2008) RNAJunction: a database of RNA junctions and kissing loops for three-dimensional structural analysis and nanodesign. *Nucleic Acids Research*, **36**, D392–7.
47. Dror, O., Nussinov, R. and Wolfson, H. (2005) ARTS: alignment of RNA tertiary structures. *Bioinformatics*, **21** (Suppl 2), ii47–53.

48. Klosterman, P.S., Tamura, M., Holbrook, S.R. and Brenner, S.E. (2002) SCOR: a Structural Classification of RNA database. *Nucleic Acids Research*, **30**, 392–4.
49. Berman, H.M., Gelbin, A. and Westbrook, J. (1996) Nucleic acid crystallography: a view from the nucleic acid database. *Progress in Biophysics and Molecular Biology*, **66**, 255–88.
50. Hastings, W., Yingling, Y.G., Chirikijian, G.S. and Shapiro, B.A. (2006) Structural and dynamical classification of RNA single-base bulges for nanostructure design. *Journal of Computational and Theoretical Nanoscience*, **3**, 63–77.
51. Mueller, F. and Brimacombe, R. (1997) A new model for the three-dimensional folding of *Escherichia coli* 16 S ribosomal RNA. I. Fitting the RNA to a 3D electron microscopic map at 20 A. *Journal of Molecular Biology*, **271**, 524–44.
52. Brown, R.A. and Case, D.A. (2006) Second derivatives in generalized Born theory. *Journal of Computational Chemistry*, **27**, 1662–75.
53. Massire, C., Jaeger, L. and Westhof, E. (1998) Derivation of the three-dimensional architecture of bacterial ribonuclease P RNAs from comparative sequence analysis. *Journal of Molecular Biology*, **279**, 773–93.
54. Jossinet, F. and Westhof, E. (2005) Sequence to Structure (S2S): display, manipulate and interconnect RNA data from sequence to structure. *Bioinformatics*, **21**, 3320–1.
55. Martinez, H.M., Maizel, J.V.J. and Shapiro, B.A. (2008) RNA2D3D: A program for generating, viewing and comparing 3-dimensional models of RNA. *Journal of Biomolecular Structure and Dynamics*, **25**, 669–83.
56. Major, F. (2003) Building three-dimensional ribonucleic acid structures. *Computing in Science and Engineering*, **5**, 44–53.
57. Kirkpatrick, S., Gelatt, C.D. Jr. and Vecchi, M.P. (1983) Optimization by simulated annealing. *Science*, **220**, 671–80.
58. Case, D.A., Cheatham, T.E. 3rd, Darden, T. *et al.* (2005) The Amber biomolecular simulation programs. *Journal of Computational Chemistry*, **26**, 1668–88.
59. Schuster, P., Fontana, W., Stadler, P.F. and Hofacker, I.L. (1994) From sequences to shapes and back: a case study in RNA secondary structures. *Proceedings in Biological Science*, **255**, 279–84.
60. Busch, A. and Backofen, R. (2006) INFO-RNA–a fast approach to inverse RNA folding. *Bioinformatics*, **22**, 1823–31.
61. Andronescu, M., Fejes, A.P., Hutter, F. *et al.* (2004) A new algorithm for RNA secondary structure design. *Journal of Molecular Biology*, **336**, 607–24.
62. Bernhart, S.H., Tafer, H., Muckstein, U. *et al.* (2006) Partition function and base pairing probabilities of RNA heterodimers. *Algorithms for Molecular Biology*, **1**, 3.
63. Brooks, B.R., Bruccoleri, R.E., Olafson, B.D. *et al.* (1983) CHARMM: A program for macromolecular energy, minimization, and dynamics calculations. *Journal of Computational Chemistry*, **4**, 187–217.
64. MacKerell, A.D. Jr., Banavali, N. and Foloppe, N. (2000) Development and current status of the CHARMM force field for nucleic acids. *Biopolymers*, **56**, 257–65.
65. MacKerell, A.D., Brooks, B., Brooks, C.L. *et al.* (1998) CHARMM: The energy function and its parameterization with an overview of the program, in *Encyclopedia of Computational Chemistry*, Vol. **1** (ed. P. v. R. Schleyer *et al.*), John Wiley & Sons, Ltd., Chichester, pp. 271–7.
66. Phillips, J.C., Braun, R., Wang, W. *et al.* (2005) Scalable molecular dynamics with NAMD. *Journal of Computational Chemistry*, **26**, 1781–802.
67. Christen, M., Hünenberger, P.H., Bakowies, D. *et al.* (2005) The GROMOS software for biomolecular simulation: GROMOS05. *Journal of Computational Chemistry*, **26**, 1719–51.
68. Ponder, J. (2006) *Tinker – Software Tools for Molecular Design – Version 4.2.*
69. Hill, D.J., Mio, M.J., Prince, R.B. *et al.* (2001) A field guide to foldamers. *Chemical Reviews*, **101**, 3893–4012.

70. Shih, W.M., Quispe, J.D. and Joyce, G.F. (2004) A 1.7-kb single-stranded DNA that folds into a nanoscale octahedron. *Nature*, **427**, 618–21.
71. Jaeger, L., Westhof, E. and Leontis, N.B. (2001) TectoRNA: modular assembly units for the construction of RNA nano-objects. *Nucleic Acids Research*, **29**, 455–63.
72. Nasalean, L., Baudrey, S., Leontis, N.B. and Jaeger, L. (2006) Controlling RNA self-assembly to form filaments. *Nucleic Acids Research*, **34**, 1381–92.
73. Heilman-Miller, S.L., Thirumalai, D. and Woodson, S.A. (2001) Role of counterion condensation in folding of the Tetrahymena ribozyme. I. Equilibrium stabilization by cations. *Journal of Molecular Biology*, **306**, 1157–66.
74. Horiya, S., Li, X., Kawai, G. *et al.* (2002) RNA LEGO: magnesium-dependent assembly of RNA building blocks through loop-loop interactions. *Nucleic Acids Research Supplement*, **2**, 41–2.
75. Pan, J., Thirumalai, D. and Woodson, S.A. (1999) Magnesium-dependent folding of self-splicing RNA: exploring the link between cooperativity, thermodynamics, and kinetics. *Proceedings of the National Academy of Sciences of the United States of America*, **96**, 6149–54.
76. Park, S.H., Yin, P., Liu, Y. *et al.* (2005) Programmable DNA self-assemblies for nanoscale organization of ligands and proteins. *Nano Letters*, **5**, 729–33.
77. Rothemund, P.W. (2006) Folding DNA to create nanoscale shapes and patterns. *Nature*, **440**, 297–302.
78. Park, S.H., Pistol, C., Ahn, S.J. *et al.* (2006) Finite-size, fully addressable DNA tile lattices formed by hierarchical assembly procedures. *Angewandte Chemie – International Edition in English*, **45**, 735–9.
79. Park, S.H., Finkelstein, G. and LaBean, T.H. (2008) Stepwise self-assembly of DNA tile lattices using dsDNA bridges. *Journal of the American Chemical Society*, **130**, 40–1.
80. Guo, F. and Cech, T.R. (2002) Evolution of Tetrahymena ribozyme mutants with increased structural stability. *Nature Structural Biology*, **9**, 855–61.
81. Bevilacqua, J.M. and Bevilacqua, P.C. (1998) Thermodynamic analysis of an RNA combinatorial library contained in a short hairpin. *Biochemistry*, **37**, 15877–84.
82. Jaeger, L. and Leontis, N.B. (2000) Tecto-RNA: one-dimensional self-assembly through tertiary interactions. *Angewandte Chemie – International Edition in English*, **39**, 2521–24.
83. Ikawa, Y., Fukada, K., Watanabe, S. *et al.* (2002) Design, construction, and analysis of a novel class of self-folding RNA. *Structure*, **10**, 527–34.
84. Brunel, C. and Romby, P. (2000) Probing RNA structure and RNA-ligand complexes with chemical probes. *Methods in Enzymology*, **318**, 3–21.
85. Ehresmann, C., Baudin, F., Mougel, M. *et al.* (1987) Probing the structure of RNAs in solution. *Nucleic Acids Research*, **15**, 9109–28.
86. Geary, C., Baudrey, S. and Jaeger, L. (2008) Comprehensive features of natural and in vitro selected GNRA tetraloop-binding receptors. *Nucleic Acids Research*, **36**, 1138–52.
87. Hansma, H.G., Oroudjev, E., Baudrey, S. and Jaeger, L. (2003) TectoRNA and 'kissing-loop' RNA: atomic force microscopy of self-assembling RNA structures. *Journal of Microscopy*, **212**, 273–9.
88. Frank, J. (2001) Cryo-electron microscopy as an investigative tool: the ribosome as an example. *BioEssays*, **23**, 725–32.
89. Tang, L., Johnson, K.N., Ball, L.A. *et al.* (2001) The structure of pariacoto virus reveals a dodecahedral cage of duplex RNA. *Nature Structural Biology*, **8**, 77–83.
90. Andersen, F.F., Knudsen, B., Oliveira, C.L. *et al.* (2008) Assembly and structural analysis of a covalently closed nano-scale DNA cage. *Nucleic Acids Research*, **36**, 1113–19.
91. Shu, D., Huang, L.P., Hoeprich, S. and Guo, P. (2003) Construction of phi29 DNA-packaging RNA monomers, dimers, and trimers with variable sizes and shapes as potential parts for nanodevices. *Journal of Nanoscience and Nanotechnology*, **3**, 295–302.

92. Khaled, A., Guo, S., Li, F. and Guo, P. (2005) Controllable self-assembly of nanoparticles for specific delivery of multiple therapeutic molecules to cancer cells using RNA nanotechnology. *Nano Letters*, **5**, 1797–808.
93. Davis, J.H., Tonelli, M., Scott, L.G. *et al.* (2005) RNA helical packing in solution: NMR structure of a 30 kDa GAAA tetraloop-receptor complex. *Journal of Molecular Biology*, **351**, 371–82.
94. Afonin, K.A. and Leontis, N.B. (2006) Generating new specific RNA interaction interfaces using C-loops. *Journal of the American Chemical Society*, **128**, 16131–7.
95. Horiya, S., Li, X., Kawai, G. *et al.* (2003) RNA LEGO: magnesium-dependent formation of specific RNA assemblies through kissing interactions. *Chemistry and Biology*, **10**, 645–54.
96. Koyfman, A.Y., Braun, G., Magonov, S. *et al.* (2005) Controlled spacing of cationic gold nanoparticles by nanocrown RNA. *Journal of the American Chemical Society*, **127**, 11886–7.
97. Gregorian, R.S. Jr. and Crothers, D.M. (1995) Determinants of RNA hairpin loop-loop complex stability. *Journal of Molecular Biology*, **248**, 968–84.
98. Seeman, N.C. (2005) From genes to machines: DNA nanomechanical devices. *Trends in Biochemical Science*, **30**, 119–25.
99. Li, H., Park, S.H., Reif, J.H. *et al.* (2004) DNA-templated self-assembly of protein and nanoparticle linear arrays. *Journal of the American Chemical Society*, **126**, 418–19.
100. Erben, C.M., Goodman, R.P. and Turberfield, A.J. (2007) A self-assembled DNA bipyramid. *Journal of the American Chemical Society*, **129**, 6992–3.
101. Goodman, R.P., Heilemann, M., Doose, S. *et al.* (2008) Reconfigurable, braced, three-dimensional DNA nanostructures. *Nature Nanotechnology*, **3**, 93–6.
102. Brody, E.N. and Gold, L. (2000) Aptamers as therapeutic and diagnostic agents. *Journal of Biotechnology*, **74**, 5–13.
103. Mosing, R.K. and Bowser, M.T. (2007) Microfluidic selection and applications of aptamers. *Journal of Separation Science*, **30**, 1420–6.
104. Joyce, G.F. (2004) Directed evolution of nucleic acid enzymes. *Annual Review of Biochemistry*, **73**, 791–836.
105. James, W. (2007) Aptamers in the virologists' toolkit. *Journal of General Virology*, **88**, 351–64.
106. Stoltenburg, R., Reinemann, C. and Strehlitz, B. (2007) SELEX–a (r)evolutionary method to generate high-affinity nucleic acid ligands. *Biomolecular Engineering*, **24**, 381–403.
107. Jaeger, L., Wright, M.C. and Joyce, G.F. (1999) A complex ligase ribozyme evolved in vitro from a group I ribozyme domain. *Proceedings of the National Academy of Sciences of the United States of America*, **96**, 14712–17.
108. Yoshioka, W., Ikawa, Y., Jaeger, L. *et al.* (2004) Generation of a catalytic module on a self-folding RNA. *RNA*, **10**, 1900–6.
109. Ikawa, Y., Tsuda, K., Matsumura, S. and Inoue, T. (2004) De novo synthesis and development of an RNA enzyme. *Proceedings of the National Academy of Sciences of the United States of America*, **101**, 13750–5.
110. Pfundheller, H.M., Sorensen, A.M., Lomholt, C. *et al.* (2005) Locked nucleic acid synthesis. *Methods in Molecular Biology*, **288**, 127–46.
111. Egli, M. (1996) Structural aspects of nucleic acid analogs and antisense oligonucleotides. *Angewandte Chemie – International Edition*, **35**, 1894–909.

9

New Paradigms in Droplet-Based Microfluidics and DNA Amplification

Michael L. Samuels[1], John Leamon[1], Jonathan Rothberg[1], Ronald Godiska[2], Thomas Schoenfeld[2] and David Mead[2]

[1]*RainDance Technologies, Lexington, MA, USA*
[2]*Lucigen Corporation, Middleton, WI, USA*

9.1 Introduction

Today, many complex problems in biology are essentially intractable without large-scale manipulation and processing. For example, whole-genome sequence analysis and whole-cell transcriptome or protein interaction analysis require $>10^6$ manipulation steps. High-throughput screens that can process 10^6 reactions are possible using traditional 384-well microtiter plate technology, but these require a room full of robotics, storage facilities and measurement systems for handling the requisite thousands of plates. Reducing the reactions to the picoliter–nanoliter scale would enable the manipulation of more than 10^6 events in a space as small as several hundred microliters, assuming that each 'nanowell' would remain separate from the others.

Essentially all of the devices that have been developed for performing submicroliter manipulations of biological matter are based on microfluidic chips with chambers, interconnecting channels and valves to control the pneumatic flow of materials. This approach has proven successful for miniaturizing DNA sequencing [1], for whole-genome amplification from single cells [2], for protein crystallization [3] and for DNA synthesis [4]. Microfluidic processing greatly reduces the analysis time and reagent consumption, and also eliminates costly macroscale robotics and laboratory apparatus. However, the throughput of current microfluidic devices is far too low for large-scale analyses.

Automation in Proteomics and Genomics: An Engineering Case-Based Approach Edited by Gil Alterovitz, Roseann Benson and Marco Ramoni

Recently, the use of mixed-phase emulsions to create nanoliter droplets has dramatically increased the number of manipulations that can be achieved. Emulsions of water in oil can be used to compartmentalize millions of discrete enzymatic reactions into individual microdroplets for protein selection and evolution [5,6]. The emulsion PCR (polymerase chain reaction), where a single nucleic acid molecule is compartmentalized in a thermostable synthetic micelle or in a water-in-oil droplet, has enabled the molecular evolution of variant DNA polymerases on a scale not previously practical [7]. A number of other droplet-based applications have emerged during the past few years, including the isolation of binding, regulatory and DNA-modifying proteins [8–10], the evolution of catalytic RNA [11], and cell-free translation [12].

Another droplet-based application is the construction of clone-free DNA libraries for sequencing [13]. Traditional clone-based libraries contain numerous gaps due to the large number of recalcitrant elements within a given genome. Examples include strong *Escherichia coli* promoter regions, toxic protein-coding sequences, AT- or GC-rich sequences and regions rich in secondary structure [14]. Clone-free sequencing strategies circumvent the challenges caused by such regions. The clone-free sequencing approach pioneered by 454 Technologies involves randomly fragmenting DNA, ligating adapters to facilitate their capture on beads (one fragment per bead), and placing one bead in one emulsion droplet containing PCR reagents [13]. The anonymous molecules on each bead are amplified using thermostable Taq DNA polymerase and primers specific to the adapters. After amplification, the emulsion is broken, the DNA denatured, and the beads containing multiple copies of a single species of DNA are distributed into the picotiter wells of a fiber optic slide. Pyrosequencing is carried out in each well using a series of enzymes and nucleotides, with the addition of each nucleotide generating light that is detected by a CCD camera. A single 8 h run of this instrument can process 400 000 wells and generate read lengths of 250 base pairs (bp), or approximately 100 million raw bases. This technology not only displaces 50 conventional capillary sequencing instruments but also eliminates the need to pick and prepare templates from 400 000 colonies. Alternative approaches achieve similarly impressive results using a planar solid-phase reactor [15].

Thermostable DNA polymerases are indispensable for next-generation and traditional Sanger sequencing, as well as for many nucleic acid amplification schemes. Notably, all commercially available thermostable DNA polymerases are derived from one of two very closely related groups of microbial polymerases, which are specific for DNA repair [16,17]. However, phage DNA polymerases represent an alternative source of improved enzymes for sequencing and amplification as they are true replicase enzymes that are significantly more diverse than those of their hosts [18]. They often possess unique and potentially more useful biochemical properties than the host repair enzymes, such as higher fidelity, higher processivity and improved nucleotide analogue incorporation [19–21]. These properties could be exploited to improve DNA amplification and sequencing in a number of ways, the best example being F/Y substitution of the microbial Taq DNA polymerase based on bacteriophage T7 DNA polymerase [22]. This modification allowed the incorporation of dideoxynucleotides that revolutionized DNA sequencing. Unfortunately, many of the unique characteristics of phage polymerases are not due to simple amino acid substitutions.

In spite of the abundance and diversity of phage in the environment [23, 24] and the potential advantages of thermostable phage DNA polymerases, only two reports have been made of DNA polymerases from high-temperature phages [25, 26]. Surprisingly, neither of these is stable enough for typical thermal cycling reactions. Traditional culture-based methods are poorly suited for identifying thermostable phage and their hosts, especially from extreme environments, and this has contributed to the lack of known thermostable phage DNAPs. A metagenomic analysis of phages from thermal aquifers has uncovered a large number of new polymerase genes, largely unrelated to any reported sequences [27]. Although the potential of these new enzymes is still being uncovered, a number of surprising new attributes have been identified.

High-throughput technologies have already increased the scale of sequencing and analysis by several orders of magnitude. For example, automated 'next-generation' DNA sequencing instruments can now decode 10^7 to 10^9 bases per run [13, 28, 15, 29]. However, the technology to perform other essential molecular and cellular biology techniques at similar high-throughput levels is currently lacking. An ideal instrument would allow an individual research scientist to manipulate millions of discrete droplets containing any combination of cells, molecules or beads. This would provide the flexibility to mix, heat, split, combine and sort samples, and to detect molecular events at the picoliter scale. In this chapter we will describe such a device – a new microfluidic platform capable of manipulating single cells or molecules. Moreover, as an example of its potential, we will discuss how this microfluidic technology could be used with novel DNA polymerases that have improved properties to provide unique capabilities to automate molecular and cellular biology tasks.

9.2 Droplet-Based Microfluidic Platform

RainDance Technologies (RDT; www.raindancetechnologies.com) has developed an instrument for the manipulation of microscopic water-in-oil droplets (Figure 9.1) whereby a disposable chip can be configured with a number of different microfluidic droplet generating, mixing and sorting elements. This allows the construction, for example, of a microfluidic fluorescence-activated cell sorting (FACS) system where the droplets can be used for the sensitive detection and sorting of fluorescently tagged markers. The microfluidic chip serves as the 'central processing unit' of the instrument, with only pressure-driven fluidics and electrical fields being required to control millions of droplets. In this way, molecules and cells can be rapidly encapsulated and mixed with other compounds, molecules and cells, such that a wide variety of microfluidic manipulations can be performed on the silicone chips, permitting complex assay designs and screening protocols.

Each droplet created by this instrument is the equivalent of a well in a microtiter plate, but is millions of times smaller (Figure 9.2). The droplets are formed by injecting water into opposing immiscible oil streams, causing the break-off of aqueous droplets in a well-defined size range, based on the nozzle design and the sample and oil flow rates. The droplets are stabilized with inert surfactants capable of withstanding thermocycling, freezing or storage at room temperature for several months, while preventing cross-contamination. The size of the droplets can be precisely varied from 5 to 500 μm (0.5 to 100.0 nl in volume) by adjusting the orifice of the generation chamber (Figure 9.3a). Each droplet can have a single molecule, bead or cell placed within it at a rate of up to 1.0×10^4 droplets per second. Premade droplet libraries can be loaded back onto a microfluidic circuit for

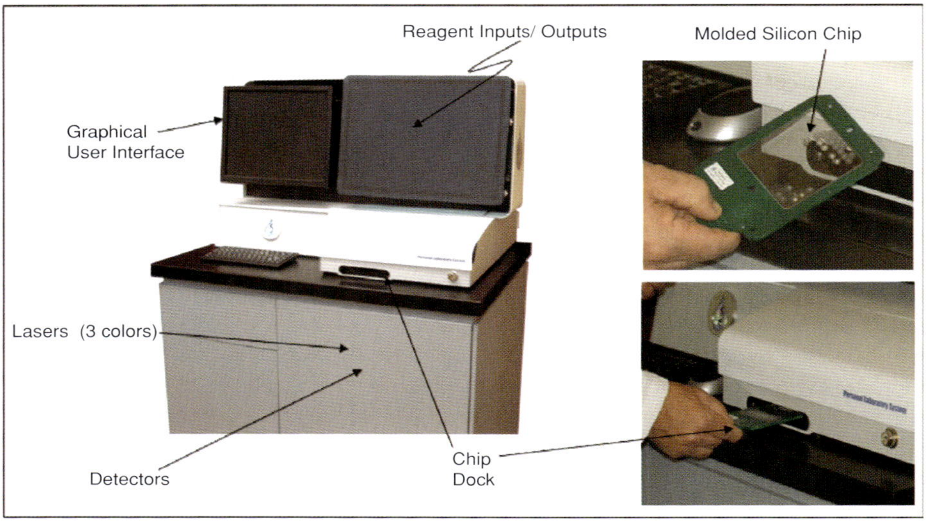

Figure 9.1 The RainDance Technologies microfluidic instrument. The overview shows the primary manipulation, detection and analytical components. A microfluidic sample chip is shown being inserted into the machine

additional manipulations after incubation or storage off-chip (Figure 9.3b). They can be mixed to create unique formulations via multiple inflow channels with variable flow rates (Figure 9.3c), or individual droplets can be combined by controlled electrical charges (Figure 9.3d). Many types of on-platform fluorescent readouts are possible [including fluorescence intensity, fluorescence polarization and Förster resonance energy transfer

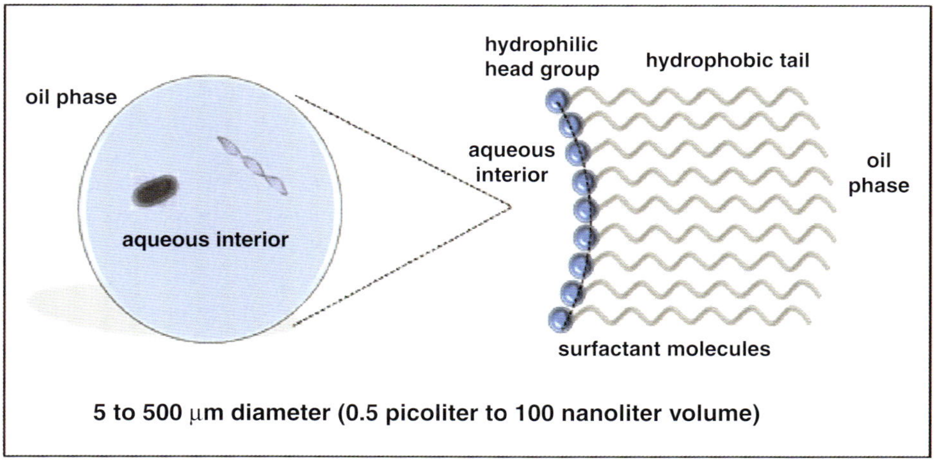

Figure 9.2 Schematic diagram of an aqueous droplet in an oil phase. The enlarged image on the right shows the structure of the surfactant interface between the aqueous interior and the external oil phase

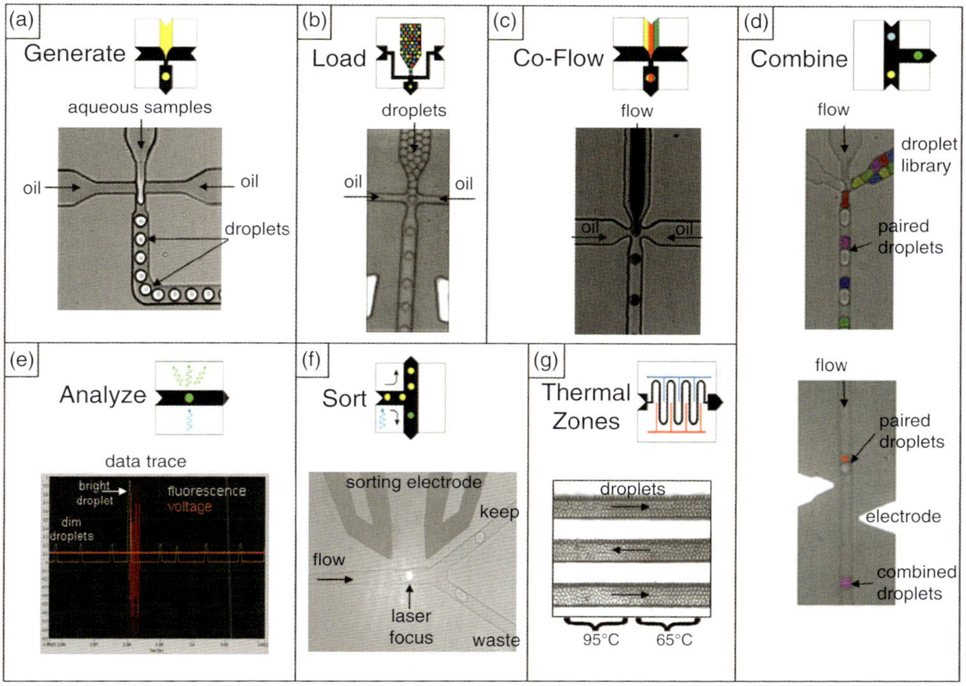

Figure 9.3 *Droplet manipulation and detection modules. (a) In a microfluidic circuit the aqueous sample stream is segregated into droplets by the force of opposing oil streams; (b) Droplet libraries can be loaded onto the chip following off-chip incubation or manipulation; (c) Multiple aqueous input streams can be infused at different flow rates to adjust component concentrations; (d) Droplets are combined by first pairing-off droplets (upper panel), which combine as they pass an electrical field (lower panel); (e,f) Detection by fluorescence or fluorescence polarization can be used to screen and sort droplets using electronic control; (g) The chip base can be heated in thermal zones for applications such as PCR, with droplets rapidly changing temperature as they pass across the chip*

(FRET)]. Quantitative detection of this fluorescence can be used to trigger the electrically controlled sorting of droplets (e.g. FACS; Figure 9.3e,f). Droplets can also be heated in thermal zones on the chip (Figure 9.3g), enabling PCR or hybridization-based applications (also see Figure 9.4). Other microfluidic elements have been designed and tested for additional applications.

The RDT instrument can assemble a broad range of components (live cells, proteins, nucleic acids, small molecule compounds, etc.) precisely into each of millions of dispersed microdroplets. The droplets can then be removed from the device for further manipulation without cross-contamination and subsequently loaded back onto a chip for analysis and processing. Single cells can either be encapsulated and screened or allowed to grow, while reactants can be added to each droplet without contamination. This process can be repeated several times such that approximately ten million cellular or molecular events can be

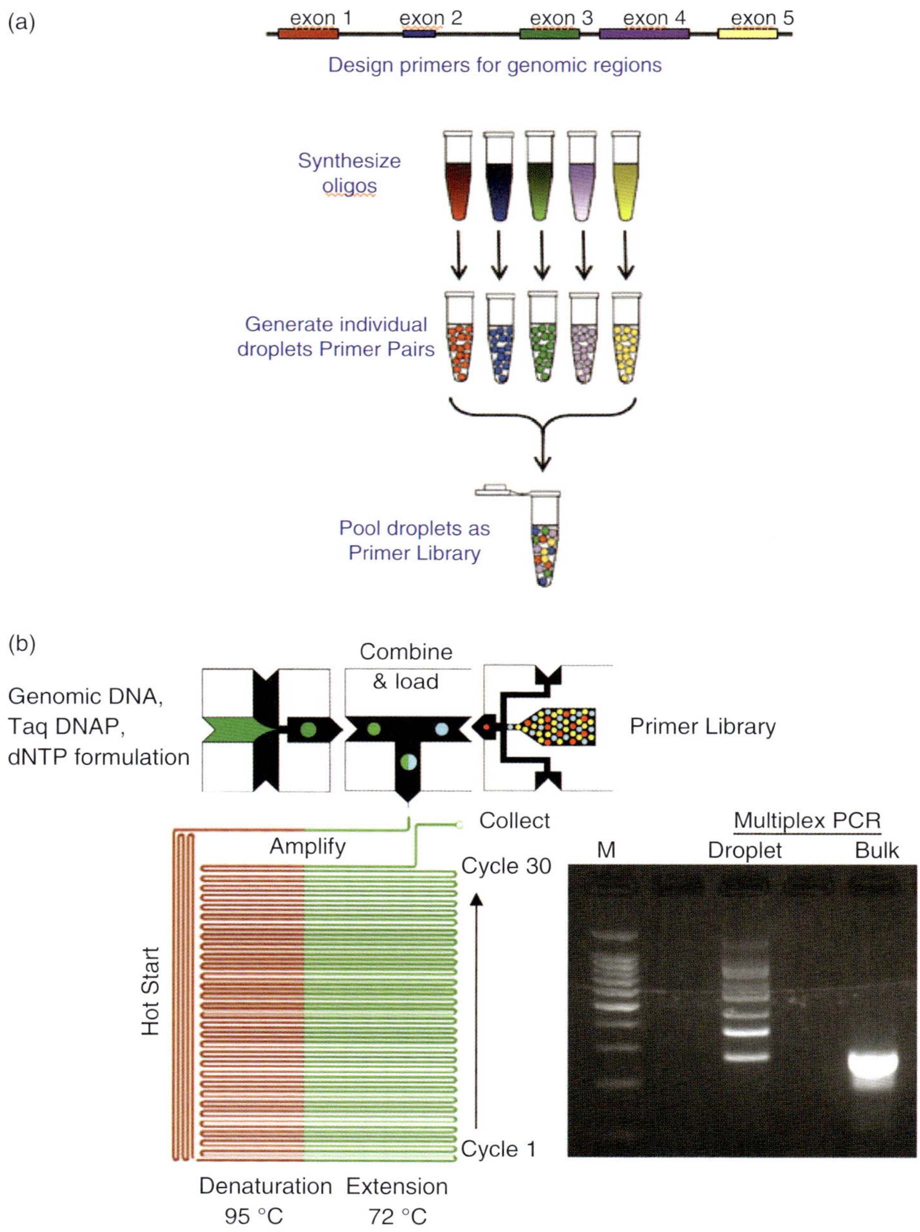

Figure 9.4 *Multiplex PCR using the RDT microfluidic chip. (a) Separate droplets of individual primer pairs are formed and merged into droplets containing the common reagents (buffer, template genomic DNA, nucleotides and DNA polymerase). The merged droplets are loaded onto an amplification chip that moves them across the temperature zones on the plate. At the end of the PCR cycle the droplets are collected, lysed and loaded on an agarose gel for electrophoretic analysis; (b) Results from a direct comparison of droplet versus bulk multiplex PCR using 10 primer pairs*

manipulated in one day, using only a few hundred microliters of solution. A few examples of the capabilities of the RDT instrument are described below.

9.3 PCR in Droplets

DNA amplification, particularly PCR, is one of the most widely used tools in molecular biology. Traditionally, one primer pair is used to produce one amplicon per 50–100 μl reaction mix. Multiplex PCR was developed to decrease costs and increase throughput by using more than one pair of primers to amplify multiple target sequences in a single reaction. Multiplex PCR is an essential cost-saving technique for large-scale genotyping, gene expression, whole-genome sequencing and the diagnosis of infectious diseases. However, the presence of multiple primers pairs and their targets, combined with differential amplification efficiencies, leads to spurious amplification that may compromise the results. Thus, true multiplexing beyond 10–20 amplicons is inherently difficult when using bulk approaches [30].

In contrast, robust multiplex PCR can be achieved using droplet-based microfluidics. Here, the primer pairs are individually packaged into single droplets, and a library of different primer pairs is merged with a series of droplets containing template DNA and the other amplification reagents (see top of Figure 9.4). Although only one primer pair is used to produce one amplicon per droplet, hundreds to thousands of different primer pair reactions can be performed in up to ten million droplets, which provides a significant increase in multiplicity. Droplet-based PCR can be performed on-chip (as shown), or alternatively the combined primers and templates can be amplified off-chip.

A direct comparison was performed using ten different primer pairs in conventional bulk multiplex PCR versus a droplet-based multiplex PCR approach. No attempt was made to optimize the simultaneous use of 20 different primers. The droplet-based multiplex reaction produced the expected ten unique bands, whereas the bulk PCR reaction generated a single broad band (Figure 9.4). This technique offers great promise for genomic and proteomic analysis; for example, it should scale uniformly to amplify the ~250 000 exons present in the human genome, enabling efficient resequencing of the annotated protein-coding portion.

The physical separation of each primer pair into separate droplets minimizes any nonspecific priming. Alternatively, the surface immobilization of oligo pairs has also been used to segregate the reactions in multiplex PCR, although the drawbacks include a loss of reactants from the surface, inefficient amplification, and considerable primer–dimer formation within pairs of primers [31–33]. Multiplex droplet PCR also saves time by only having to validate that a given primer pair produces the correct amplicon, without having to balance the amplification from competing primers.

Single cell genetic analysis from bacterial and mammalian cells has been demonstrated using a droplet-based microfluidic device similar to that described above (see Figure 9.5) [34]. In these experiments, a dilution series of human lymphocyte or *E. coli* cells were mixed with beads conjugated to a reverse PCR primer and a bulk solution containing a fluorescent forward primer, as well as the other amplification reagents (mammalian single cell, shown in Figure 9.5a). As a result, emulsion droplets of approximately 2.5 nl diameter were formed, the temperature was cycled 40 times and the beads recovered for fluorescence

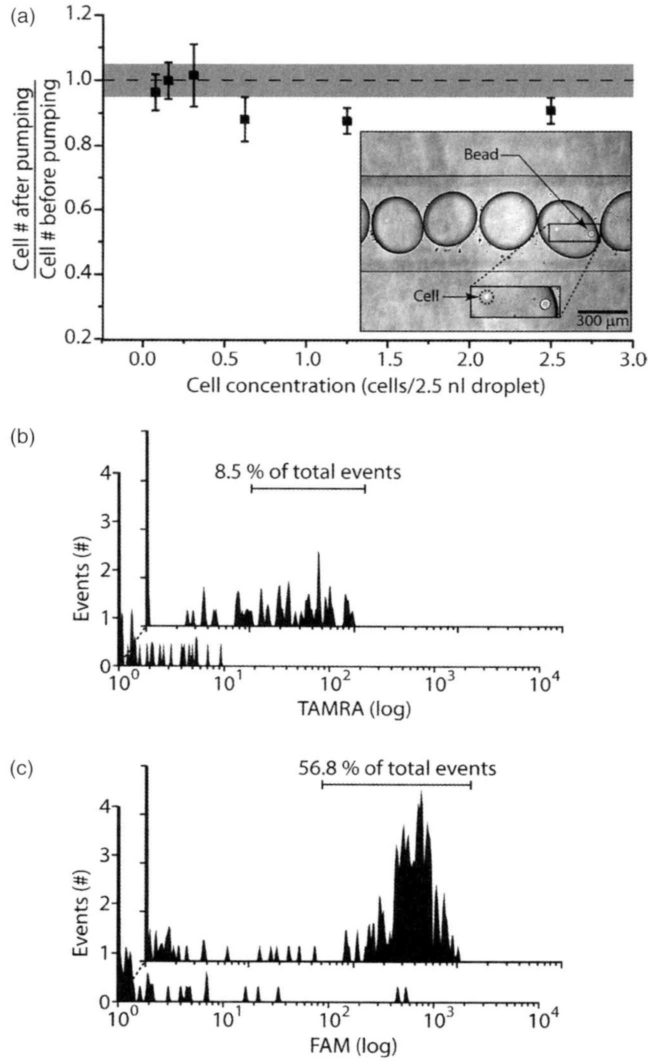

Figure 9.5 *Single-cell genetic analysis. (a) Comparison of the numbers of mammalian cells (human lymphocyte cell line) before and after pumping at different cell concentrations. The inset shows an optical micrograph of an emulsion droplet containing a single bead and a single mammalian cell; (b) Flow cytometry analysis of beads from emulsion droplet bead PCR, starting with 0.1 human lymphocyte cell per droplet (upper) and 0 human lymphocyte cell per droplet (lower). Agarose beads are conjugated with reverse primer targeting the human* GAPDH *gene, while the corresponding forward primer is labeled with TAMRA; (c) Flow cytometry analysis of beads from emulsion droplet bead PCR, starting with 1* Escherichia coli *K12 cell per droplet (upper) and 0* E. coli *K12 cell per droplet (lower). Reverse-primer targeting the* gyr B *gene of* E. coli *is linked to agarose beads, and the forward primer is labeled with FAM. With permission from Dr Richard Mathies at UC Berkley*

analysis using a flow cytometer with a 488 nm excitation source. A Poisson distribution predicted that 9.5% and 63% of the processed beads should fluoresce, using an average concentration of 0.1 and 1 cell per 2.5 nl droplet, respectively. The mammalian gene amplification experiment shown in Figure 9.5b showed 8.5% of the total bead population to be strongly fluorescent when using 0.1 cell per droplet, while the bacterial *E. coli* experiment in Figure 9.5c showed that 57% of the beads fluoresced at the one cell per droplet level. These results were consistent with single cell amplification of single copy genes.

9.4 Transcription and Translation in Droplets

Genomic sequence information provides detailed information about local and global regulatory sequences, coding regions and homology relationships. However, functional analysis ultimately requires the expression and characterization of the protein. Standard methods for functional analysis include heterologous protein expression in microbial hosts, or *in vitro* transcription and translation (IVT) systems [35]. Classic examples of IVT systems include wheat germ or rabbit reticulocyte extracts, which contain all of the protein and RNA components required for IVT, as well as fully characterized reconstituted systems [36].

A massively parallel approach to the functional analysis of gene libraries can be achieved by using the microfluidic capabilities of the RDT instrument. The encapsulation of individual genes (either cDNA or genomic) inside droplets containing IVT reagents is easily achievable using limiting dilutions, and rare or enhanced functional proteins can be detected after transcription and translation if a fluorescence assay is available to monitor the protein's function. The catalytic efficiency of several enzymes has been improved by using a bulk double emulsion-based system (water-in-oil-in-water droplets) [5, 6], while further studies using microfluidic water-in-oil approaches have demonstrated the functional analysis of beta-galactosidase activity using a limiting dilution of the *lac*Z gene in droplets (RDT, unpublished data). Ultimately, the development of IVT-based functional screens in droplets should greatly accelerate the discovery of gene functions, by eliminating host transformation steps and allowing for functional screening without sequence information.

9.5 Screening Libraries of Host Cells for Secreted Enzyme Activity and Evolution in Droplets

Currently, there is a great need for improving the catalytic properties of enzymes for both industrial and academic purposes. Present molecular biology practices enable the cloning of genes that encode the desired enzyme, with a variety of mutational techniques permitting the generation of a large number ($>10^6$) of enzyme variants. In addition, transformation of the mutant library into a range of host organisms enables screening for improved protein function. Enzyme evolution strategies are based on successive rounds of improved enzyme selection, followed by additional rounds of mutagenesis and further screening. However, a key bottleneck in this process is the limited number of screening techniques that allow sufficient throughput to examine protein function on a clone-by-clone basis. Today, the standard procedures in industry rely on placing individual clones on agar plates for halo-based assays, these being typically scored qualitatively by color or zonal clearing. In this way it is possible to resolve approximately 1000 individual colonies on large agar plates, while 1000 plates are required to screen a million-member library. Secondary screening for enzyme variants is onerous, mandating expensive robotic infrastructure and material handling.

RDT's droplet-based microfluidic platform provides the workflow and screening methodology to screen up to ten million events per day (see Figure 9.6). For this, a mutant gene library containing millions of transformed cells is produced using standard molecular biology techniques. To avoid capturing more than one cell per droplet, the library is diluted such that only one in ten droplets is expected to contain a cell (Poisson distribution regime). The droplets provide both the medium for cell growth and an encapsulated space to contain the secreted enzyme. Fluorogenic substrates can be coencapsulated with the cells for detection and quantitation of the enzyme, followed by sorting of the fluorescent cell droplets on the RDT instrument.

Bacteria that express both a green fluorescent protein (GFP) and a secreted enzyme were used to test the expression and sorting capabilities (Figure 9.6). Single bacteria expressing GFP were seen in droplets generated at limiting dilution, after which the droplets were collected and incubated overnight at 37 °C to allow for bacterial growth and enzyme secretion. After incubation, the droplets were reinjected for analysis and sorting on-chip.

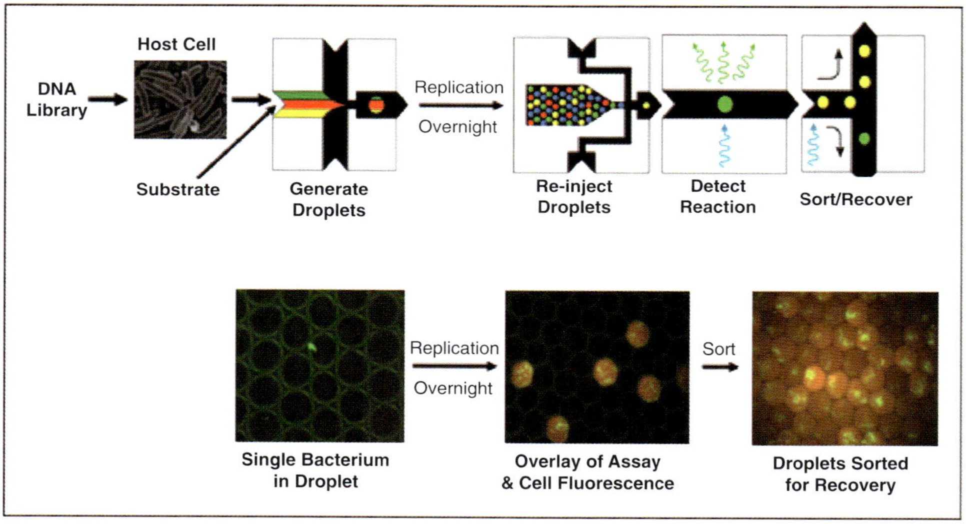

Figure 9.6 *Massively parallel screening of microdroplet libraries containing mutant enzymes. A mutant gene pool is created, transformed into a host cell, and dispersed into a droplet library containing a fluorogenic substrate. Limiting dilution is used initially to generate droplets containing single host cells. Those droplets containing host cells can be enriched from empty droplets by FACS. The mutant enzyme library can be screened for catalytic activity by fluorescent probes and sorted for recovery and analysis*

Although the fluorogenic substrate for the secreted enzyme was present in all droplets, the red assay signal was seen only in those droplets that contained bacteria (bacteria are seen as yellow, when overlaying the red assay and green cell images).

When using this technique, the current sorting throughput is approximately 10^6 droplets per hour. Quantitative analysis enables sorting at any desired threshold stringency, and the entire screen requires less than 1 ml of bacterial culture. Positively sorted cells can be recovered onto agar plates for subsequent validation and further rounds of mutagenesis and screening.

The success of next-generation sequencing clearly demonstrates the advantages of a high-throughput format, not only with regards to the amount of data obtained but also in the scope of possible research projects. Whilst the RDT instrument is expected to have a similar impact on other conventional molecular biology reactions, additional advances can be made by combining such instrumentation with improved enzymes. In the next section we will discuss how a search for novel DNA polymerases has yielded enzymes which not only have unique activities but may also be combined with microfluidics to achieve additional advances.

9.6 The Ideal DNA Polymerase

DNA polymerases are key reagents in nucleic acid amplification, sequencing and genotyping applications. Indeed, these enzymes – and the associated technology – comprise a market which currently is approaching US$ 2 billion per year. To date, more than 100 DNA polymerases have been cloned, expressed and characterized, and approximately 20 have been commercialized. All of the thermostable polymerases utilized *in vitro* fall into one of two closely related groups based on amino acid alignment, namely bacterial enzymes or archaeal enzymes [18]. These enzymes function as DNA repair enzymes, but are not true DNA replicases (the latter are complex, multigene, multisubunit proteins that are prohibitively expensive to produce commercially).

Taq DNA polymerase derivatives, ThermoSequenase, or AmpliTaq FS, have certain important attributes, including thermostability and a reasonable incorporation of dye terminators [37, 38], although they retain many of the deficiencies inherent in the parent enzyme. They have a low processivity, a lack of strand displacement activity, a high error rate, a high level of slippage, a relatively low affinity for DNA-primer templates, problems with certain sequence contexts, and detectable discrimination against nucleotide analogues [16, 39–41]. These restrictions become especially apparent when certain difficult sequences are encountered (G/C- or A/T-rich, direct or inverted repeats), and when limiting amounts of template are sequenced. The low fidelity of *Taq* DNA polymerase results in mutant amplification products, which complicates the cloning of genes from rare samples [42]. The efficient extension of mispaired bases [43] can also impair cloning and complicate the results of single nucleotide discrimination assays. The inconsistent addition of a nontemplated nucleotide to amplification products results in a measurable source of error in genotyping studies [44], and this same activity can also interfere with certain cloning and mutagenesis applications [45, 46]. Genetic analysis of the small nucleotide repeats associated with a number of human diseases is compromised by the PCR stutter and slippage-induced expansion artifacts [40, 47–49].

Significant efforts to ameliorate these activities have met with only partial success. One promising development was the introduction of proofreading thermostable enzymes such as *Pfu* [50, 51] DNA polymerases. However, these enzymes unfortunately have strong exonuclease activities that can rapidly degrade the amplification primers unless they are chemically modified [52]. In addition, *Pfu* induces mutations at levels considered excessive for genetic analysis [53]. Whilst *Taq* DNA polymerase and other available enzymes are certainly adequate for many applications, they all have activities that will compromise certain results.

An alternate source of DNA polymerases is bacteriophage or archaeaphage (collectively, 'phage'). In contrast to the complex cellular DNA replicases, phage replicases are often simple, single-protein enzymes that are amenable to cloning, overexpression and production in large quantities. DNA replicases have been cloned from a handful of nonthermal-stable phages (e.g. bacteriophages T4, T7 and phi29). These phage DNA polymerases are superior to their bacterial counterparts in areas crucial to DNA sequencing and amplification, such as improved nucleotide analog incorporation, read length, copy accuracy and strand displacement activity. However, the phage enzymes are not suitable for high-throughput DNA sequencing or amplification, as they are not thermostable. The primary biochemical attributes of the most widely used commercial DNA polymerases are summarized in Table 9.1.

The 'ideal' DNA polymerase would possess attributes that are optimal for many different applications. For example, amplifying or sequencing templates with strong secondary structure would require an enzyme with high strand displacement activity and processivity; a repetitive template would require high processivity and no replication slippage or terminal transferase activity; amplifying DNA templates >50 kb would require high affinity for primed templates and high processivity. The bottom portion of Table 9.1 shows the activities that facilitate the efficient amplification or sequencing of a given type of template. An enzyme that incorporates all of these qualities would be ideal for conventional and next-generation sequencing platforms, as it would greatly reduce the gaps in the draft sequence that require finishing. Although an ideal enzyme has not been identified to date, there is ample evidence to suggest that improved enzymes exist among the thermophilic phage DNA polymerases.

9.7 The Physiological Role and Characteristics of Phage Versus Bacterial DNA Polymerases

The physiological role of phage DNA polymerases is fundamentally different from that of commercially available bacterial or archaeal polymerases. The current reagent DNA polymerases (including *Taq*) are DNA *repair* enzymes that primarily fill in short gaps, whereas the phage enzymes are *replicases* that are under selective pressure to rapidly replicate entire genomes with high fidelity. Phage polymerases are typically highly processive, have high rates of extension, high affinities for templates and nucleotides, and are able to deal with torsional constraints inherent in replicating long sequences. Microbial replicases employ various accessory proteins, such as processivity factors, fidelity factors, helicases and primase, in order to function this effectively.

Table 9.1 *Biochemical properties of known DNA polymerases and optimal activities of an 'ideal' DNA polymerase for various applications. A blank space indicates that the value has not been measured*

Known DNA polymerase	Processivity	K_m DNA	Error rate ($\times 10^{-6}$)	Strand displacement	Thermo-stable	5′-3′ exo	3′-5′ exo	Slippage	Stutter
ThermoSequenase AmpliTaq FS	42	2 n*M*	285	−	+	+	−	+	+
Taq	42	2 n*M*	285	−	+	+	−	+	+
Vent	7	0.1 n*M*	57	+	+	−	+	+	−
Bst	>1000			+	−	+	−	−	
T7	>1000	18 n*M*	15	+	−	−	+	+	
T4	12		<1	−	−	−	+	−	
phi29	>70 000		<1	+	−	−	+	−	
E. coli Pol I	20	5 n*M*	9	−	−	+	+	+	+
Klenow	12		40	+	−	−	+	+	+
Ideal polymerases for specific applications									
>50 kb PCR	High	Low	<1	+++	+		−	− − −	−
Secondary structure, repetitive sequence	High	Low	<1	++	+		−	− − −	−
Homopolymers, GC-/AT-rich, trace amount of template	High	Low	<1	+	+		−	−	−
Nucleotide discrimination	High	Low	<1	+	+		−	−	−

The T4 phage replicates its genome ten times faster than does its *E. coli* host, while T7 incorporates 300 nucleotides per second, six times faster than Pol I [54]. The T7, T4, T5 and phi29 DNA polymerases all have extraordinarily high levels of processivity, with phi29 polymerase, in particular, having the highest measured level of processivity measured to date (>70 kb) [55]. In fact, phi29 polymerase can amplify a 20-kb template 1000-fold in 1 h [55]. Importantly, the phi29 and T5 polymerases are highly processive without the aid of host proteins [54], suggesting a high affinity for the template.

Interestingly, the DNA polymerases of phages T5 and phi29 are able to strand displace without the aid of helicases or other accessory proteins [55,56], a property which is rare among bacterial or archaeal polymerases. The processivity and strand displacement are inversely related to slippage [47]; accordingly, phi29 and T4 DNAPs have no detectable slippage during extension through repetitive DNA *in vitro*, unlike available thermostable DNA polymerases. Lacking terminal transferase activity, the phage DNA polymerases do not produce 'stutter' bands [44].

DNA polymerases are specialized for various types of templates: repair polymerases most efficiently use nicked double-stranded (ds) DNA templates, but they are less effective in extending long stretches of primed single-stranded (ss) DNA [54]. Phage polymerases, including those of T4, T7 and phi29, are highly efficient at extending long stretches of primed ssDNA. The efficient primer extension of single-stranded templates is a key requirement for sequencing and amplification. In some cases, phage DNA polymerases, in addition to any proofreading activities, have a higher discrimination than Pol I at the initial incorporation step [57,58], further increasing their fidelity.

9.8 Diversity among Phage DNA Polymerases

Whereas the commercially available thermophilic bacterial or archaeal DNA polymerases fall into either of two groups, the polymerases of phage T4, T5, T7, Spo1, Spo2, PRD1, phi29 and M2 are distinct [18]. The phage polymerases as a group are also far more diverse than the bacterial and archaeal DNA polymerases. Surprisingly, T4 DNA polymerase appears more similar to eukaryotic than prokaryotic enzymes [19].

New DNA polymerases traditionally have been identified by isolating the abundant repair polymerases from cultured microbes. However, due to the similarity of these enzymes and difficulties in culturing extremophiles [59], it is unlikely that novel enzyme activities will be discovered in this way. Uncultured thermostable phage clearly represent a large, untapped, currently inaccessible resource of diversity, although accessing these enzymes will require novel approaches be developed.

9.9 Phage Metagenomics of Thermal Aquifers

In order to circumvent the difficulties of culturing thermophilic phage, a metagenomics approach was initiated. Phage particles were isolated from hot springs (74–93 °C) in Yellowstone National Park and purified from microbial cells [27]. Representative phage particles were imaged using transmission electron microscopy (TEM) (Leo 912AB, operating at 80 KV) (Figure 9.7). Direct phage enumeration by epifluorescence microscopy [60]

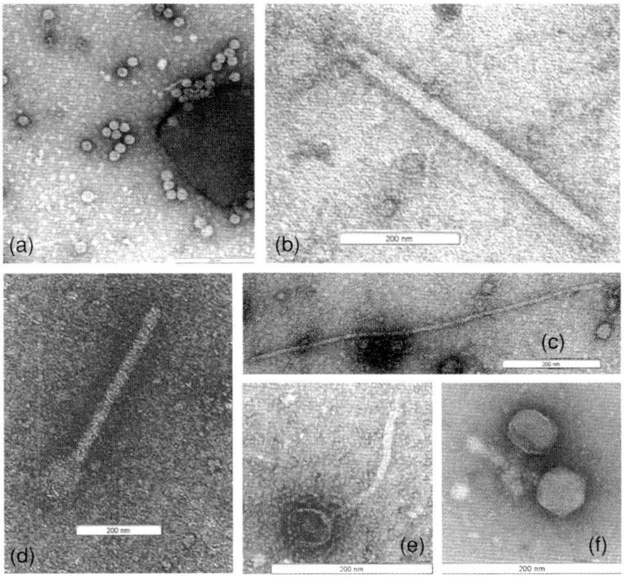

Figure 9.7 *TEM images of phage-like particles directly isolated from Yellowstone National Park hot springs. The scale bar in each figure represents 200 nm. (Image courtesy of Sue Brumfield and Mark Young, Montana State University)*

showed phage abundances ranging from 10^5 to 10^6 particles per milliliter, which was consistent with previous studies of thermophilic viruses and lower than concentrations in typical temperate waters [61].

The phage nucleic acid was extracted and physically sheared to 3–6 kb using a HydroShear device (Genomic Solutions, MI, USA). The ends were made blunt using the DNATerminator end repair kit (Lucigen, WI, USA), and the fragments ligated to double-stranded asymmetric linkers and PCR-amplified. The amplification products were cloned into the transcription-free pSMART vector (Lucigen) and used to transform *E. coli* 10G cells (Lucigen). In collaboration with the Department of Energy's Joint Genome Institute (Walnut Creek, CA, USA), a total of approximately 29,000 sequence reads was determined (~30 Mb total).

The longest contig from these reads was 16.5 kb, assembled at 50% identity, which included 187 reads. GeneMark [62] predicted 26 open reading frames (ORFs) of greater than 100 nucleotides, including an apparent replication operon. The genes with the strongest similarity to these ORFs encode primase, uracil DNA glycosylase, Family B DNA polymerase, nucleotide excision repair nuclease (*dnaG*, *udg*, *polB* and ERCC4 genes, respectively) and homologues to a zinc finger-like protein and a transposase. Homologues of these ORFs belong to crenarchaeal DNA replication/repair complexes [63–65]. Sequences from three discrete clones homologous to the *polB* gene in this contig have been expressed in *E. coli* as functional thermostable DNA polymerases (data not shown).

In total, the ~30 Mb of sequence data contained several hundred apparent *pol* gene homologues, 59 of which appeared full length. Genes with similarity to essentially every type of known DNA polymerase type were identified, with BLASTx E values as low as 10^{-140} being seen, indicating a very high degree of similarity. The similarity was strongest in the highly conserved catalytic domains of the polymerase genes, ten of which have been expressed to produce thermostable DNA polymerase, while seven were completely sequenced.

The predicted amino acid sequences were compared, using clustalW [66], to those of commonly used thermostable DNA polymerases, including *Taq*, *Tth*, *Bst*, Vent and *Pfu* (Figure 9.8). Also included were representatives of each of the viral DNA polymerase families. By this analysis, the diversity of the PyroPhage DNA polymerases appeared to be much higher than that of the available thermostable DNA polymerases. *Taq* and *Tth* were seen to be 85% identical, while Vent and *Pfu* were 75% identical. PyroPhage 4110, 2323 and 2783 formed a clade of >90% identity. Likewise, PyroPhage 3173 was 48% identical to 488 and 82% identical to 967. Otherwise, the PyroPhage enzymes were very distinct (<20% identical) from all known thermostable DNA polymerases, and also from one

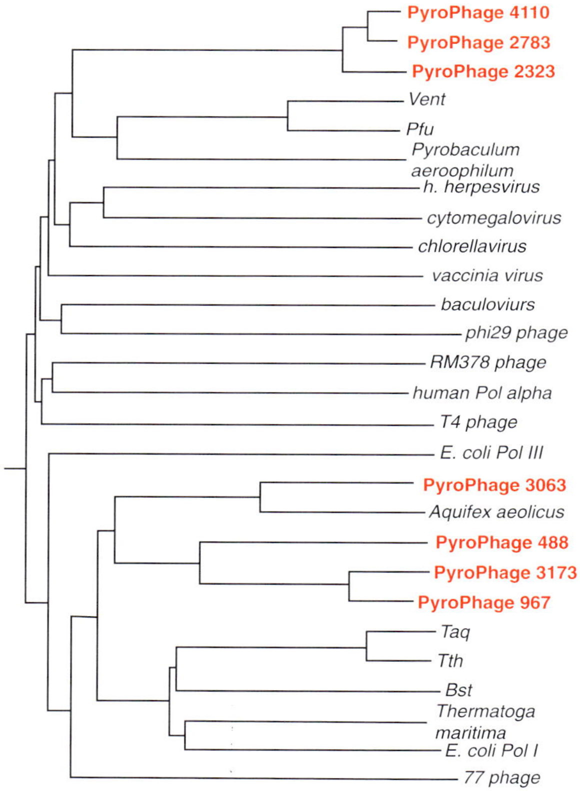

Figure 9.8 *Amino acid alignment of microbial and phage DNA polymerases. The seven complete PyroPhage genes are shown in red. Selected viral and cellular DNA polymerases include several commercial thermostable DNA polymerases (Taq, Tth, Bst, Vent and Pfu)*

another. PyroPhage 3063 showed the closest association to a microbial enzyme, *Aquifex aeolicus* DNAP [67], at 63% amino acid identity.

The degree of molecular diversity suggested a substantial biochemical diversity. PyroPhage 3173 DNAP has been most fully studied, and it does indeed have several novel properties that may make it especially useful for a variety of applications.

9.10 Biochemical Characteristics of PyroPhage 3173 DNA Polymerase

PyroPhage 3173 DNA polymerase has a unique combination of characteristics (see Table 9.2 and below). For example, it is the only known phage DNA polymerase that is thermostable to 95 °C, and it effectively amplifies most templates up to 4 kb using PCR. Perhaps, due to its high strand-displacement activity, PyroPhage DNA polymerase is more effective than current PCR enzymes in amplifying certain difficult templates, in particular, templates containing a high GC content, repetitive sequences and secondary structure (Figure 9.9). The wild-type version of PyroPhage 3173 DNA polymerase demonstrates a strong proofreading activity, while its fidelity is among the highest measured for enzymes used in PCR (Figure 9.10). Following site-directed mutation of 3173 to disable the 3′–5′ exonuclease, the fidelity was comparable to that of nonproofreading microbial enzymes.

9.11 Reverse Transcription

The detection and amplification of RNA, rather than DNA, is vital for many types of analyses, as the characterization of RNA provides additional insight into gene structure and expression. In addition, several phage genomes consist only of RNA. The conventional method for RNA amplification is that of reverse-transcription PCR (RT-PCR), where the RNA is first copied to cDNA by a viral reverse transcriptase. The cDNA is subsequently amplified in a separate PCR step using a standard thermostable DNA polymerase and PCR buffer. This two-step method is acceptable for microliter-scale reactions, although a

Table 9.2 *Biochemical characteristics of PyroPhage 3173 DNA polymerase*

3173 DNA polymerase	Wild-type	Exo-minus
3′-5′ exonuclease	Strong	None
5′-3′ exonuclease	None	None
Strand displacement	Strong	Strong
Extension from nicks	Strong	Strong
$T_{1/2}$ @ 95°	10 min	10 min
K_m dNTPs	40 μ*M*	40 μ*M*
K_m DNA	5.3 n*M*	5.3 n*M*
Processivity	n.d.	47 nt
Fidelity	8×10^4	1.5×10^4
3′ ends of amplicons	Blunt	Single base A and G overhangs
Template	DNA or RNA	DNA or RNA

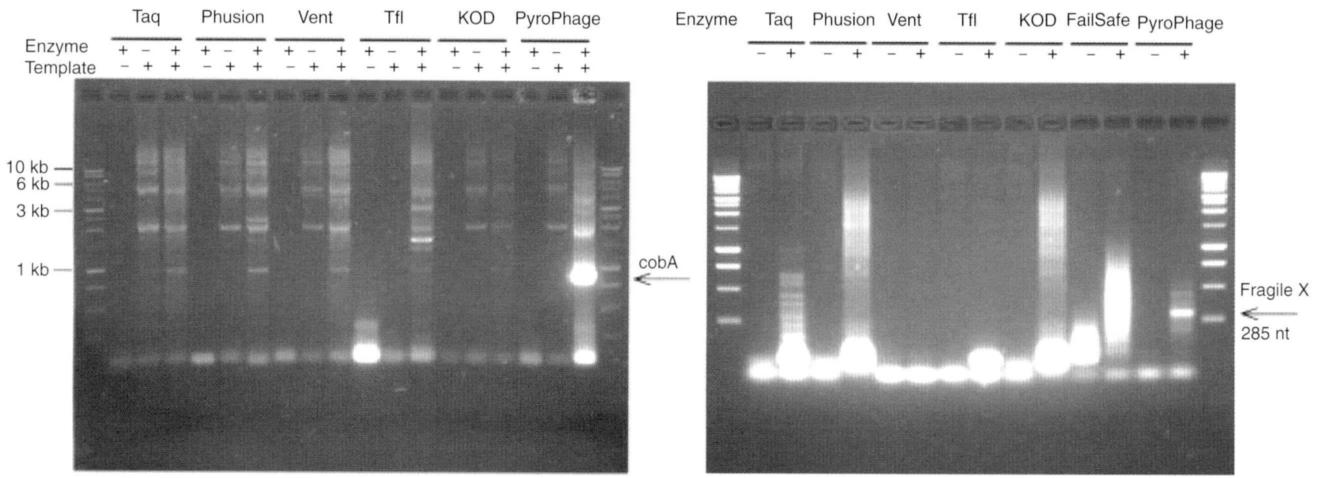

Figure 9.9 *PCR amplification of difficult templates. PCR amplification of the* Bacillus *cobA gene (left) and the human Fragile X gene (right) using PyroPhage 3173 (Exo Minus) or the indicated DNA polymerase*

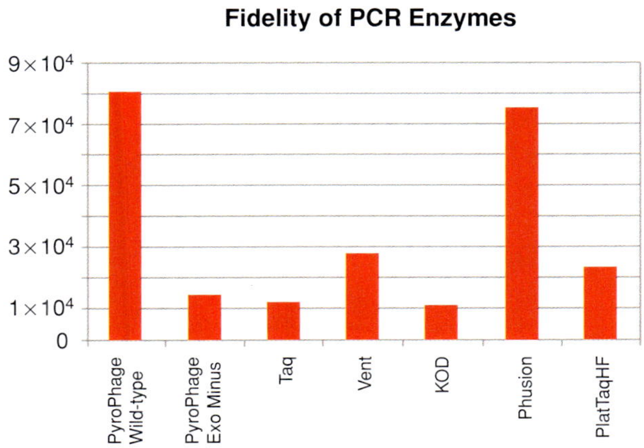

Figure 9.10 *PyroPhage 3173 fidelity. Fidelity measurements, shown as the ratio of correct to incorrect nucleotide incorporations, are based on the LacI^q forward mutation assay (Hogrefe). PyroPhage 3173 DNAP (Wild-Type or Exo Minus) was compared to various commercially available enzymes*

single-tube reaction would be preferable. For high-throughput applications, a single-enzyme, single-tube method of RT–PCR would be very valuable.

The RT–PCR activity of 3173 DNAP was tested by attempting to amplify the mouse actin gene from liver RNA. A single band of expected size was detected, indicating that this enzyme could efficiently amplify RNA into dsDNA (Figure 9.11).

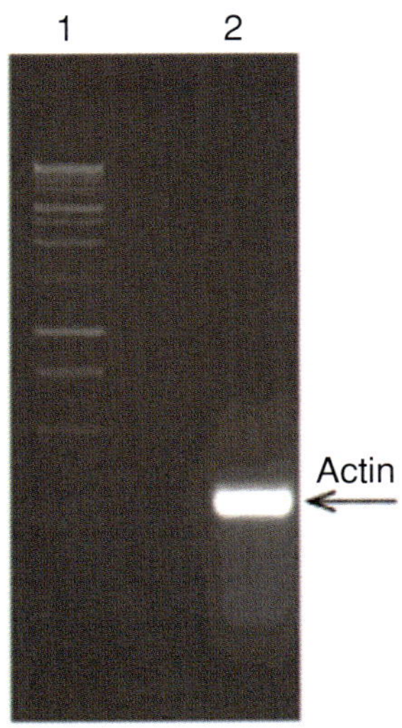

Figure 9.11 *RT–PCR activity of PyroPhage 3173 DNA polymerase. Mouse liver RNA was purified and used as a template for single-tube RT–PCR with 3173 DNAP and actin-specific primers. A strong band at the expected size of 283 bp was detected*

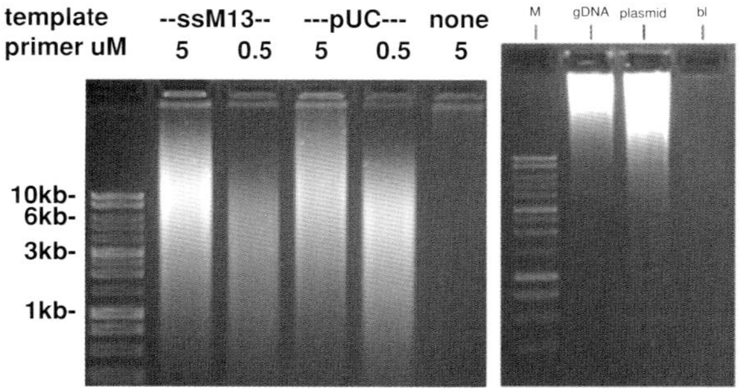

Figure 9.12 *Isothermal whole-genome amplification. PyroPhage 3173 DNA polymerase was used to amplify 10 ng each of circular ssDNA (ssM13) and dsDNA (pUC) in the presence of the indicated concentrations of random decamer primers (left panel). The same enzyme was used to amplify* E. coli *gDNA and a supercoiled plasmid in the absence of exogenous primers, but in the presence of the site-directed nicking enzyme, Nt.BstNBI (right panel). In both cases, the reactions were incubated for 2 h at 55 °C, and one-tenth of the total product was resolved on the gel. A negative control without template is shown*

9.12 Isothermal DNA Amplification using PyroPhage 3173 DNA Polymerase

PyroPhage 3173 DNA polymerase (wild-type and exonuclease-minus) is effective in the isothermal replication of linear, circular or supercoiled DNA, either single- or double-stranded (Figure 9.12, left panel). In general, a greater than 10 000-fold amplification is regularly achieved in 2 h at 55 °C. A typical DNA synthesis was initiated using exogenous primers; in addition, this enzyme initiated synthesis efficiently from genomic or plasmid DNA that had been nicked using single-strand nicking enzymes (Figure 9.12, right panel). In the absence of template DNA, no amplification product was observed in these assays. In conjunction with its thermostability, this enzyme can therefore perform single-cell DNA amplification following just a heat lysis step for template preparation, thus significantly simplifying the process of genomic amplification and reducing the possibility of sample contamination. This attribute is most important for single-cell DNA amplification.

9.13 Single-Cell Genomics

Determining the genomic sequence of microbes is complicated by technical challenges, with only a small minority of microbes capable of being cultured for the isolation of large amounts of genomic DNA (gDNA). The metagenomic sequencing of a mixed uncultivated community provides a snapshot of relatively small fragments that cannot be readily related to any given species. Assembling a contiguous genome from these fragments requires very deep sequencing [68], and is computationally and realistically impractical. The bias of oversampling abundant species and heterogeneity within species also complicates this approach. Single-cell genomics – the sequence analysis of single individual genomes – could overcome these limitations by providing insight into rare species, and into the genetic heterogeneity of the population, the composition of phage and viral populations, and the presence of eukaryotic and archaeal microorganisms. It could also be used to demonstrate the complete sequences of large operons and neighboring genes from long contiguous stretches of DNA.

Several recent reports have outlined different strategies for partially sequencing the genome of a single cell [15, 2, 69–71]. The genomic content of a prokaryotic or archaeal cell is approximately 2 fg of DNA [72]; because current sequencing technologies require microgram quantities of template DNA, an amplification process must faithfully replicate this single DNA molecule 5×10^9-fold, ideally with high fidelity, low bias and complete coverage. The technique of multiple displacement amplification (MDA) is indispensable for generating sufficient nucleic acid from limiting amounts of sample [73, 74]. The MDA reaction combines random primers, a strand-displacing DNA polymerase and genomic DNA, in an isothermal reaction. Due to an efficient strand displacement by the enzyme, replicons from upstream primers displace ssDNA from downstream regions; the displaced ssDNA can then hybridize to additional random primers, with the amplification process repeating itself. In this way, MDA is able to generate a series of staggered duplications of the original template.

The efficient isolation of individual cells from other microorganisms and extracellular DNA is critical for single-cell amplification. Single prokaryotic cells have been isolated by micromanipulation [75], serial sample dilution [15], microfluidic chambers [76] and FACS [77]. A nanoscale device for microfluidic sample processing of single cells is an important recent advance [2, 76]. Here, a microfluidic chip using 60-nl chambers was used to carry out MDA reactions on eight isolated bacteria. However, because the small initial reaction volume yielded only a few nanograms of material, a second amplification was employed to generate the microgram amounts of DNA required for sequencing.

The most commonly used MDA enzymes are phi29 DNA polymerase [78] and *Bst* DNA polymerase. *Bst* DNA polymerase has a strong constitutive strand-displacement activity [79], it does not suffer from replication slippage (as do most polymerases) [40], it has a high affinity for DNA-primer complexes [80], and it demonstrates high processivity [81]. The strand displacement activity of *Bst* DNA polymerase can extend primed templates beyond 50 Kb [79, 82, 83], and can amplify DNA by as much as 10^{12}-fold [79, 84, 85].

Bacteriophage phi29 DNAP is the only other enzyme with many of these properties [47, 86]. Phi29 DNA polymerase has been used to amplify plasmid DNA from single colonies 10 000-fold via multiply primed rolling circle amplification (RCA) [87]. RCA has also been used to amplify trace amounts of human and bacterial genomic DNA [88–90] and DNA from single cells [78]. Unfortunately, amplification by phi29 DNA polymerase leads to a significant bias in whole-genome sequence analysis, as demonstrated by the sequence analysis of microbial genomes. An AT-rich template showed a 19-fold bias towards particular regions, whereas a GC-rich genome showed an over 100-fold bias [73]. Other pitfalls included excessive genomic sequence gaps, chimeras and nonspecific amplification due to primer dimers [15].

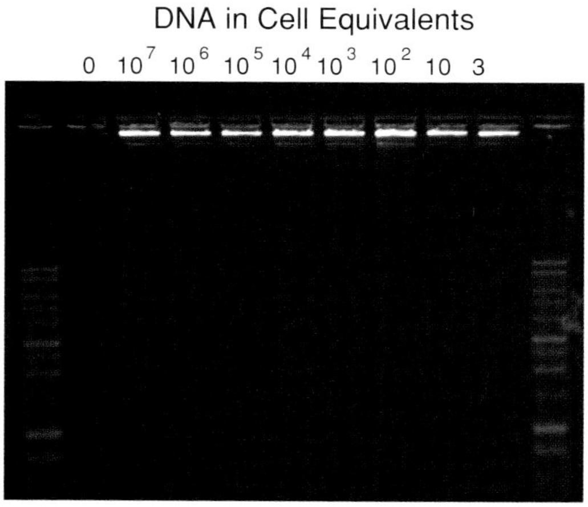

Figure 9.13 *Strand displacement amplification using PyroPhage 3173 DNA polymerase. A dilution series of* E. coli *DNA was incubated at 70 °C for 16 h in a 10-μl MDA reaction*

9.14 PyroPhage 3173 DNA Polymerase for Single-Cell Genomics

The use of a DNA polymerase that is thermostable and capable of strand displacement amplification could overcome several of these problems. Both, phi29 and *Bst* DNA polymerases are readily inactivated above 65 °C. PyroPhage 3173 DNAP may be suitable for single-cell amplification, as it is capable of withstanding thermal conditions for lysis of cells, and its sensitivity is sufficient for single cell amplification (Figure 9.13).

Primer auto-amplification was reduced to undetectable levels by using a small reaction volume and limited amounts of primer. These conditions also were correlated with a very high molecular weight of the amplified DNA (Figure 9.13). Compared to the broad smear typically seen with phi29 amplifications, the high-molecular-weight DNA may contain longer stretches with fewer branches. To date, we have cloned and sequenced DNA amplified by PyroPhage 3173 using strand displacement, and found no sequence differences compared to unamplified DNA. However, pretreatment of the reagents with thermolabile nucleases was essential to eliminate contaminating DNA.

9.15 Thermophilic Phage DNA Polymerases and Cell-Free Droplet-Based Biology

Clone-free metagenomics, massively parallel single-cell genomics, whole-genome multiplex PCR, single-cell transcriptional analysis and additional improvements to DNA sequencing are just a few of the applications that will be enabled by these new technologies. The enzymes may provide DNA amplification tools that simplify single-cell genomics, while the droplet-based platform may be used to manipulate single cells and their genomes. A schematic example workflow for automating single-cell, whole-genome amplification using a microfluidic droplet instrument is shown in Figure 9.14.

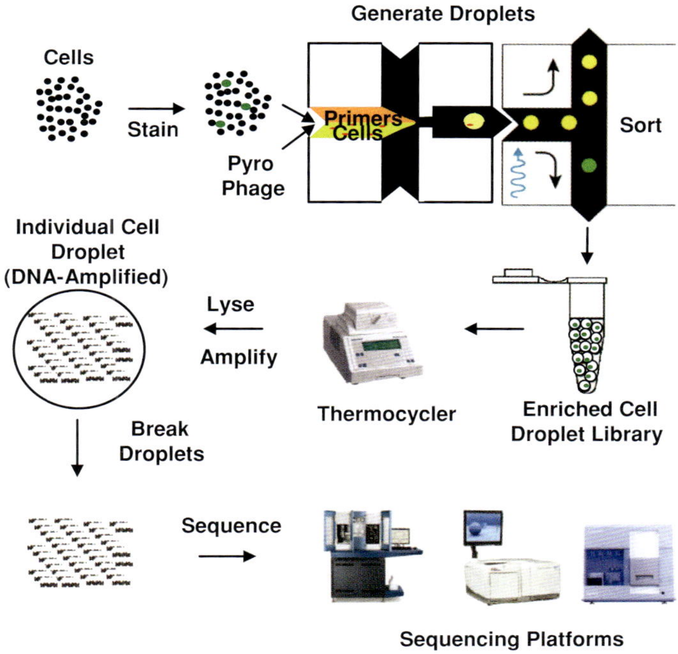

Figure 9.14 Workflow for droplet-enabled, single-cell, whole-genome amplification using PyroPhage 3173 DNA polymerase. By eliminating the requirement for a separate addition step after thermal lysis, a simplified workflow is provided utilizing a homogeneous droplet. This example includes sorting of prestained cells for enrichment of a rare subpopulation for genome amplification

Eliminating the need for a host cell for transformation, cloning, expression and screening of enzymes represents a new frontier in automating molecular and cellular biology. An important advantage of this approach is the removal of numerous incompatibilities between the host cell and the gene, such as toxicity of the encoded ORFs, codon bias, instability due to structure-rich sequences, selection against large genes or membrane proteins, and the insolubility of heterologous proteins *in vivo*.

An example of the utility of this technology is sequencing the 'second' human genome, consisting of the trillions of uncultivated microbes harbored by our bodies. This complex community is estimated to contain about 100 times as many genes as the human genome [91]. Although the mammalian microbial community has a profound impact on the metabolism and immune status of the host [92–96], surprisingly little is known about them. The emerging picture is that of the human as a 'supraorganism' which is influenced by the amalgam of microbial and host cellular metabolic features. Understanding the connection between the human microbiome and human health could transform biomedical research over the next decade, leading to new therapies and diagnostics.

The migration away from cell-based cloning to amplification-based cloning presents a number of new challenges. First, the highly efficient replication mechanism inherent in microbial hosts must be replaced by a reliable, unbiased, high-fidelity *in vitro* replication system capable of amplifying large fragments. Second, miniaturization of the reaction volume to nanoliters or picoliters presents its own set of challenges. The emulsion droplets being used for PCR in the current high-throughput sequencing platforms are limited to amplifying small fragments, with the efficiency of amplification dropping off dramatically above 1 kb [13]. The ability to precisely control the size of the droplet will be critical in optimizing the amplification of large DNA fragments, as well as whole genomes. It also requires the efficient replication of templates that typically are refractory to synthesis (e.g. those with regions of secondary structure or high GC content).

Phage replicases hold great promise for advanced applications, compared to the more commonly used microbial DNA polymerases. Phage DNA polymerases are much more molecularly diverse than microbial enzymes, and their remarkable biochemical characteristics enable several moderate temperature applications. Viral enzymes such as retroviral reverse transcriptase and DNA polymerases from phages T7, Phi29 and T4 have been indispensable for DNA and RNA amplification and analysis.

Unfortunately, even the known viral enzymes have limitations, foremost among which is the absence of a thermostable phage DNA polymerase. Many of the most important amplification and thermocycle sequencing applications rely on thermal denaturation, and therefore thermostable DNA polymerases are essential. Even isothermal applications that do not depend on thermal denaturation can be improved at higher temperatures, resulting in higher stringency and suppression of background artifacts. The compelling attributes found in phage DNA polymerases, in combination with thermostability, should prove especially advantageous when addressing these needs. Serious technical challenges associated with the discovery of new thermophilic phage DNA polymerases by traditional microbial culturing have been overcome using a metagenomic approach. The combination of these enzymes and their derivatives with advances in the microfluidic handling of large numbers of nanoliter droplets should lead to extraordinary advances for cell-free systems.

To summarize, the droplet-based microfluidic instrument developed by RDT has been used to demonstrate a new level of high-throughput biology in the few examples presented here. Yet, the exploitation of the many capabilities of droplet-based biology will require optimization of existing technologies and the evolution of new molecules and tools.

Acknowledgments

The authors thank Palani Kumaresan and Richard Mathies for kindly supplying Figure 9.5.

References

1. Blazej, R.G., Kumaresan, P. and Mathies, R.A. (2006) Microfabricated bioprocessor for integrated nanoliter-scale Sanger DNA sequencing. *Proceedings of the National Academy of Sciences of the United States of America*, **103**, 7240–5.
2. Marcy, Y., Ouverney, C., Bik, E.M. *et al.* (2007A) Dissecting biological "dark matter" with single-cell genetic analysis of rare and uncultivated TM7 microbes from the human mouth. *Proceedings of the National Academy of Sciences of the United States of America*, **104**, 11889–94.

3. Anderson, M.J., DeLabarre, B., Raghunathan, A. *et al.* (2007) Crystal structure of a hyperactive *Escherichia coli* glycerol kinase mutant Gly230 → Asp obtained using microfluidic crystallization devices. *Biochemistry*, **46**, 5722–31.
4. Huang, Y., Castrataro, P., Lee, C.C. and Quake, S.R. (2007) Solvent resistant microfluidic DNA synthesizer. *Lab on a Chip*, **7**, 24–6.
5. Tawfik, D.S. and Griffiths, A.D. (1998) Man-made cell-like compartments for molecular evolution. *Nature Biotechnology*, **16**, 652–6.
6. Griffiths, A.D. and Tawfik, D.S. (2006) Miniaturizing the laboratory in emulsion droplets. *Trends in Biotechnology*, **24**, 395–402.
7. Ghadessy, F.J., Ong, J.L. and Holliger, P. (2001) Directed evolution of polymerase function by compartmentalized self-replication. *Proceedings of the National Academy of Sciences of the United States of America*, **98**, 4552–7.
8. Sepp, A., Tawfik, D.S. and Griffiths, A.D. (2002) Microbead display by in vitro compartmentalisation: selection for binding using flow cytometry. *FEBS Letters*, **532**, 455.
9. Bernath, K., Hai, M., Mastrobattista, E. *et al.* (2004) In vitro compartmentalization by double emulsions: sorting and gene enrichment by fluorescence-activated cell sorting. *Analytical Biochemistry*, **325**, 151–7.
10. Zheng, Y. and Roberts, R.J. (2007) Selection of restriction endonucleases using artificial cells. *Nucleic Acids Research*, **35**(11), e83.
11. Agresti, J.J., Kelly, B.T., Jaschke, A. and Griffiths, A.D. (2005) Selection of ribozymes that catalyse multiple-turnover Diels–Alder cycloadditions by using in vitro compartmentalization. *Proceedings of the National Academy of Sciences of the United States of America*, **102**, 16170–5.
12. Dittrich, P.S., Jahnz, M. and Schwille, P. (2005) A new embedded process for compartmentalized cell-free protein expression and on-line detection in microfluidic devices. *ChemBioChem*, **6**, 811–14.
13. Margulies, M., Egholm, M., Altman, W.E. *et al.* (2005) Genome sequencing in microfabricated high-density picolitre reactors. *Nature*, **437**, 376–80.
14. Godiska, R., Patterson, M., Schoenfeld, T. and Mead, D.A. (2005) Beyond pUC: Vectors for cloning unstable DNA, in *DNA Sequencing: Optimizing the Process and Analysis* (ed. J. Kieleczawa), Jones and Bartlett Publishers, Sudbury, MA.
15. Zhang, K., Martiny, A.C., Reppas, N.B. *et al.* (2006) Sequencing genomes from single cells by polymerase cloning. *Nature Biotechnology*, **24**, 680–6.
16. Perler, F.B., Kumar, S. and Kong, H. (1996) Thermostable DNA polymerases. *Advances in Protein Chemistry*, **48**, 377–435.
17. Hogrefe, H.H., Cline, J., Lovejoy, A.E. and Nielson, K.B. (2001) DNA polymerases from hyperthermophiles. *Methods in Enzymology*, **334**, 91–116.
18. Braithwaite, D.K. and Ito, J. (1993) Compilation, alignment, and phylogenetic relationships of DNA polymerases. *Nucleic Acids Research*, **214**, 787–802.
19. Karam, J.D. (ed.) (1994) *Molecular Biology of Bacteriophage T4*, American Society for Microbiology, Washington.
20. Tabor, S. and Richardson, C.C. (1987) DNA sequence analysis with a modified bacteriophage T7 DNA polymerase. *Proceedings of the National Academy of Sciences of the United States of America*, **8414**, 4767–71.
21. Tabor, S. and Richardson, C.C. (1995) A single residue in DNA polymerases of the *Escherichia coli* DNA polymerase I family is critical for distinguishing between deoxy- and dideoxyribonucleotides. *Proceedings of the National Academy of Sciences of the United States of America*, **9214**, 6339–43.
22. Reeve, M.A. and Fuller, C.W. (1995) A novel thermostable polymerase for DNA sequencing. *Nature*, **376**, 796–7.

23. Angly, F.E., Felts, B., Breitbart, M. *et al.* (2006) The marine viromes of four oceanic regions. *PLoS Biology*, **411**, e368.
24. Suttle, C.A. (2007) Marine viruses–major players in the global ecosystem. *Nature Reviews Microbiology*, **510**, 801–12.
25. Hjörleifsdottir, S.H., Hreggvidsson, G.O., Fridjonsson, O.H. *et al.* (2002) U. S. Patent 6,492,161.
26. Naryshkina, T., Liu, J., Florens, L. *et al.* (2006) Thermus thermophilus bacteriophage phiYS40 genome and proteomic characterization of virions. *Journal of Molecular Biology*, **364**, 667–77.
27. Schoenfeld, T., Patterson, M., Richardson, P.M. *et al.* (2008) Assembly of viral metagenomes from Yellowstone hot springs. *Applied Environmental Microbiology*, **74**, 4164–74.
28. Bentley, D.R. (2006) Whole-genome resequencing. *Current Opinion in Genetic Development*, **16**, 545–52.
29. Rusk, N. and Kiermer, V. (2008) Primer: Sequencing – the next generation. *Nature Methods*, **5**, 15.
30. Rachlin, J., Ding, C., Cantor, C. and Kasif, S. (2005) Computational tradeoffs in multiplex PCR assay design for SNP genotyping. *BMC Genomics*, **6**, 102.
31. Adessi, C., Matton, G., Ayala, G. *et al.* (2000) Solid phase DNA amplification: characterization of primer attachment and amplification mechanisms. *Nucleic Acids Research*, **28**, e87.
32. Shapero, M.H., Leuther, K.K., Nguyen, A. *et al.* (2001) SNP genotyping by multiplexed solid-phase amplification and fluorescent minisequencing. *Genome Research*, **11**, 1926–34.
33. Pemov, A., Modi, H., Chandler, D.P. and Bavykin, S. (2005) DNA analysis with multiplex microarray-enhanced PCR. *Nucleic Acids Research*, **33**, e11.
34. Kumaresan, P., Yang, C.J., Cronier, S.A. *et al.* (2008) High-throughput single copy DNA amplification and cell analysis in engineered nanoliter droplets. *Analytical Chemistry*, **80**, 3522–9.
35. Jagus, R. and Beckler, GS. (2003) Overview of eukaryotic in vitro translation and expression systems. Current Protocols in Cell Biology, **Chapter 11**, Unit 11.1.
36. Hoffmann, M., Nemetz, C., Madin, K. and Buchberger, B. (2004) Rapid translation system: a novel cell-free way from gene to protein. *Biotechnology Annual Review*, **10**, 1–30.
37. Peterson, M.G. (1988) DNA sequencing using Taq polymerase. *Nucleic Acids Research*, **1622**, 10915.
38. Slatko, B.E. (1994) Thermal cycle dideoxy DNA sequencing. *Methods in Molecular Biology*, **31**, 35–45.
39. Chen, J., Sahota, A., Stambrook, P.J. and Tischfield, J.A. (1991) Polymerase chain reaction amplification and sequence analysis of human mutant adenine phosphoribosyltransferase genes: the nature and frequency of errors caused by Taq DNA polymerase. *Mutation Research*, **2491**, 169–76.
40. Viguera, E., Canceill, D. and Ehrlich, S.D. (2001) In vitro replication slippage by DNA polymerases from thermophilic organisms. *Journal of Molecular Biology*, **3122**, 323–33.
41. Ji, J., Clegg, N.J., Peterson, K.R. *et al.* (1996) In vitro expansion of GGC:GCC repeats: identification of the preferred strand of expansion. *Nucleic Acids Research*, **2414**, 2835–40.
42. Flaman, J.M., Frebourg, T., Moreau, V. *et al.* (1994) A rapid PCR fidelity assay. *Nucleic Acids Research*, **22**(15), 3259–60.
43. Huang, M.M., Arnheim, N. and Goodman, M.F. (1992) Extension of base mispairs by Taq DNA polymerase: implications for single nucleotide discrimination in PCR. *Nucleic Acids Research*, **2017**, 4567–73.
44. Smith, J.R., Carpten, J.D., Brownstein, M.J. *et al.* (1995) Approach to genotyping errors caused by nontemplated nucleotide addition by Taq DNA polymerase. *Genome Research*, **53**, 312–17.
45. Clark, J.M. (1988) Novel nontemplated nucleotide addition reactions catalyzed by procaryotic and eucaryotic DNA polymerases. *Nucleic Acids Research*, **1620**, 9677–86.
46. Mead, D.A., Pey, N.K., Herrnstadt, C. *et al.* (1991A) A universal method for the direct cloning of PCR amplified nucleic acid. *Biotechnology*, **9**, 657–63.

47. Canceill, D., Viguera, E. and Ehrlich, S.D. (1999) Replication slippage of different DNA polymerases is inversely related to their strand displacement efficiency. *Journal of Biological Chemistry*, **27439**, 27481–90.
48. Virtaneva, K., Paulin, L., Krahe, R. *et al.* (1998) The minisatellite expansion mutation in EPM1: resolution of an initial discrepancy. Mutations in brief no. 186. Online. *Human Mutation*, **123**, 218.
49. Walsh, P.S., Fildes, N.J. and Reynolds, R. (1996) Sequence analysis and characterization of stutter products at the tetranucleotide repeat locus vWA. *Nucleic Acids Research*, **2414**, 2807–12.
50. Lundberg, K.S., Shoemaker, D.D., Adams, M.W. *et al.* (1991) High-fidelity amplification using a thermostable DNA polymerase isolated from *Pyrococcus furiosus*. *Gene*, **1081**, 1–6.
51. Mattila, P., Korpela, J., Tenkanen, T. and Pitkänen, K. (1991) Fidelity of DNA synthesis by the *Thermococcus litoralis* DNA polymerase – an extremely heat stable enzyme with proofreading activity. *Nucleic Acids Research*, **19**, 4967–73.
52. Skerra, A. (1992) Phosphorothioate primers improve the amplification of DNA sequences by DNA polymerases with proofreading activity. *Nucleic Acids Research*, **2014**, 3551–4.
53. Andre, P., Kim, A., Khrapko, K. and Thilly, W.G. (1997) Fidelity and mutational spectrum of Pfu DNA polymerase on a human mitochondrial DNA sequence. *Genome Research*, **78**, 843–52.
54. Kornberg, A. and Baker, T. (1992) *DNA Replication*, 2nd edn, W. H Freeman and Co., New York.
55. Meijer, W.J., Horcajadas, J., Salas, M. (2001) Phi29 family of phages. *Microbiology and Molecular Biology Reviews*, **652**, 261–87.
56. Andraos, N., Tabor, S. and Richardson, C.C. (2004) The highly processive DNA polymerase of bacteriophage T5. Role of the unique N and C termini. *Journal of Biological Chemistry*, **279**(48), 50609–18.
57. Patel, S.S., Wong, I. and Johnson, K.A. (1991) T7 DPOL is kinetically distinct from E. coli pol I Pre-steady-state kinetic analysis of processive DNA replication including complete characterization of an exonuclease-deficient mutant. *Biochemistry*, **302**, 511–25.
58. Bebenek, A., Dressman, H.K., Carver, G.T. *et al.* (2001) Interacting fidelity defects in the replicative DNA polymerase of bacteriophage RB69. *Journal of Biological Chemistry*, **276**(13), 10387–97.
59. Robb, F.T. and Place, A.R. (1995) Thermophiles, in *Archaea: A Laboratory Manual*, 1st edn (eds F.T. Robb, K.R. Sowers, H.J. Shreier, S. DasSarma, and E.M. Fleischmann) Cold Spring Harbor Laboratory Press, Cold Spring Harbor, NY.
60. Noble, R.T. and Fuhrman, J.A. (1998) Use of SYBR Green I for rapid epifluorescence counts of marine viruses and bacteria. *Aquatic Microbial Ecology*, **14**, 113–18.
61. Breitbart, M., Salamon, P., Andresen, B. *et al.* (2002) Genomic analysis of uncultured marine viral communities. *Proceedings of the National Academy of Sciences of the United States of America*, **99**, 14250–5.
62. Lukashin, A. and Borodovsky, M. (1998) GeneMark.hmm: new solutions for gene finding. *Nucleic Acids Res.*, **26**, 1107–15.
63. Barry, E.R. and Bell, S.D. (2006) DNA replication in the archaea. *Microbiology and Molecular Biology Reviews*, **704**, 876–87.
64. Dionne, I. and Bell, S.D. (2005) Characterization of an archaeal family 4 uracil DNA glycosylase and its interaction with PCNA and chromatin proteins. *Biochemical Journal*, **387**, 859–63.
65. Roberts, J.A., Bell, S.D. and White, M.F. (2003) An archaeal XPF repair endonuclease dependent on a heterotrimeric PCNA. *Molecular Microbiology*, **482**, 361–71.
66. Thompson, J.D., Higgins, D.G. and Gibson, T.J. (1994) CLUSTAL W: improving the sensitivity of progressive multiple sequence alignment through sequence weighting, position-specific gap penalties and weight matrix choice. *Nucleic Acids Research*, **22**, 4673–80.

67. Chang, J.R., Choi, J.J., Kim, H.K. and Kwon, S.T. (2001) Purification and properties of *Aquifex aeolicus* DNA polymerase expressed in *Escherichia coli*. *FEMS Microbiology Letters*, **201**, 73–7.
68. Venter, J.C., Remington, K., Heidelberg, J.F. *et al.* (2004) Environmental genome shotgun sequencing of the Sargasso Sea. *Science*, **304**, 66–74.
69. Raghunathan, A., Ferguson, H.R. Jr., Bornarth, C.J. *et al.* (2005) Genomic DNA amplification from a single bacterium. *Applied Environmental Microbiology*, **71**, 3342–7.
70. Geigl, J.B. and Speicher, M.R. (2007) Single-cell isolation from cell suspensions and whole genome amplification from single cells to provide templates for CGH analysis. *Nature Protocols*, **2**, 3173–84.
71. Kvist, T., Ahring, B.K., Lasken, R.S. and Westermann, P. (2007) Specific single-cell isolation and genomic amplification of uncultured microorganisms. *Applied Microbiology and Biotechnology*, **74**, 926–35.
72. Bakken, L.R. and Olsen, R.A. (1989) DNA content of soil bacteria of different cell size. *Soil Biology and Biochemisrty*, **21**, 789–93.
73. Pinard, R., de Winter, A., Sarkis, G.J. *et al.* (2006) Assessment of whole genome amplification-induced bias through high-throughput, massively parallel whole genome sequencing. *BMC Genomics*, **7**, 216.
74. Spits, C., Le Caignec, C., De Rycke, M. *et al.* (2006) Whole-genome multiple displacement amplification from single cells. *Nature Protocols*, **1**, 1965–70.
75. Frohlich, J. and Konig, H. (1999) Rapid isolation of single microbial cells from mixed natural and laboratory populations with the aid of a micromanipulator. *Systematic and Applied Microbiology*, **22**, 249–57.
76. Marcy, Y., Ishoey, T., Lasken, R.S. *et al.* (2007B) Nanoliter reactors improve multiple displacement amplification of genomes from single cells. *PLoS Genetics* **3**, 1702–8.
77. Stepanauskas, R. and Sieracki, M.E. (2007) Matching phylogeny and metabolism in the uncultured marine bacteria, one cell at a time. *Proceedings of the National Academy of Sciences of the United States of America*, **104**, 9052–7.
78. Lasken, R.S. (2007) Single-cell genomic sequencing using multiple displacement amplification. *Current Opinion in Microbiology*, **10**, 510–16.
79. Thomas, D. C., Nardone, G.A. and Randall, S.K. (1999) Amplification of padlock probes for DNA diagnostics by cascade rolling circle amplification or the polymerase chain reaction. *Archives of Pathology and Laboratory Medicine*, **123**, 1170–6.
80. Mead, D., McClary, J.A., Luckey, J.A. *et al.* (1991B) Bst polymerase permits rapid sequence analysis from nanogram amounts of template. *BioTechniques*, **11**, 76–87.
81. McClary, J., Ye, S.Y., Hong, G.F. and Witney, F. (1991) Sequencing with the large fragment of DNA polymerase I from *Bacillus stearothermophilus* J. *DNA Sequencing and Mapping*, **1**, 173–80.
82. Faruqi, A.F., Hosono, S., Driscoll, M.D. *et al.* (2001) High-throughput genotyping of single nucleotide polymorphisms with rolling circle amplification. *BMC Genomics*, **2**, 4.
83. Voisey, J., Hafner, G.J., Morris, C.P. *et al.* (2001) Isothermal amplification and multimerization of DNA by Bst DNA polymerase. *Biotechniques*, **30**, 852–6.
84. Lizardi, P.M., Huang, X., Zhu, Z. *et al.* (1998) Mutation detection and single-molecule counting using isothermal rolling-circle amplification. *Nature Genetics*, **19**, 225–32.
85. Zhang, D.Y., Brandwein, M., Hsuih, T. and Li, H.B. (2001) Ramification amplification (RAM): a novel isothermal DNA amplification method. *Molecular Diagnosis*, **6**, 141–50.
86. Blanco, L., Lazaro, J.M., de Vega, M. *et al.* (1994) Terminal protein-primed DNA amplification. *Proceedings of the National Academy of Sciences of the United States of America*, **91**, 12198–202.
87. Dean, F.B., Nelson, J., Giesler, T.L. and Lasken, R.S. (2001) Rapid amplification of plasmid and phage DNA using Phi 29 DNA polymerase and multiply-primed rolling circle amplification. *Genome Research*, **11**, 1095–9.

88. Dean, F.B., Hosono, S., Fang, L. *et al.* (2002) Comprehensive human genome amplification using multiple displacement amplification. *Proceedings of the National Academy of Sciences of the United States of America*, **99**, 5261–6.
89. Lage, J.M., Leamon, J.H., Pejovic, T. *et al.* (2003) Whole genome analysis of genetic alterations in small DNA samples using hyperbranched strand displacement amplification and array-CGH. *Genome Research*, **13**, 294–307.
90. Detter, J.C., Jett, J.M., Lucas, S.M. *et al.* (2002) Isothermal strand-displacement amplification applications for high-throughput genomics. *Genomics*, **80**, 691–8.
91. Backhed, F., Ley, R.E., Sonnenburg, J.L. *et al.* (2005) *Science* **307**, 1915–20.
92. Backhed, F., Ding, H., Wang, T. *et al.* (2004) The gut microbiota as an environmental factor that regulates fat storage. *Proceedings of the National Academy of Sciences of the United States of America*, **101**, 15718–23.
93. Dumas, M.E., Barton, R.H., Toye, A. *et al.* (2006) Metabolic profiling reveals a contribution of gut microbiota to fatty liver phenotype in insulin-resistant mice. *Proceedings of the National Academy of Sciences of the United States of America*, **103**, 12511–16.
94. Turnbaugh, P.J., Ley, R.E., Mahowald, M.A. *et al.* (2006) An obesity-associated gut microbiome with increased capacity for energy harvest. *Nature*, **444**, 1027–31.
95. Kitano, H. and Oda, K. (2006) Robustness trade-offs and host-microbial symbiosis in the immune system. *Molecular Systems Biology*, **2**, 2006–22.
96. Nicholson, J.K., Holmes, E. and Wilson, I.D. (2005) Gut microorganisms, mammalian metabolism and personalized health care. *Nature Reviews Microbiology*, **3**, 431–8.

10

Synthetic Networks

Jongmin Kim

CbsBioscience Inc., Daejeon, Korea

10.1 Introduction

In order to understand and utilize the diverse functionality displayed by biological organisms, it is first necessary to comprehend the regulatory network underlying such complex behavior. Fortunately, biological networks share certain properties of engineered networks [1], and thus are potentially amenable to automated design and characterization. Recent advances in both metabolic and genetic engineering have made feasible the investigation of novel biological functionality through the design and implementation of synthetic biological networks. Well-characterized 'parts' would be essential for streamlining synthetic network design processes, such that complex functionality can be created without reinventing all details of the molecules involved. One example of such an effort towards standardized parts for abstraction is the Registry of Standard Biological Parts (http://parts.mit.edu). Another important research venue is mathematical modeling including quantitative analysis, which allows for the circuit behavior to be explored with uncertain parameter sets and external disturbances. Today, several software tools are available to aid biochemical kinetic simulations [2]. In this chapter, the current understanding of cellular networks, synthetic network construction and the remaining challenges towards automating biochemical processes using synthetic circuitry are reviewed.

10.2 Cellular Network: Functional Design

Cells live in a complex environment and can sense many different signals, whether physical, chemical or biological. Cells also have the ability to process information for survival

Automation in Proteomics and Genomics: An Engineering Case-Based Approach Edited by Gil Alterovitz, Roseann Benson and Marco Ramoni

and reproduction, such as detecting nutrients and avoiding harmful chemicals, by using functional circuits composed of many interacting molecular species [1]. Hence, information processing through regulatory networks lies at the heart of all living systems. By taking a 'top-down' view of protein–protein interactions, signaling pathways and gene regulatory pathways, the basic architecture of biological networks has been analyzed [3]. The network description of cellular circuits allows the application of tools and concepts which have been developed in fields such as graph theory, physics, sociology and engineering [4]. Remarkably, biological networks share the design principles of engineered networks, namely modularity, robustness and recurring circuit elements. A module in a network is a set of nodes that have strong interactions and a common function [1]. Modules in engineering – and presumably also in biology – have special features that make them easily embedded in almost any system. The robustness of a cellular network design requires that the design must function under plausible fluctuations and interferences due to the components and to the environment [5]. Recurring network motifs for signal processing tasks, such as filtering out input noise, accelerating throughput of the network or temporal programming, can be found in biological networks [6]. The fact that a biological organism must function and compete for resources imposes severe constraints on the regulatory network design, which could have shaped the biological networks with characteristics analogous to human-engineered networks. These design principles of cellular networks will help delineate system architecture with limited data, such that researchers can focus on modular and robust patterns. Indeed, some of these patterns are already known as network motifs.

10.2.1 Network Motifs

Alon and colleagues studied the transcription network of *Escherichia coli* to identify meaningful patterns on the basis of statistical significance. The transcription network was compared to an ensemble of randomized networks, with similar characteristics such as the same number of nodes and edges but with random connections between nodes and edges. Patterns that occur in the real network significantly more often than in randomized networks were termed network motifs [6, 7]. One network motif is that of negative autoregulation, where a protein product binds to its own promoter and represses its own transcription. Negative autoregulation has two useful features – the speed-up of response time and robustness to fluctuation. The response time, which is defined as the time to reach halfway between the initial and final levels in a dynamic process, depends simply on degradation and dilution rates in unregulated transcription and translation processes. In order to achieve the same steady-state value, negative autoregulation employs a stronger promoter than its unregulated counterpart; therefore, the initial build-up of signals is fast with negative autoregulation, cutting down the response time. Moreover, the steady-state protein level is stable with negative autoregulation, albeit with fluctuations in the production rate. An important three-node motif – termed the feedforward loop – is defined by a transcription factor X that regulates a second transcription factor Y, such that both X and Y jointly regulate an operon Z (Figure 10.1a). Most of the feedforward loops are coherent; that is, the direct regulation of X on Z and indirect regulation of X on Z through Y are of the same sign. Mathematical analysis suggests that the coherent feedforward loop can act as a persistence detector, rejecting short pulses of activation signals from the general transcription factor responses. Consider the case where both X and Y transcription factors are required for the

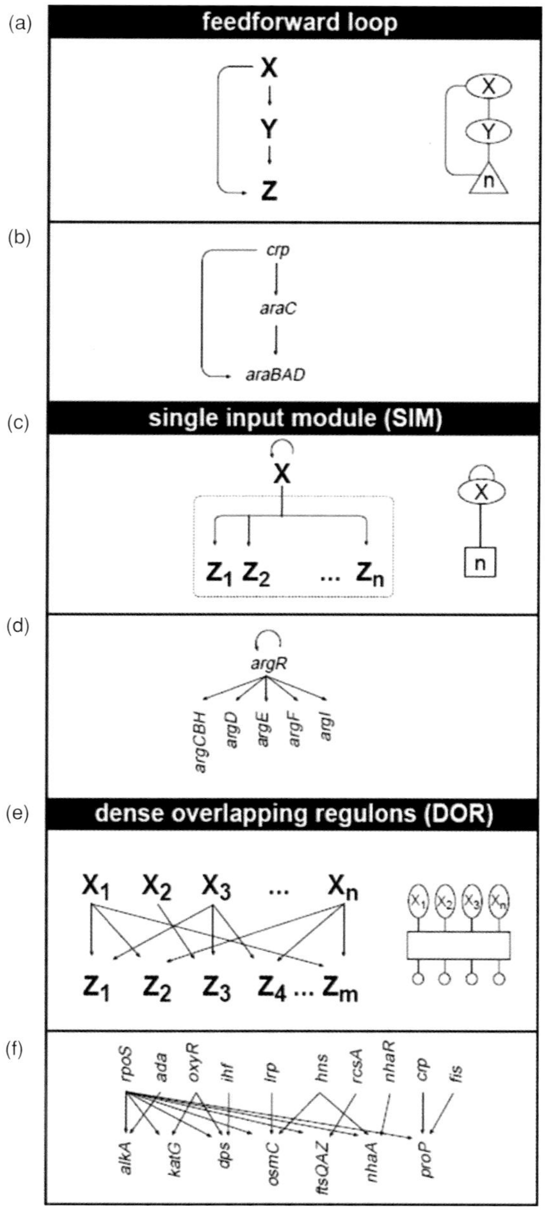

Figure 10.1 *Network motifs found in the* E. coli *transcriptional network. (a) Feedforward loop: a transcription factor X regulates a second transcription factor Y, and both jointly regulate one or more operons $Z_1 \ldots Z_n$; (b)* L*-arabinose utilization network; (c) SIM motif: a single transcription factor, X, regulates a set of operons $Z_1 \ldots Z_n$; (d) Arginine biosynthesis network; (e) DOR motif: a set of operons $Z_1 \ldots Z_m$ are regulated by a combination of a set of input transcription factors, $X_1 \ldots X_n$; (f) Stationary phase response network. (Reprinted by permission from Macmillan Publishers Ltd. Ref. [6])*

activation of Z in a feedforward loop. Upon arrival of activation signal for X, the activation of Z is delayed because Y takes time to build up to a threshold level. Thus, if the activation signal for X has a short duration, Y cannot reach the threshold level needed to activate Z. Response to signals such as nutrients that activate X incurs production cost for the final enzyme Z, but no significant benefit can be gleaned if the nutrients disappear by the time enzyme Z level is sufficiently high. A cost–benefit analysis indicates that a coherent feedforward loop offers more benefit over the simple regulation of X and Y on Z under a fluctuating environment where transient activation signal is common [8].

Two other larger motifs are called the single-input module (SIM) and the dense overlapping regulon (DOR). The SIM network motif – a simple pattern in which one regulator controls a group of genes – can generate temporal programs of expression, in which genes are turned on one by one in a defined order (Figure 10.1c). In contrast, the DOR network motif is a layer of overlapping interactions between operons and a group of input transcription factors, in which the signal inputs are integrated and the output genes are under a combinatorial control (Figure 10.1e). Other network motifs appear in a developmental transcription network [9], such as a positive feedback loop and a long cascade. A positive feedback loop can serve as a memory, locking in the cell fate if an early developmental signal ever reaches a threshold level. Long cascades are uncommon in sensory information processing due to significant delays, but prove useful in developmental timing that spans several cell generations. Transcription regulatory networks operate on the timescale of tens of minutes to hours, whereas signal transduction networks rely on protein–protein interactions to process sensory signals on the timescale of seconds to minutes. A more complete picture of cellular networks requires an analysis of the interaction of different network components operating at different timescales and searching for novel regulatory mechanisms operating on such interfaces.

Network motifs provide a powerful tool to understand cellular organization from a functional point of view, bypassing the biochemical details. The spontaneous evolution of modularity and network motifs has been demonstrated in computational evolution models of electronic circuits and neural networks [10]. Many such models use networks in a population explored by means of mutations, crossover and duplication to be selected for a defined goal. The evolved systems typically result in intricately wired nonmodular solutions because these are more optimized than their human-engineered counterparts. A lack of modularity has been cited as one of the reasons why computational evolution can generate design patterns for simple tasks, but cannot be scaled-up to more complex tasks. If the network evolution is constrained to fulfill modularly varying goals, then the achieved architecture is built of more computational units solving subproblems; this framework has an increased modularity but is suboptimal. Modularity decreases quickly when the network is trained on a single goal or nonmodularly varying goals. Kashtan and Alon [10] have suggested that modularity allows a higher adaptability to be achieved, and is therefore a characteristic that a biological network must have in order to evolve in a constantly changing environment that requires a certain set of basic functions in different combinations.

10.2.2 Network Architecture

An alternative approach is to abstract features from the overall architecture of cellular networks. The architecture of a network places boundaries on its performance capabilities, and also explains its possible evolutionary path [3]. Clearly, cellular networks differ from regular networks, where nearest neighbors are linked in a regular fashion, or from random networks, where randomly selected nodes are joined together. In cellular networks, a few nodes have a large number of connections, while most of the nodes have relatively few connections – this is a feature of a 'scale-free' network. 'Scale-free' means that the number of molecules (N) with a given number of connections (k) falls off as a power law, $N(k) \sim k^{-g}$, where no characteristic peak value can be found. In a scale-free network, the average distance between any two nodes is almost as small as the random network, while the extent to which neighbors of a node are themselves connected (known as its clustering coefficient) is almost as large as in a regular network. Protein–protein interaction maps have the features of a scale-free network, with their degree sequences (number of edges per node) often following a long-tailed distribution [11]. However, the fact that a network has scale-free properties is of limited use, since power laws occur widely in nature, possibly with different mechanistic origins. Thus, a much closer examination of small-scale networks, such as subnetworks or molecular complexes, should complement the top-down network description [3].

It has been suggested that biological networks have additional constraints that are beyond simple scale-free networks [12]. Networks that are simple connection networks, such as the Internet, are able to grow in an unconstrained way, whereas regulatory networks – such as genetic regulatory networks in biology – must be able to operate in a globally responsive way. In order to maintain global connectivity, the number of connections must be scaled quadratically with the network size. As a consequence, the need for an increased number of connections at the regulatory level naturally imposes a limit on the size of the network and its complexity [12]. Although dedicated hierarchies could solve such a scalability problem, each level of regulatory hierarchy will introduce time delays and increase stochastic noise [13]. Regulatory proteins scale almost quadratically with genome size in prokaryotes [14], and the extrapolation of this relationship suggests that prokaryotes have reached their complexity limit by their reliance on a protein-based regulatory architecture. Eukaryotes have a far more developed RNA processing and signaling system than prokaryotes, which appears to be linked to a more sophisticated pathway of gene regulation. Recently it was suggested that, in addition to being a digital storage medium, noncoding RNA themselves are actually transmitting digital signals [15]. In contrast, regulatory proteins act mainly as analogue components because their signals are transmitted as their concentrations. Following the comparison with electronic circuits, it is possible that the cellular network complexity limit was lifted by the use of both digital and analogue signals.

In summary, biological networks present different features at different scales, behaving like scale-free networks on a large scale, and consisting of recurring network motifs and basic functionalities on a smaller scale. Network motifs found in transcriptional networks illustrate that the network design has functional consequences. Other modalities of regulatory strategies such as RNA processing and post-translational modifications, although sophisticated regulatory examples are known, have not been discussed here. Investigating the cellular networks at different levels of complexity starting from basic network motifs merits future research efforts that would lead to an understanding of the complexity of regulation strategies and provide useful insights.

10.3 Synthetic Approaches to Understand Cellular Networks

A network description in an abstract sense is not enough to understand cellular networks with quantitative details and to construct predictive models. Rather, the investigation of detailed kinetics and reaction mechanisms among the constituent macromolecules is required. The reductionist approach attempts to explain the behavior of cellular networks in terms of the behavior of the components. Despite many molecular components of biological organisms being identified and characterized using genetic and biochemical techniques, it is still not possible to predict system behavior, except in the simplest systems. This indicates that the great complexity of cellular network hinders the prediction of system behavior from characterized components, and that alternative approaches for understanding cellular network behavior and design principles may be necessary.

Synthetic biology provides an alternative to the study of cellular networks, by constructing increasingly complex analogues of natural circuits. This is a 'bottom-up' approach that attempts to test the sufficiency of mechanistic models by actively synthesizing them: this allows insights to be gained that observation and analysis alone do not provide [16]. A synthetic biology approach shares the spirit of engineering community in that a successful model system should operate upon synthesis. For engineering purposes, parts are most suitable when they contribute independently to the whole. This 'independence property' allows one to predict the behavior of an assembly by characterizing parts. In terms of satisfying independence property, the DNA molecules described by the Watson–Crick model stand out because each nucleotide pair contributes independently to the stability of a duplex, to a good approximation [17]. However, the DNA molecule is rather an exception than the rule; for instance, the behavior of a protein is generally not a function of the behavior of its constituent amino acids.

Although amino acids may be a poor unit for the application of independence property, natural folded proteins can be treated as interchangeable parts. Several synthetic networks constructed by rearranging the regulatory components in a cell have been characterized, including autoregulators [18, 19], feedforward cascades [13, 20], bistable memory element [21] and oscillators [22, 23]. In order for this type of network design to lead to an improved understanding of naturally occurring networks, detailed studies of the synthetic systems are needed [16], for example, through a systematic examination of the effects of parameter variations with quantitative modeling and analysis [24]. Some example networks and their design principles will be discussed.

10.3.1 Synthetic Networks *In Vivo*

A bistable memory was constructed by Gardner *et al.* [21] by employing a mutual repression system which used two genes that each coded for a transcriptional repressor of the other gene. These authors used combinations of the lac repressor (LacI), tetracycline repressor (TetR) and the temperature-sensitive lambda repressor (cI). An external stimulus inhibits the activity of a specific repressor and pushes the system to one steady state. For the mutual repression system shown in Figure 10.2a, isopropyl-β-D-thiogalactopyranoside (IPTG) inhibits the lac repressor, while a high temperature inhibits the cI repressor. Thus, the addition of IPTG pushed the system to a lac-off/lambda-on state and a concomitant increase in the green fluorescent protein (GFP) signal. This system demonstrated hysteresis,

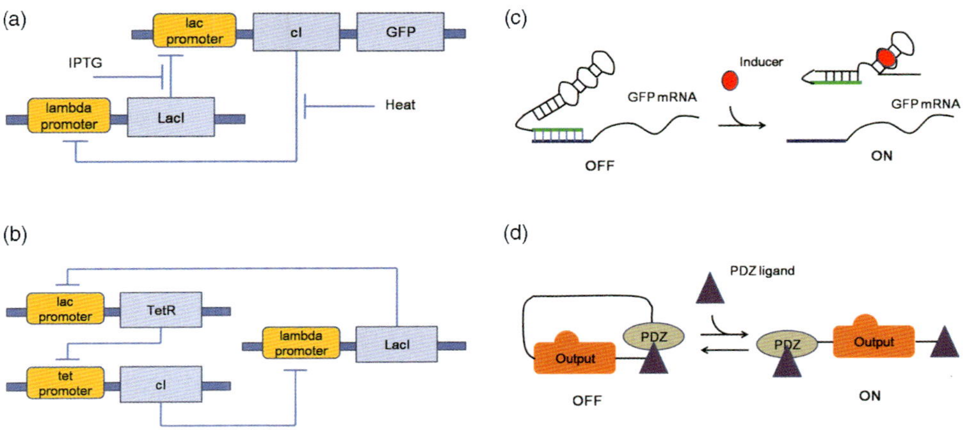

Figure 10.2 *Engineered* in vivo *networks. (a) A genetic toggle switch uses a mutual repression motif. Two genes, lac repressor and lambda repressor, repress the expression of the other gene. Transient exposure to either heat or IPTG will shift the steady-state of the system to the expression of only one repressor; (b) A circular arrangement of repressors comprises the genetic ring oscillator. Oscillatory output was observed via GFP expression regulated by the tetracycline repressor; (c) The RNA 'anti-switch' relies on ligand-binding regions of RNA that, when bound to ligand, induce changes in RNA structure. When bound to an inducer ligand, the anti-switch hides an antisense region of RNA that hybridizes to the 5′-UTR of target mRNA, encompassing the translational start site; (d) An allosteric switch based on the natural N-WASP allosteric switch. A PDZ-binding domain is used with a C-terminal PDZ ligand, resulting in autoinhibition of N-WASP output domain. When an exogenous PDZ ligand is added, the intramolecular PDZ interaction is disrupted, and the output domain stimulates actin polymerization. (Reprinted by permission from Macmillan Publishers Ltd. Refs. [21, 22, 26, 27])*

such that once the switch was flipped toward one steady state it remained there, even in the absence of the original stimulus. Several plasmid constructs with different promoters and ribosome-binding sequences were shown to be bistable, except for one construct. Thus, bistability can be achieved for a wide range of parameter space, if two repressor strengths are balanced. Furthermore, a toggle switch design can be embedded in a larger system. Kobayashi *et al.* [25] used a lac repressor/lambda repressor toggle switch as a memory subsystem within the DNA damage sensor. The lambda repressor is naturally cleaved upon DNA damage and induction of the SOS response, leading to a lac-on/lambda-off state. The engineered cells also contained the *traA* gene, which activates biofilm formation under the control of lambda repressor. Consequently, exposure of the cells to DNA-damaging agents resulted in biofilm formation.

The first synthetic oscillator was a ring oscillator constructed by Elowitz and Leibler [22], where three repressors (the lac, lambda and tetracycline repressors) regulated the expression of the next repressor in the cycle (Figure 10.2b). A GFP reporter protein under the control of tetracycline repressor was used to monitor periodic changes of output. An important part of the design process was a rough quantitative model of the system to explore parameter spaces. A tightly regulated promoter and a shorter protein half-life improved the performance in the mathematical analysis, which was implemented in the experimental design. The authors described a single plasmid construct, which suggests that the approximate calculation used to design the ring oscillator was enough to achieve oscillatory behavior in engineered *E. coli* cells. Interestingly, the oscillation period showed much more variability than did natural oscillators, with only 40% of the cells exhibiting oscillation. These findings suggested that the stability properties observed in wild-type circadian oscillators might result from the coupling of these clocks to other cellular processes. Alternatively, the architecture of the oscillator itself may dictate the stability of oscillation. In fact, the models of circadian oscillators fall in the class of relaxation oscillators [28], where a positive feedback loop and a negative feedback loop operate with slow and fast time scales. The synthetic oscillator design of Elowitz and Leibler does not fall into this category, but is a phase oscillator [29]. The oscillator design of Atkinson *et al.* [23] involved a positive autoregulatory circuit linked to a repressor module, analogous to the relaxation oscillator model of Barkai and Leibler [28]. Atkinson and colleagues used the components of a nitrogen-regulated response system for the activation signal and LacI for the inhibitory signal. This design did not involve a degradation sequence, as was used by Elowitz and Leibler [22], to shorten the protein lifetime, and the experiments were performed in a continuous bioreactor under constant cell density. Surprisingly, this oscillator displayed oscillation dynamics at population level, despite the oscillation being damped. Through mathematical analysis, the authors suggested a variety of parameter changes, such as messenger RNA stability and protein stability, to achieve sustained oscillation. Yet, an experimental exploration of such parameter change was not achieved and the mechanism for synchronization was unclear [30].

RNA molecules play important and diverse regulatory roles in the cell by virtue of their interaction with other nucleic acids, proteins and small molecules. For instance, diverse *cis* and *trans* gene regulation by noncoding RNA molecules such as microRNAs [31] and antisense RNAs [32] have been characterized in natural organisms. Researchers have engineered RNA molecules with new biological functions realized in bacteria and yeast [26, 33]. Isaacs *et al.* [33] achieved the repression of a target gene by forming a hairpin structure in the 5′ untranslated region (UTR) of the mRNA (*cis*-regulator), sequestering the ribosome-binding sequence. The expression of a targeted *trans*-RNA activator allowed translation from modified mRNA by exposing the ribosome-binding sequence. Bayer and Smolke [26] developed RNA regulatory molecules that have an aptamer domain to recognize specific effector molecules and an antisense domain to control gene expression, analogous to naturally found riboswitches (Figure 10.2c). The specific and dose-dependent switching responses of these regulatory RNA molecules have been demonstrated; for example, theophylline and tetracycline were each used to control the expression of GFP and yellow fluorescent protein (YFP) reporter proteins, without significant crosstalk. The stem stability of the designed RNA regulators turned out to be an important parameter that shifted switching thresholds. These results point to an intriguing possibility where designed RNA switches can be employed as cellular sensors and effectors to create programmable cells [34]. However, the engineered synthetic RNA regulation systems mainly demonstrated switching behavior rather than general network construction; consequently, quantitative models for the dynamics of RNA regulators need to be developed.

The signal transduction cascades composed of multiple proteins with enzymatic and structural interactions mediate many cellular functions and interactions with the environment. The interaction domains within signaling proteins can be rearranged to create novel interactions. For example, when Dueber *et al.* [27] described the modular reprogramming of an allosteric protein signaling switch in yeast, their hybrid protein was constructed with an N-WASP-regulated actin polymerization output domain, a PDZ domain and a PDZ ligand (Figure 10.2d). This synthetic design has autoinhibitory architecture because the binding of a PDZ domain and a PDZ ligand blocks actin polymerization output, analogous to its natural counterpart GTPase-binding domain that represses actin polymerization. An external supply of PDZ ligand releases this autoinhibition in a dose-dependent manner. Furthermore, a library of hybrid proteins was created using PDZ- and SH3- binding domains with a variety of ligand affinities. Exploiting novel protein–protein interactions in addition to transcriptional regulation will enlarge the design space of synthetic networks.

10.3.2 Synthetic Networks *In Vitro*

An *in vitro* reconstruction with known components offers a unique opportunity to investigate how system behavior derives from reaction mechanisms. The first nontrivial system behavior created by an *in vitro* chemical system was the Belousov–Zhabotinsky oscillator [35], although it was difficult to see how these reaction mechanisms could support a wide variety of chemical logic, as is found in biochemistry. An excellent example of *in vitro* reconstruction using biochemical components is the cyanobacterial circadian clock, the operation of which has been shown to be independent of transcription and translation [36]. Operating and characterizing biochemical circuits in a cell-free system present some challenges, partly due to the complexity of synthesis machinery. A reconstituted cell-free transcription–translation system requires almost 100 purified components [37] or poorly characterized cell extracts [38]. Yet, several research groups were able to successfully construct a variety of interesting circuits within cell-free transcription–translation systems. For instance, Noireaux *et al.* [38] constructed transcriptional activation and repression cascades, where the protein product of each stage activated or inhibited the following stage. Isalan *et al.* [39] constructed a transcription–translation network that emulated *Drosophila* embryonic patterns and, by utilizing regulatory interaction mediated by previously characterized zinc-finger proteins, different network connections were tested. The patterning behavior was qualitatively correct and more mutual repression led to an overall lower activity, but with sharper patterns. Moreover, the addition of a protease stabilized the pattern over time. Thus, these bare-bone *in vitro* systems can be used to illustrate design principles, although further refinement of model systems and quantitative characterization would be required.

Nucleic acid-based networks greatly reduce the complexity of the production machinery. For example, feedback circuits modeled after predator–prey dynamics have been constructed as a much simpler *in vitro* system containing only three enzymes – T7 RNA polymerase, M-MLV reverse transcriptase and *E. coli* RNase H [40, 41]. The reaction scheme is based on self-sustained sequence replication, an isothermal amplification scheme for the coupled amplification of both DNA and RNA oligomers [42]. Mathematical modeling suggests that coupling prey and predator cycles (where the prey cycle provides a primer for the predator cycle) with an appropriate flow rate in a chemostat can lead to oscillation. Yet, a quantitative agreement of models and experiments was not achieved, possibly because of unmodeled dead-end side reactions and further couplings of reaction rates by the common use of enzymes. Kim *et al.* [43] presented an alternative approach which relied on the transcription and degradation of RNA signals rather than replication and dilution. These authors constructed and analyzed feedforward circuits and a bistable mutual repression circuit with reasonable agreement to a mathematical model. However, it remains to be seen whether such nucleic acid-based networks can be utilized for regulating cellular behavior.

10.4 Challenges in Synthetic Networks

10.4.1 Saturation of Degradation Machinery *In Vivo*

Predictions about network behaviors through computational modeling and analytical theory is central to computational and systems biology. Many models of biological systems use simplifying assumptions [22, 44, 45] such as no spatial dependency of molecular species and no crosstalk between promoters. It is a widely accepted abstraction to view translation, transcription and degradation as composite processes, neglecting the detailed underlying reactions; however, these simplifying assumptions turned out to be inappropriate in some cases.

The transcriptional regulatory networks of Guet *et al.* [46] used three repressors – the lac repressor, lambda repressor and tetracycline repressor – with combinatorially assigned promoters; this allowed for a total of 27 different network topologies. The output of the network was monitored using GFP under control of the lambda repressor. Experimentally, GFP outputs were measured under four conditions: (1) without effector; (2) with IPTG, which inhibits LacI; (3) with anhydrotetracycline (aTc), which inhibits TetR; and (4) with both effectors. Kim and Tidor [47] studied the behavior of these combinatorial circuits by assuming a monotonic dependency of transcription, translation and degradation reactions to substrates and effectors, without detailed functional description or parameterization. Thus, without any detailed measurements of regulatory functions, it was possible to predict – for certain network topologies – the network output as upregulation, downregulation, or no change. Interestingly, two networks of equivalent topology (but with interchanged regulatory elements) showed different behavior in the study conducted by Guet and coworkers. According to the model, the addition of IPTG to the first network led to an increased production of both LacI and TetR, as the effect of LacI autorepression was decreased. Consequently, the model predicted that the cI level would decrease and the GFP output level would increase, in contrast to the experimental observations (Figure 10.3a). However, the addition of aTc in network 2 showed an increase of GFP output level, as predicted by the model (Figure 10.3b).

After ruling out some of the potential weakness of their model, such as not accounting for cell growth and stochastic noise, Kim and Tidor proposed that the saturation of degradation machinery could be one possible mechanism to reconcile the experimental results and model predictions. As all three repressors of the synthetic network were known to carry ssrA tags, they would be degraded by a special cellular machinery, the Clp system [48]. Because the components of Clp system are at fairly low cellular concentrations, this degradation machinery could be saturated. In network 1, IPTG released the LacI repression on both LacI and TetR production, which in turn reached high cellular concentrations and

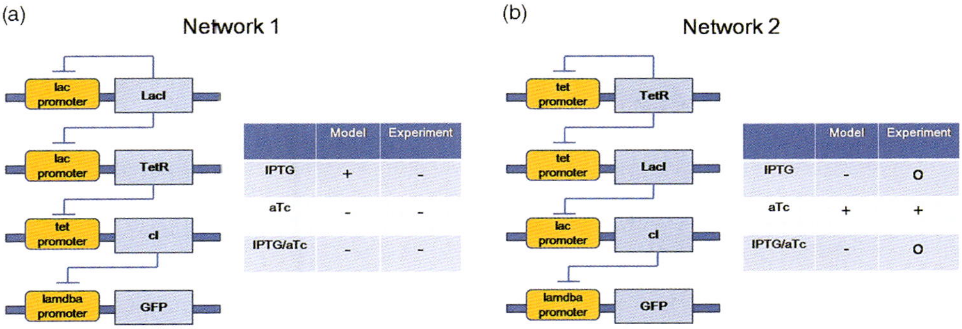

Figure 10.3 Experimental and predicted behavior of synthetic repressor networks consisting of lac, lambda and tetracycline repressors. The two networks shown in (a) and (b) have identical topologies but with lac repressor and tetracycline repressor interchanged. In the rows titled IPTG, aTc and IPTG/aTc, the GFP level changes are shown as + (increase), – (decrease), 0 (no change). (Reproduced by permission of Cold Spring Laboratory Press. Ref. [47])

outcompeted cI for degradation. Consequently, under saturating degradation conditions cI could potentially accumulate, even with basal expression rates. A direct measurement of cellular repressor levels could answer the question of the validity of this scenario. The results of this study show that care must be taken for a seemingly general assumption such as the monotonic dependency of production and degradation functions on substrates and effectors, particularly with synthetic networks that introduce new components and novel interactions among the cellular machinery.

10.4.2 Saturation of Production Machinery *In Vitro*

Noireaux and colleagues [38] characterized the cell-free genetic circuits constructed in a transcription–translation extract by engineering transcriptional activation and repression cascades in which the protein product of each stage was the input required to drive or block the following stage. The protein expression reactions were carried out in batch mode, without any continuous exchange of nutrients and byproducts. In order to boost protein production, 5′-polyguanylic acid was used to increase the mRNA lifetime [49] from 20–30 min to 2 h. At the same time, both the creatine phosphate concentration (for ATP regeneration) and the magnesium concentration were adjusted to optimal levels.

A single-level cascade was constructed as a T7-luc plasmid composed of T7 RNA polymerase promoter site and firefly luciferase gene. Upon the addition of T7 RNA polymerase, this single-level cascade began to accumulate luciferase protein after 15 min, reaching a maximum concentration of 500 n*M* after 6 h. A two-stage cascade, constructed with the plasmids T7-SP6RNAP and SP6-luc, used SP6 RNA polymerase produced from T7-SP6RNAP plasmid to drive the production of luciferase output from the luciferase gene downstream of SP6 polymerase promoter (Figure 10.4a). The two-stage cascade started to produce luciferase after a 1 h delay, such that the final luciferase level was 100 n*M* – fivefold less than for a single-stage cascade. A three-stage cascade constructed with the plasmids T7-SP6RNAP, SP6-rpoF and Ptar-luc using *E. coli* sigma factor F from the *rpoF*

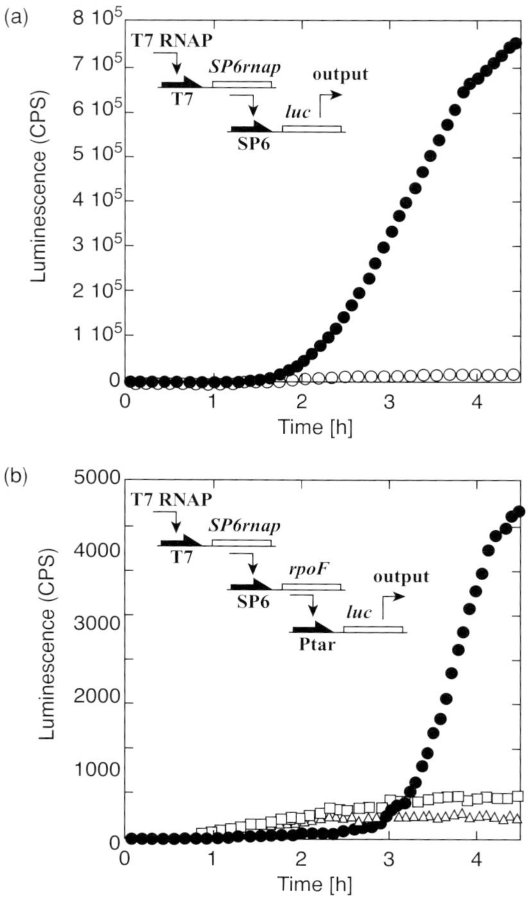

Figure 10.4 *Two- and three-stage cascades. (a) Kinetics of expression of the cascade with T7 RNA polymerase, and both T7-SP6rnap and SP6-luc plasmids (filled circles) or SP6-luc plasmid only (open circles); (b) Kinetics of expression of the cascade with T7 RNA polymerase and all three plasmids (filled circles) or two plasmids, SP6-rpoF and Ptar-luc (open squares) or Ptar-luc only (open triangles). (Copyright National Academy of Sciences, U.S.A. Ref. [38])*

gene as a new relay signal (Figure 10.4b) produced luciferase after about a 3 h delay, reaching a final concentration of only 1 n*M* after 6 h. Interestingly, substantial time delays and dramatic decreases in output were observed with each additional stage.

A detailed characterization of the two-stage cascade with various RNA polymerase and plasmid concentrations revealed that the translation machinery was saturated for the combinations of polymerase and plasmid concentrations which resulted in high transcript concentrations. Above the first-stage transcription rate, that maximized luciferase production, the overproduced first-stage mRNA occupied translation machinery and inhibited luc mRNA translation. In contrast, luciferase production did not show saturation for similar RNA polymerase and plasmid concentrations if short-lifetime mRNAs without polyG modification were transcribed. The results of the study indicated that a conventional approach of maximizing single-protein synthesis in cell-free systems must be reconsidered for *in vitro* gene circuits. The authors suggested that a rapid turnover of mRNA might avoid saturation of the translation machinery and that implementing gene autoregulation would prevent overproduction.

In a follow-up study of the cell-free expression system, Noireaux and Libchaber [50] employed the phospholipid encapsulation of synthesis machinery to construct a vesicle bioreactor. Without access to nutrients outside, the vesicle bioreactor could not prolong the expression of reporter proteins by more than 5 h. In order to solve the material and energy limitation, the α-hemolysin pore protein from *Staphylococcus aureus* was expressed inside the vesicle to create a selective permeability for nutrients. Subsequently, the vesicle bioreactor thus created could take up nutrients from a feeding solution containing amino acids and nucleic acids, and maintained protein expression for up to four days. This study proved to be an important step towards the synthesis of a minimal, self-reproducing cell.

10.4.3 Saturation in a Mutual Repression Circuit

The saturation of production and degradation machinery has a significant impact on the network dynamics. Take an example of a mutual repression circuit where two repressors, X and Y, downregulate the synthesis rates of each other (Figure 10.5a). By assuming equivalence of the two repressors, the behavior of the circuit can be understood using the following dimensionless model (Equation 10.1):

$$\begin{aligned}\frac{\mathrm{d}x}{\mathrm{d}t} &= \frac{\alpha}{1+y^n} - x,\\ \frac{\mathrm{d}y}{\mathrm{d}t} &= \frac{\alpha}{1+x^n} - y,\end{aligned} \tag{10.1}$$

where x and y are the concentrations of the repressors, α is the effective synthesis rate of repressors, and n is the cooperativity of repressor binding. The repressor binding to promoter is fast compared to transcription, translation and degradation processes. Therefore, it is assumed that the promoter–repressor binding is already at steady state when considering repressor production and degradation dynamics. Thus, the fraction of active gene x with an unoccupied promoter region can be described by $1/(1+y^n)$, and similarly for the fraction of active gene y. With the repressor cooperativity >1 and for a large synthesis rate, the two nullclines ($\mathrm{d}x/\mathrm{d}t = 0$ and $\mathrm{d}y/\mathrm{d}t = 0$) were seen to intersect at three points, producing one unstable and two stable steady states [21]. The nullclines for the circuit with cooperativity of two and maximum production rate of five indicates such bistable behavior (Figure 10.5b).

Consider the case where the production machinery is saturated for the mutual repression circuit. Assuming that α is the maximum synthesis rate for the system, and that the sharing of synthesis machinery is strictly between two repressor genes with unoccupied promoters,

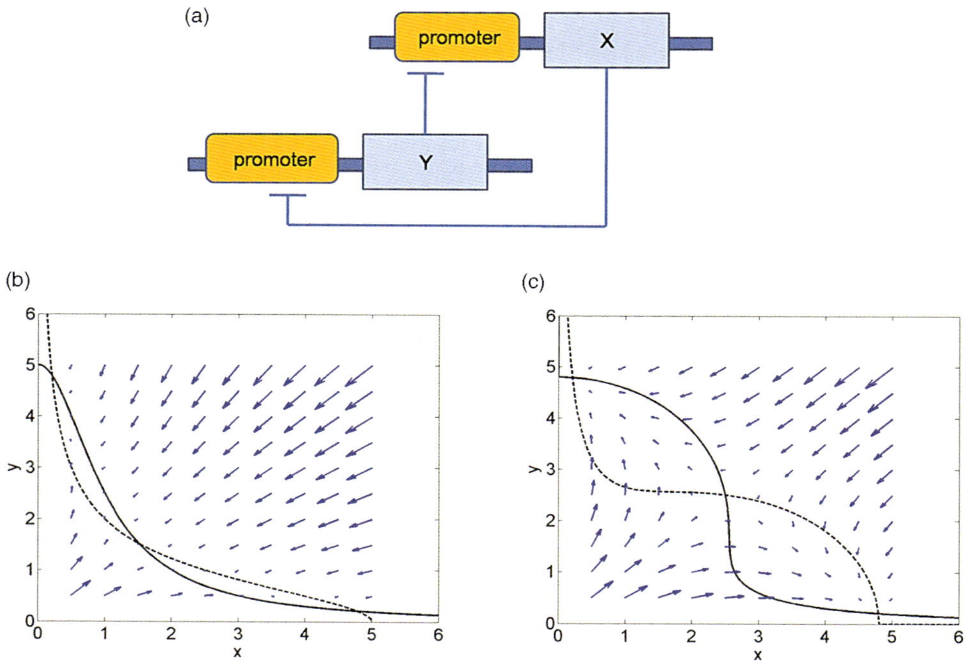

Figure 10.5 Dynamics of a mutual repression system. (a) A mutual repression system constructed from two repressors, X and Y, that repress the expression of each other; (b, c) Dynamics of mutual repression system without saturation of synthesis machinery (b) or with saturated synthesis machinery (c). Nullclines are drawn for both $dx/dt = 0$ (dotted line) and $dy/dt = 0$ (solid line) with vector flow (arrows). The parameters are $\alpha = 5$ and $n = 2$

the behavior of the circuit can be described using the following dimensionless model (Equation 10.2):

$$\begin{aligned}\frac{dx}{dt} &= \alpha \cdot \frac{\frac{1}{1+y^n}}{\frac{1}{1+y^n} + \frac{1}{1+x^n}} - x = \alpha \cdot \frac{1+x^n}{2+x^n+y^n} - x,\\ \frac{dy}{dt} &= \alpha \cdot \frac{\frac{1}{1+x^n}}{\frac{1}{1+x^n} + \frac{1}{1+y^n}} - y = \alpha \cdot \frac{1+y^n}{2+x^n+y^n} - y.\end{aligned} \tag{10.2}$$

The nullclines for the circuit intersect at three points with a cooperativity of two and a maximum synthesis rate of five, analogous to the previous example (Figure 10.5c). However, the circuit dynamics around the unstable steady state is different: the approach towards the unstable steady state is slower, while the exit from the unstable steady state is faster than the previous example. The production of repressor X, in effect, inhibits the production of repressor Y because the two promoters compete for the same synthesis machinery. Thus, it is expected that saturated production leads to bistability even when the repressor cooperativity is relatively low. For example, bistability is achieved for the repressor cooperativity of 1.4 and a maximum synthesis rate of five with saturated synthesis

(Equation 10.2), but bistability is not achieved for the same parameters in the other model (Equation 10.1).

On the other hand, saturation of the degradation machinery would be detrimental to the bistability of a mutual repression circuit because the accumulation of one repressor would allow an accumulation of the other repressor. In natural organisms, it is rarely the case that a few proteins dominantly occupy the synthesis and degradation machinery. However, for synthetic networks *in vivo* or *in vitro*, inducing the overproduction of network elements can lead to the saturation of such machinery. Hence, saturation effect must be carefully modeled, depending on the context, and can potentially be exploited for circuit operation.

10.4.4 Waste Product in an *In Vitro* Oscillator

Kim and colleagues developed an experimental analogue to a genetic regulatory circuit that uses only T7 RNA polymerase and *E. coli* RNase H in addition to synthetic DNA templates regulated by RNA transcripts [43]. A synthetic template – a gene analogue – consists of a regulatory domain, a promoter and an output domain. Each synthetic template requires a DNA oligonucleotide activating signal that complements the promoter region for a strong transcription of its output. The addition of an RNA inhibitor complementary to the DNA-activating signal hybridizes to – and consequently eliminates – the DNA-activating signal from the target synthetic template and greatly reduces transcription rates. At the same time, the degradation of RNA signals by RNase H releases the DNA signals from a functionally inert DNA–RNA hybrid state. Thus, the difference of activating and inhibitory signals determines the transcription speed of outputs. Consequently, a sigmoidal response curve with adjustable thresholds is achieved through a competitive binding of nucleic acid species.

A two-node oscillator was constructed as follows. An RNA activator (rA) activates the production of an RNA inhibitor (rI) by regulating a synthetic template (gene I), while the RNA inhibitor, in turn, inhibits the production of RNA activator by controlling gene A (Figure 10.6a). These two genes form a negative feedback loop and can potentially show oscillatory behavior. By measuring RNA signals, up to six oscillation cycles were observed before the production rate could no longer be sustained due to exhaustion of the NTP fuel (Figure 10.6c). Interestingly, the concentration of rI was seen to build up after each cycle, although it was expected that the RNA inhibitor signal would oscillate around a fixed threshold, the concentration of DNA-activating signal. One hypothesis was that the short fragments of rI generated by degradation process might interfere with the correct hybridization reaction of rI signals to its regulatory target, gene A, and therefore, more signals would be needed to overcome the interference. The short fragments of rI produced by RNase H processing would encompass the toehold binding sequence of rI because RNase H cannot process several bases on the 5′ side of the RNA strand on an RNA/DNA hybrid substrate [51]. Thus, the short fragment of rI could block the (otherwise freely available) toehold region that was essential for providing a fast kinetic pathway [52]. The concentration of short degradation products estimated from the gel showed a linear build-up over time (Figure 10.6b). Intriguingly, subtracting a fraction of short products from rI signal resulted in an oscillation around a fixed threshold (Figure 10.6c). A mathematical model taking account of the interference from short products was able to reproduce these experimental observations qualitatively. Taken together, the *in vitro* oscillator

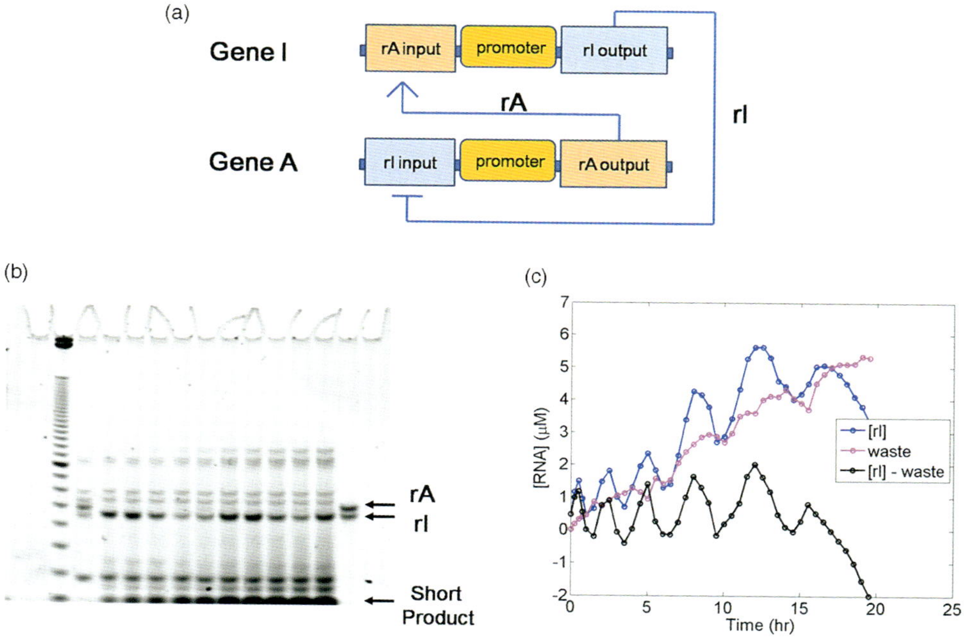

Figure 10.6 *A synthetic* in vitro *oscillator. (a) The synthetic oscillator is composed of two gene analogues, an activator and a repressor; (b) Gel measurement of oscillator outputs up to 4 h. The leftmost lane contains size markers, while the rightmost lane contains purified rA and rI products. It is assumed that the band of ~35 nucleotides in the gel is representative of accumulating short products; (c) The rI signal, the short product level normalized to be of similar scale to rI, and rI signal minus the normalized short product level are shown*

demonstrated sustained oscillations and was robust to the build-up of interfering signals to some extent. However, for a sustained and reliable operation of oscillators the incomplete degradation products need to be further processed, ultimately to the mononucleotide level.

Lessons can be learned from the degradation machinery of natural organisms. For example, *E. coli* has a high-molecular-weight complex called the degradosome which consists of RNase E, polynucleotide phosphorylase (PNPase), an ATP-dependent helicase, RhlB and enolase, a glycolytic enzyme [53]. When the decay of mRNA is initiated via endonucleolytic cleavage by RNase E, the newly formed 3′ end can be attacked by PNPase, which performs processive exonucleolytic digestion. The ATP-dependent RNA helicase in the degradosome presumably helps the degradation by unwinding RNA structures that impede the cleavage by RNase E and PNPase. The concerted action of these enzymes would explain the observation that, once initiated, the decay of mRNA proceeds without any accumulation of the decay intermediates. Although many mRNAs are subject to alternative decay processes, the existence of a highly orchestrated multienzyme complex such as the degradosome indicates that a complete degradation of messages without byproducts is an essential regulatory step.

10.5 The Minimal Cell

On a larger scale of synthetic efforts, the assembly of a type of cell – that is, a self-replicating, membrane-encapsulated collection of biomolecules – would be the next major challenge [54]. However small, a cellular gene set must be self-sufficient in the sense that cells generally import metabolites, but not functional macromolecules. *Mycoplasma genitalium*, a parasitic bacterium with a small genome size, is recognized as an attractive model in the search for the minimal genome. After comparing the 468 predicted *M. genitalium* protein sequences with the 1703 *Haemophilus influenzae* protein sequences, Mushegian and Koonin [55] suggested 256 genes as a minimal genome set, including 234 *M. genetalium* genes. Most of the proteins encoded by genes from the minimal set suggested by these authors had eukaryotic or archaeal homologues, whereas the key proteins of DNA replication did not, which led these authors to speculate that the last common ancestor had an RNA genome. The estimated gene number could be further reduced by eliminating cofactors and regulatory genes, and by applying the parsimony principle [56].

A recent estimate suggested that the minimal genome would comprise 151 genes, 38 RNAs and 113 proteins [54]. Lipids alone have been shown to be sufficient for the formation of rudimentary membranous compartments capable of both the transmembrane transport of small molecules and autocatalytic fission [57]. A bare-bones genome would perform basic DNA replication, transcription and translation processes, in which alternative approaches for essential mechanisms such as the adaptation of rolling circle amplification for DNA replication were employed to reduce the number of genes. A surprisingly large fraction (96%) of the minimal gene set is devoted to translation mechanisms, including ribosome components, a set of transfer RNAs (tRNAs), a set of translational initiation, elongation and release factors, and a few chaperones. In light of this, the simplest approach for creating a minimal cell may be to evolve an RNA polymerase made exclusively from RNA that would replace all of the protein components of the *in vitro* replicating and evolving systems [57]. An exciting development in this direction is the templated assembly of RNA products catalyzed by ribozymes [58]; these ribozymes used nucleoside triphosphates and the coding information of an RNA template to extend an RNA primer by the successive addition of up to 14 nucleotides, with high accuracy. These findings support the 'RNA-world' hypothesis regarding the early evolution of life – the main tenet of which is that ribozymes would have been far easier to duplicate than proteinaceous enzymes. Given that most of the minimal gene set is devoted to translation, a nucleic acid-based artificial cell would certainly be attractive, justifying a search for different sets of ribozymes through *in vitro* evolution approaches.

Estimates of the minimal genome typically do not include catabolism (nucleases and proteases), the active conversion or removal of waste products (energy-regenerating enzymes and membrane transporters) and regulatory feedback. It is unclear whether a minimal cell could sustain growth and replication without such regulatory mechanisms. At any rate, a much simpler purified system based on a real cell would be easier to model and understand, and it could certainly answer questions that cannot be answered *in vivo*, such as which set of macromolecules would be sufficient for a functional cellular subsystem [54]. The iterative synthetic process in which the performance of an *in vitro* model system is continuously improved may, in time, culminate in viable minimal cells as complex analogues of cells.

10.6 Conclusions

Today, synthetic biology provides the ability to study cellular regulation and behavior using *de novo* networks, with future applications of synthetic systems extending also to the fields of medicine and biotechnology. Yet, challenges remain that call for novel approaches and creative solutions. Synthetic networks *in vivo* have recycled previously used parts because a single point mutation may alter the *in vivo* activity of the network, and it is difficult to predict how redesigned molecules such as synthetic promoters would behave [59]. Mutations and the loss of synthetic network control can be a serious problem, especially when a large population of cells is considered. A 'population control' circuit [60] has been described which utilized a bacterial quorum-sensing system linked to a cell death signal to regulate the cell density of an *E. coli* population. Here, the steady-state cell density in the regulated cell culture was about tenfold lower than that of the control culture. However, due to the disadvantage in growth rate, cells that acquired mutations to disrupt the synthetic circuit control easily outgrew the regulated cells. A microfluidic microreactor was used to alleviate this problem by greatly reducing the population size [61], and allowed the synthetic circuit behavior to be monitored over hundreds of hours. Engineered cells would retain the synthetic network design that conferred a selective advantage in cellular growth rate, allowing further observation and analysis. For *in vitro* networks, the lack of any complex feedback regulation for the production and degradation machinery can lead to a high variability and a lack of robustness in their performances. As observed previously, dead-end side reactions, the saturation of the enzyme machinery and interference from incomplete products must be correctly addressed for successful *in vitro* network construction. Further developments of *in vitro* networks, accompanied by effective encapsulation in membranous compartments and ensuing growth and fission, will provide a good starting point for a minimal cell.

These synthetic approaches have successfully demonstrated several interesting networks, and have provided valuable engineering tools to study motifs, modularity and the robustness of cellular networks. Nonetheless, the development of new frameworks for regulatory costs, trade-offs and energy consumption of network structures remains a major problem, the solution of which could eventually lead to the construction of viable minimal cells.

References

1. Hartwell, L.H. Hopfield, J.J., Leibler, S. and Murray, A.W. (1999) From molecular to modular cell biology. *Nature*, **402**, C47–52.
2. Alves, R., Antunes, F. and Salvador, A. (2006) Tools for kinetic modeling of biochemical networks. *Nature Biotechnology*, **24**, 667–72.
3. Bray, D. (2003) Molecular networks: The top-down view. *Science*, **301**, 1864–5.
4. Alon, U. (2003) Biological networks: The tinkerer as an engineer. *Science*, **301**, 1866–7.
5. Savageau, M.A. (1971) Parameter sensitivity as a criterion for evaluating and comparing the performance of biochemical systems. *Nature*, **229**, 542–4.
6. Shen-Orr, S.S., Milo, R., Mangan, S. and Alon, U. (2002) Network motifs in the transcriptional regulation network of *Escherichia coli*. *Nature Genetics*, **31**, 64–8.

7. Milo, R. *et al.* (2002) Superfamilies of evolved and designed networks. *Science*, **303**, 1538–42.
8. Dekel, E., Mangan, S. and Alon, U. (2005) Environmental selection of the feed-forward loop circuit in gene-regulation networks. *Physical Biology*, **2**, 81–8.
9. Davidson, E.H., Rast, J.P., Oliveri, P., Ransick, A. *et al.* (2002) A genomic network for development. *Science*, **295**, 1669–79.
10. Kashtan, N. and Alon, U. (2005) Spontaneous evolution of modularity and network motifs. *Proceedings of the National Academy of Sciences of the United States of America*, **102**, 13773–8.
11. Strogatz, S.H. (2001) Exploring complex networks. *Nature*, **410**, 268–76.
12. Mattick, J.S. and Gagen, M.J. (2005) Accelerating networks. *Science*, **307**, 856–8.
13. Hooshangi, S., Thiberge, S. and Weiss, R. (2005) Ultrasensitivity and noise propagation in a synthetic transcriptional cascade. *Proceedings of the National Academy of Sciences of the United States of America*, **102**, 3581–6.
14. van Nimwegen, E. (2003) Scaling laws in the functional content of genomes. *Trends in Genetics*, **19**, 479–84.
15. Mattick, J.S. and Makunin, I.V. (2006) Non-coding RNA. *Human Molecular Genetics*, **15**, R17–29.
16. Benner, S.A. and Sismour, A.M. (2005) Synthetic biology. *Nature Reviews in Genetics*, **6**, 533–43.
17. SantaLucia, J. Jr. (1998) A unified view of polymer, dumbbell, and oligonucleotide DNA nearest-neighbor thermodynamics. *Proceedings of the National Academy of Sciences of the United States of America*, **95**, 1460–5.
18. Becskei, A. and Serrano, L. (2000) Engineering stability in gene networks by autoregulation. *Nature*, **405**, 590–3.
19. Becskei, A., Seraphin, B. and Serrano, L. (2001) Positive feedback in eukaryotic gene networks: Cell differentiation by graded to binary response conversion. *The EMBO Journal*, **20**, 2528–35.
20. Basu, S., Mehreja, R., Thiberge, S., Chen, M.T. *et al.* (2004) Spatiotemporal control of gene expression with pulse-generating networks. *Proceedings of the National Academy of Sciences of the United States of America*, **101**, 6355–60.
21. Gardner, T.S., Cantor, C.R. and Collins, J.J. (2000) Construction of a genetic toggle switch in *Escherichia coli*. *Nature*, **403**, 339–42.
22. Elowitz, M.B. and Leibler, S. (2000) A synthetic oscillatory network of transcriptional regulators. *Nature*, **403**, 335–8.
23. Atkinson, M.R., Savageau, M.A., Myers, J.T. and Ninfa, A.J. (2003) Development of genetic circuitry exhibiting toggle switch or oscillatory behavior in *Escherichia coli*. *Cell*, **113**, 597–607.
24. Ozbudak, E.M., Thattai, M., Lim, H.N., Shraiman, B.I. *et al.* (2004) Multistability in the lactose utilization network of *Escherichia coli*. *Nature*, **427**, 737–40.
25. Kobayashi, H., Kaern, M., Araki, M. *et al.* (2004) Programmable cells: Interfacing natural and engineered gene networks. *Proceedings of the National Academy of Sciences of the United States of America*, **101**, 8414–19.
26. Bayer, T.S. and Smolke, C.D. (2005) Programmable ligand-controlled riboregulators of eukaryotic gene expression. *Nature Biotechnology*, **23**, 337–43.
27. Dueber, J.E., Yeh, B.J., Chak, K. and Lim, W.A. (2003) Reprogramming control of an allosteric signaling switch through modular recombination. *Science*, **301**, 1904–8.
28. Barkai, N. and Leibler, S. (2000) Circadian clocks limited by noise. *Nature*, **403**, 267–8.
29. Schibler, U. and Naef, F. (2005) Cellular oscillators: rhythmic gene expression and metabolism. *Current Opinion in Cell Biology*, **17**, 223–9.
30. Wong, W.W. and Liao, J.C. (2006) The design of intracellular oscillators that interact with metabolism. *Cellular and Molecular Life Sciences*, **63**, 1215–20.
31. Carrington, J.C. and Ambros, V. (2003) Role of microRNAs in plant and animal development. *Science*, **301**, 336–8.

32. Kramer, C., Loros, J.J., Dunlap, J.C. and Crosthwaite, S.K. (2003) Role for antisense RNA in regulating circadian clock function in *Neurospora crassa*. *Nature*, **421**, 948–52.
33. Isaacs, F.J., Dwyer, D.J., Ding, C. *et al.* (2004) Engineered riboregulators enable post-transcriptional control of gene expression. *Nature Biotechnology*, **22**, 841–7.
34. Isaacs, F.J., Dwyer, D.J. and Collins, J.J. (2006) RNA synthetic biology. *Nature Biotechnology*, **24**, 545–54.
35. Zaikin, A.N. and Zhabotinsky, A.M. (1970) Concentration wave propagation in 2-dimensional liquid-phase self-oscillating system. *Nature*, **225**, 535–7.
36. Nakajima, M., Imai, K., Ito, H. *et al.* (2005) Reconstitution of circadian oscillation of cyanobacterial KaiC phosphorylation in vitro. *Science*, **308**, 414–15.
37. Shimizu, Y., Inoue, A., Tomari, Y. *et al.* (2001) Cell-free translation reconstituted with purified components. *Nature Biotechnology*, **19**, 751–5.
38. Noireaux, V., Bar-Ziv, R. and Libchaber, A. (2003) Principles of cell-free genetic circuit assembly. *Proceedings of the National Academy of Sciences of the United States of America*, **100**, 12672–7.
39. Isalan, M., Lemerle, C. and Serrano, L. (2005) Engineering gene networks to emulate *Drosophila* embryonic pattern formation. *PLoS Biology*, **3**, e64.
40. Wlotzka, B. and McCaskill, J.S. (1997) A molecular predatory and its prey: Coupled isothermal amplification of nucleic acids. *Chemistry and Biology*, **4**, 25–33.
41. Ackermann, J., Wlotzka, B. and McCaskill, J.S. (1998) In vitro DNA-based predator-prey system with oscillatory kinetics. *Bulletin of Mathematical Biology*, **60**, 329–53.
42. Guatelli, J.C., Whitfield, K.M., Kwoh, D.Y. *et al.* (1990) Isothermal, in vitro amplification of nucleic acids by a multienzyme reaction modeled after retroviral replication. *Proceedings of the National Academy of Sciences of the United States of America*, **87**, 1874–8.
43. Kim, J., White, K.S. and Winfree, E. (2006) Construction of an in vitro bistable circuit from synthetic transcriptional switches. *Molecular Systems Biology*, **2**, 68.
44. Arkin, A., Ross, J. and McAdams, H.H. (1998) Stochastic kinetic analysis of developmental pathway bifurcation in phage λ-infected *E. coli* cells. *Genetics*, **149**, 1633–48.
45. Thattai, M. and van Oudenaarden, A. (2001) Intrinsic noise in gene regulatory networks. *Proceedings of the National Academy of Sciences of the United States of America*, **98**, 8614–19.
46. Guet, C.C., Elowitz, M.B., Hsing, W. and Leibler, S. (2002) Combinatorial synthesis of genetic networks. *Science*, **296**, 1466–70.
47. Kim, P.M. and Tidor, B. (2003) Limitations of quantitative gene regulation models: A case study. *Genome Research*, **13**, 2391–5.
48. Keiler, K.C., Waller, P.R.H. and Sauer, R.T. (1996) Role of a peptide tagging system in degradation of proteins synthesized from damaged messenger RNA. *Science*, **271**, 990–3.
49. Shen, X.-C. *et al.* (1999) Poly[G] improved protein productivity of cell-free translation by inhibiting mRNase in wheat germ extract. *Journal of Biotechnology*, **75**, 221–8.
50. Noireaux, V. and Libchaber, A. (2004) A vesicle bioreactor as a step toward an artificial cell assembly. *Proceedings of the National Academy of Sciences of the United States of America*, **101**, 17669–74.
51. Lima, W.F. and Crooke, S.T. (1997) Cleavage of single strand RNA adjacent to RNA-DNA duplex regions by *Escherichia coli* RNase H1. *Journal of Biological Chemistry*, **272**, 27513–16.
52. Yurke, B. and Mills A.P. Jr. (2003) Using DNA to power nanostructures. *Genetic Programming and Evolvable Machines*, **4**, 111–22.
53. Grunberg-Manago, M. (1999) Messenger RNA stability and its role in control of gene expression in bacteria and phages. *Annual Review of Genetics*, **33**, 193–227.
54. Forster, A.C. and Church, G.M. (2006) Towards synthesis of a minimal cell. *Molecular Systems Biology*, **2**, 45.

55. Mushegian, A.R. and Koonin, E.V. (1996) A minimal gene set for cellular life derived by comparison of complete bacterial genomes. *Proceedings of the National Academy of Sciences of the United States of America*, **93**, 10268–73.
56. Benner, S.A., Ellington, A.D. and Tauer, A. (1989) Modern metabolism as a palimpsest of the RNA world. *Proceedings of the National Academy of Sciences of the United States of America*, **86**, 7054–8.
57. Szostak, J.W., Bartel, D.P. and Luisi, P.L. (2001) Synthesizing life. *Nature*, **409**, 387–90.
58. Johnston, W.K., Unrau, P.J., Lawrence, M.S. *et al.* (2001) RNA-catalyzed RNA polymerization: Accurate and general RNA-templated primer extension. *Science*, **292**, 1319–25.
59. Ventura, B.D., Lemerle, C., Michalodimitrakis, K. and Serrano, L. (2006) From in vivo to in silico biology and back. *Nature*, **443**, 527–33.
60. You, L., Cox, R.S. 3rd, Weiss, R. and Arnold, F.H. (2004) Programmed population control by cell-cell communication and regulated killing. *Nature*, **428**, 868–71.
61. Balagadde, F.K., You, L., Hansen, C.L. *et al.* (2005) Long-term monitoring of bacteria undergoing programmed population control in a microchemostat. *Science*, **309**, 137–40.

Section 4
Integration

11

Molecular Modeling of CYP Proteins and its Implication for Personal Drug Design

Jing-Fang Wang, Cheng-Cheng Zhang, Jing-Yi Yan, Kuo-Chen Chou and Dong-Qing Wei

Department of Bioinformatics and Biostatistics, College of Life Sciences and Technology, Shanghai Jiao Tong University, China

11.1 An Introduction to CYPs

Cytochrome P450s (CYPs), which belong to a superfamily of hemoproteins, can be found in virtually all types of organism, including Bacteria, Eukaryotes and even Archaea [1]. In animals, CYPs are located in either the endoplasmic reticulum or the inner membrane of the mitochondrion. Cytochrome P450s – so named because the heme pigment that they contain absorbs light at a wavelength of 450 nm when complexed with carbon monoxide – are mainly membrane-associated.

During drug metabolism, both Phase I and Phase II reactions occur:

- In Phase I reactions, polar groups may be introduced or unmasked, leading to more water-soluble metabolites, such that drugs are either activated or inactivated. In humans, the CYPs are the most important enzymes responsible for Phase I drug metabolism.
- Phase II reactions usually include detoxication processes, where mainly conjugation reactions take place.

J.-F. Wang and C.-C. Zhang contributed equally to this chapter.

Automation in Proteomics and Genomics: An Engineering Case-Based Approach Edited by Gil Alterovitz, Roseann Benson and Marco Ramoni

The common reaction of CYPs in Phase I metabolism is that of a monooxygenase (Equation 11.1); such a reaction generally makes the substrates more water-soluble such that they can be excreted in the urine:

$$RH + O_2 + 2H^+ + 2e^- \rightarrow ROH + H_2O \text{ (where R is a substrate)} \quad (11.1)$$

CYPs are found throughout the body, the highest concentrations being associated with liver proteins, and have specialized roles in controlling the levels of endogenous compounds; examples include vitamin D metabolism, cholesterol synthesis and hormone synthesis and breakdown. CYP enzymes are also involved with vascular autoregulation, especially in the brain, and are vital to the synthesis of cholesterol, steroids and arachidonic acid metabolites. They also clear metabolic products from the body, an example being the breakdown product of hemoglobin, bilirubin.

11.1.1 CYP Nomenclature

In order to explain the system of nomenclature for CYPs, we will take CYP3A4 as an example. Virtually all enzymes of the CYPs are designated with the root 'CYP', followed by an Arabic numeral for the gene family (CYP3), a capital letter for the subfamily (CYP3A), and another Arabic number for a particular gene (CYP3A4). The enzymes in the same family share at least 40% amino acid identity, whereas enzymes in the same subfamily share at least 55% amino acid identity. There is, however, no correlation between nomenclature and enzyme function.

11.1.2 CYP Isoforms and Single Nucleotide Polymorphism (SNP)

To date, more than 7700 distinct CYP sequences have been identified. Many antidepressant and antipsychotic drugs are metabolized by CYP2D6 or CYP2C19, while CYP3A4 alone metabolizes more than 50% of all drugs; clearly, a mutation of the latter enzyme could result in clinical disaster. As CYP1A2, 2C9, 2C19, 2D6, 2E1 and 3A4 metabolize more than 90% of all known drugs, members in the CYP1, 2, 3 families have achieved most recognition among biochemists.

CYPs are also notable for their pharmacogenetic characteristics [2]. Typically, humans carry a series of CYP gene alleles, which have only minimal variation in terms of their genetic sequence that is attributable to nucleotide changes or polymorphisms (SNP). These polymorphic variations may lead to inter-individual and within-population differences in the tolerance to toxins and drugs and, as CYPs are specific with regards to which drugs they clear from the body (and which they activate), SNP variations in different CYP genes may lead to different effects. Consequently, because of these possible variations the possible design of 'personal' drugs for individuals with the same illness is a daunting task.

11.2 Computational Methods

Protein structure modeling is of major for understanding and explaining CYP functions. In the case of CYPs for which details of crystal structures are available, a convenient approach is to perform docking studies and molecular dynamic simulations in order to model the drug–protein and protein–protein interactions. However, for those CYPs lacking any high-resolution structure data, it is essential that these structures are predicted by using computational methods, the target being to deduce the three-dimensional (3-D) protein structure from the amino acid sequence.

Currently, three main (theoretical) methods are available for protein structure prediction, namely homology modeling, fold recognition (threading) and *ab initio*/*de novo* methods.

- *Homology modeling*, which is the oldest and most widely used method for protein structure prediction, requires a known protein structure, the sequence of which must show about 30% similarity to that of the protein under study. In recent years, the method's accuracy has rapidly improved due to an availability of expanding protein structure databases and structure analysis algorithms.
- The *fold recognition* method serves as a supplement to homology modeling for detecting remote homologues, and has also improved rapidly in recent years. The method focuses on the element of folds, and can be applied to both 2-D and 3-D structures, and considerable progress has been made in all aspects of protein-fold predictions.
- The *ab initio*/*de novo* method is based on the force fields of atomic details, such that the protein structure is built using sequence-only information. Although advances in simulation methodology and forces that drive protein folding have had a major impact on this method, it is very time-consuming, for two reasons: first, there are often too many conformations to sample; and second, the time scale of protein motion is measured in seconds, whereas atomic motion is measured in femtoseconds.

Since 1994, the performance of these methods has been assessed biannually using critical assessment of structure prediction (CASP) experiments. (Additional CASP information is available at http://www.predictioncenter.org/), while the overlap between the three methods has increased dramatically.)

11.2.1 Homology Modeling

For homology modeling the most frequently used techniques are segment matching or coordinate reconstruction, both of which require a crystal structure as a template. This type of approach is based on the realization that most hexapeptide segments of protein or enzymes structures can be clustered into only 100 structurally different classes. Thus, comparative models can be prepared by using a subset of atomic positions from a selected template structure as guiding positions, as well as by identifying and assembling short, all-atom segments, which are suitable for the guiding positions. The template structure should be homologous to the target protein, and preferably have a high structural resolution. However, in recent years remote homologue detection has become the major goal of this method. From a practical viewpoint, the entire homology modeling process comprises four steps.

11.2.1.1 Homology Modeling: Step 1

In this first step, the target protein is broken into a series short sequence segments, and a known structure having a good sequence identity with the target sequences is searched. Considerable increases in sensitivity over traditional pairwise alignment methods have led to sequence profiling methods becoming the optimal approach in homologue detection in recent years. The Basic Local Alignment Search Tool (BLAST) is an algorithm for comparing the amino acid sequences of different proteins, or the nucleotides of DNA sequences. BLAST compares a query sequence with a database of sequences to identify similar subsequences to those in the query, using an heuristic approach that approximates the Smith–Waterman algorithm for high-scoring sequence alignments. Its relatively good accuracy and speed make BLAST the most popular bioinformatics search tool. A more recently developed version – PSI-BLAST – is used to identify any distant relatives of a protein. PSI-BLAST operates in stepwise fashion: first, a list is created of all closely related proteins, after which these proteins are combined into a profile which is an average sequence. A query is then run against the database, using this profile to identify a larger group of proteins, and another profile is then constructed using this larger group of proteins. The whole process is then repeated. Consequently, PSI-BLAST is much more sensitive at detecting distant homologues than is the standard BLAST. Another method worthy of mention here is the hidden Markov model (HMM). This is similar to PSI-BLAST, but an initial HMM is created from a single given query sequence, after which a database of potential homologues of the query sequence is constructed by searching a large protein database using WU-BLAST. New sequences with good local alignment scores to the HMM from the database of potential homologues are selected, and a new HMM and a new multiple alignment for the query sequence is created. This step is repeated several times, whereupon the final HMM can be used to search a selected database for homologues of the query sequence. Both, PSI-BLAST and HMM methods have greatly improved the accuracy of sequence alignments and increased the ability of remote homologue detection. Unfortunately, there is an inherent shortcoming in sequence alignment technology, as a good linear sequence alignment does not directly reflect good 3-D thermodynamic relationships; however, this difficulty is the subject of ongoing research.

11.2.1.2 Homology Modeling: Step 2

In step 2, the segments are matched according to the template protein or, in other words, target–template alignment. Differences in the target and template structures in certain regions can result in alignment mistakes, particularly where the sequence similarity is not sufficiently high to identify structurally equivalent residues. However, three strategies have greatly improved the probability of generating correct target–template alignments. First, a 3D-shotgun builds multiple models from fragments of the initial models and constructs a final model based on a measure of structural similarity. The fragments used to build into the final model are the most often observed. Second, the final model is evaluated using Verify3D, where low-scoring regions of the alignments are shifted to obtain better scoring models. In a third approach – the Robetta method – alternate alignments are constructed by sampling different parameters that reflect various measurements of similarity, and by enumerating suboptimal alignments directly.

11.2.1.3 Homology Modeling: Step 3

In step 3, in order to construct a model, the coordinates of the matched segments are fitted into the growing target under the monitor, in order to avoid any van der Waals overlap until all atomic coordinates of the target have been gained. The methods used in this step involve the assembly of rigid bodies, segment matching or coordinate reconstruction, satisfaction of spatial restraints, loop modeling and side-chain modeling. Several programs are available to construct a 3-D model, including SWISS-MODEL, MODELLER, HHpred and 3D-JIGSAW. Although each of these programs produces models that are as similar as possible to the templates, they are inadequate for the important features of a model that are structurally distinct from their templates; examples include different conformations of the side chains and loops between the secondary structure elements and between the target and template structures. For side-chain conformation calculations, a rotamer library generated from a database of known structures is used to observe the relationship between side-chain conformation and backbone conformation. The side-chain torsional angles for the preferred conformations of a specific side chain define the rotamer library. The accuracy of side-chain modeling is close to experimental level in many cases. For loop modeling, the programs construct the loop model in an open conformation, where one end of the loop is disconnected to the succeeding residue. The loop is then connected using different algorithms. The process is then repeated using various starting conformations, and the resulting conformations evaluated using the energy function. Although, in terms of computer and server capabilities, the computational demands are heavy, the suggestion is that remarkably accurate results can be obtained when extensive sampling and conformational energy evaluation are combined. Both, side-chain and loop modeling should be based on the correct backbone conformation, as different orientations and secondary structure element numbers are recognized as potential fundamental problems of the two methods. Indeed, attempts to resolve such problems constitute an expanding area of research.

11.2.1.4 Homology Modeling: Step 4

By repeating Steps 1 to 3 ten times to generate an average model and minimize global energy, the final structure can be created and a model can be assessed. Although, to date, there are no reliable procedures to assess a model, many programs are available to observe whether the model possesses good stereochemistry and overall conformational energy. A scoring function, Very3D, is commonly used to assess models; this evaluates how well residues in the segments of the model fit into the environments. Another strategy is to measure the conformational stability of all atoms under molecular mechanics force fields in an aqueous solvent; the case study in Section 11.3.5 is an example of such an investigation.

11.2.2 Fold Recognition

In addition to homology modeling, fold recognition (threading) represents another type of comparative protein modeling. This concentrates on the element of folds rather than amino acid sequences in the homology modeling technique. There are fewer classes of folds than sequences, as explained by computer scientist Adam Zemla:

> *Because there are 20 different amino acids, a medium-size protein with 300 amino acids would theoretically have 20300 possibilities in sequence. In nature, not all combinations of amino acids can exist. Scientists estimate that the number of different protein sequences is close to a few million.*

Fold recognition can be used to supplement homology modeling to detect remote homologues when proteins share the same fold category and no relationship of the sequences can be detected. Bioinformatics methods have been developed to identify the fold category to which a protein belongs, and can be divided into two groups: (i) sequence-based methods; and (ii) structure-based methods.

- *Sequence-based methods* use the amino acid sequence or predicted secondary structure information for alignments, and determine whether proteins share the same fold, or not. PSI-BLAST and HMM techniques (see above), along with genetic algorithms and support vector machines techniques, are applied to obtain amino acid sequences or secondary structure information. Interestingly, one group [3] developed a fold recognition approach based on secondary structure information and solvent accessibility that outperformed methods which take information on the 3-D structure (fold) into account.
- In *structure-based methods* (threading), energy functions are used to evaluate how well a probe sequence matches a target known 3-D protein structure (a fold). Recently, the Skolnick team have developed and successfully applied threading methods in CASP experiments [4, 5].

The major challenges which face fold recognition methods include the acquisition of a better understanding of the intermolecular forces and folding mechanism–solvation interactions. Currently, computer algorithms of fold recognition are able to predict accurately the structure of small proteins, which in turn will shed light on the predictions for larger, more complicated proteins.

11.2.3 *Ab initio/de novo* Methods

The *ab initio/de novo* method is usually based on force fields of atomic details, which builds the protein structure by using sequence-only information. The most outstanding feature of this method is the ability to render the backbone of protein flexible in the protein structure prediction process. In many computational methods the backbone of the protein is kept rigid; this means that the 3-D coordinates of all the α-carbon atoms are fixed in order to reduce to a remarkable extent the computation steps, the search space and, consequently, the time required for reaching the minimum energy state. Keeping the backbone of proteins rigid is a dubious practice, however, as actual proteins exhibit great flexibility. For example, there are eight CYP2A6 crystal structures available, and the backbone of each varies substantially one from another. A case in point [6] shows that a CYP3A4 backbone undergoes a dramatic shift when a ligand is introduced to the active site of the protein.

One way to allow for backbone flexibility is through an ensemble of related backbone conformations close to the protein, and which are generated randomly using genetic algorithms and Monte Carlo sampling. The sequences will then be designed for these conformations using rigid backbones assumptions, with the lowest energy backbone-sequence combination being selected [7]. A natural way to allow for backbone flexibility is to make every position in the protein variable, using an integer linear optimization technique with a distance-dependent force field in the sequence selection stage, as introduced in Ref. [8].

Unfortunately, the *ab initio*/*de novo* method is too time-consuming for two reasons:

1. There are too many conformations to sample.
2. The time scale of protein motion is on the order of seconds, while atomic motion is measured in femtoseconds.

In addition, only small proteins with less than 100 residues have been successfully predicted, and improvements of better force fields, scoring functions and more accurate search methods are needed for predicting larger proteins. This will, in turn, bring about vast opportunities, making the *de novo* method an active area of research.

11.3 Crystal Structures of CYPs

Although, both nuclear magnetic resonance (NMR) and X-ray crystallography are capable of providing high-resolution structural information, the CYP proteins usually contain more than 400 residues and hence are too large for NMR methods to be applied. Consequently, in recent years X-ray crystallography has been the main technique for obtaining structural information for the CYPs.

As CYPs are mainly membrane-associated proteins, it is necessary to modify the N-terminal transmembrane domain or to truncate the N-terminal hydrophobic domain in order to increase the solubility of the enzymes, thus facilitating crystallization. In the case of CYP2C9, the N-terminal transmembrane domain (residues 1–29) was replaced by a hydrophilic polypeptide sequence MAKKTSSKGR and a histidine tag introduced in the C-terminus. It is generally accepted that such changes do not alter the enzyme function, because the kinetic parameters of the mammalian CYPs remain unaltered. A recent study on the heterologous expression of CYP1A2 without the conventional N-terminal modification in *Escherichia coli* further confirmed this conclusion [9].

At the time of writing this chapter (March 2008), a total of 30 crystal structures of eight CYP enzymes (CYP 1A2, 2A6, 2B4, 2C5, 2C8, 2C9, 2D6 and 3A4) have been published (see Table 11.1). As a milestone in CYPs structure studies, the first crystal structure of a mammalian CYP – the rabbit CYP2C5 – was reported in 2000 [10]. Since that time, many computational tools – including homology modeling, docking, molecular dynamic simulations and quantitative structure–activity relationships (QSARs) – have been used to analyze or predict structural information based on crystal structures. With the publication of the CYP1A2 crystal structure in 2007, the structures of all of the major drug-metabolizing enzymes had been obtained, while homology modeling methods were able to predict the 3-D structures of all other members in the CYP1, 2 and 3 families [11, 12]. This, in turn, makes personalized drug design more feasible.

As crystal structures provide only static structural information, they are insufficient for the study of enzyme flexibility. Crystal structures do, however, facilitate molecular dynamic simulations such that, when combined with experimental techniques, such as high-pressure UV/visible spectroscopy, the mechanism of the CYPs' metabolism and details of their flexibility and stability CYPs should be acquired.

Table 11.1 *The crystal structures of mammalian CYPs*

CYP	PDB code	Published year	Substrate	Organism	Resolution (Å)	Reference
2C5	1DT6	2000	Free	Rabbit	3.00	[10]
2C5	1N6B	2003	Dimethyl sulfaphenazole	Rabbit	2.30	[13]
2C5	1NR6	2003	Diclofenac	Rabbit	2.10	[14]
2C8	1PQ2	2004	Palmitic acid[a]	Human	2.70	[15]
2C8	2NNI	2007	Montelukast	Human	2.80	To be published
2C8	2NNJ	2007	Felodipine	Human	2.28	To be published
2C8	2NNH	2007	9-*cis* retinoic acid ×2	Human	2.60	To be published
2C9	1OG2	2003	Free	Human	2.60	[16]
2C9	1OG5	2003	*S*-warfarin	Human	2.55	
2C9	1R9O	2004	Flurbiprofen	Human	2.00	[17]
2D6	2F9Q	2005	Free	Human	3.00	[18]
3A4	1TQN	2004	Free	Human	2.05	[19]
3A4	1W0E	2004	Free	Human	2.80	[20]
3A4	1W0F	2004	Progesterone	Human	2.65	
3A4	1W0G	2004	Metyrapone	Human	2.73	
3A4	2J0D	2006	Erythromycin	Human	2.75	[6]
3A4	2V0M	2007	Ketoconazole	Human	3.80	
2B4	1SUO	2004	4-(4-Chlorophenyl)imidazole	Rabbit	1.90	[21]
2B4	2BDM	2005	Bifonazole	Rabbit	2.30	[22]
2B4	2Q6N	2007	1-(4-Cholorophenyl)imidazole	Rabbit	3.20	[23]
2A6	1Z10	2005	Coumarin	Human	1.90	[24]
2A6	1Z11	2005	Methoxsalen	Human	2.05	
2A6	2FDU	2006	*N*,*N*-Dimethyl(5-(pyridin-3-yl)furan-2-yl)methanamine	Human	1.85	[25]
2A6	2FDV	2006	*N*-Methyl(5-(pyridin-3-yl)furan-2-yl)methanamine	Human	1.65	
2A6	2FDW	2006	(5-(Pyridin-3-yl)furan-2-yl)methanamine	Human	2.05	
2A6	2FDY	2006	Adrithiol	Human	1.95	[26]
2A6	2PG5	2007	Free[b]	Human	1.95	[27]
2A6	2PG6	2007	Free[b]	Human	2.53	
2A6	2PG7	2007	Free[b]	Human	2.80	
1A2	2HI4	2007	Alpha-naphthoflavone	Human	1.95	

[a] Outside the active site of the enzyme.

[b] Mutants of CYP2A6: 2PG5: N297Q; 2PG6: L240C/N297Q; 2PG7: N297Q/I300V.

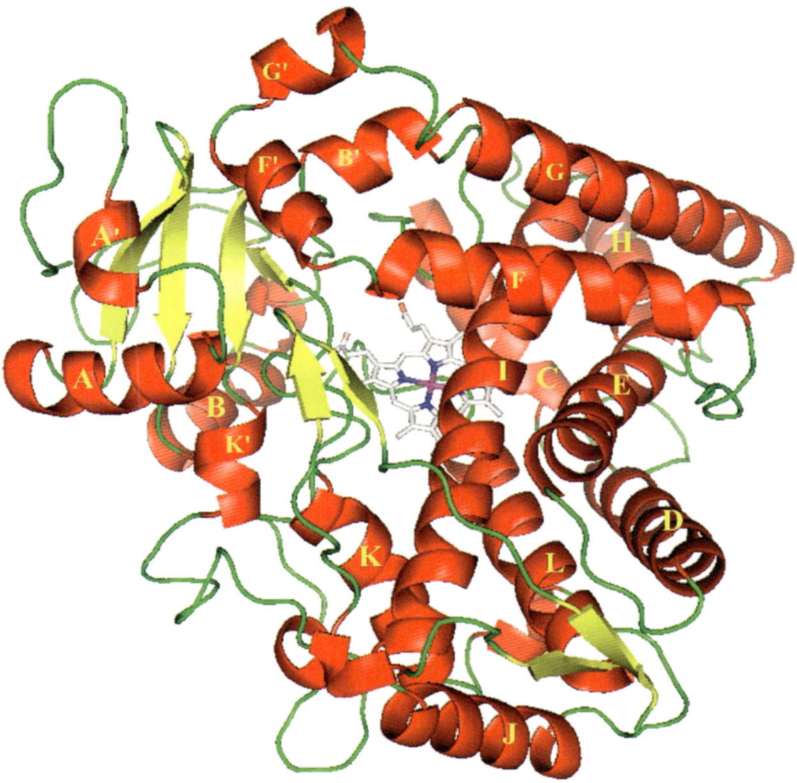

Figure 11.1 *A typical structure of CYPs, monomer of CYP2C9 (PDB code: 1OG2) with labeled secondary elements, where the red, yellow and green colors represent helices, β sheets and loops, respectively, and the molecule-with-stick representation is the heme cofactor*

11.4 Common Features of CYPs

In recent years, the structural features of several mammalian CYPs have been identified (a typical CYP structure is shown in Figure 11.1). According to the CATH classification,[1] the structures of mammalian CYPs are mainly α-helical with an orthogonal bundle architecture [28]. The F and G helices and F/G-loop along with the B/C-loop form the access to the active site of CYPs [29]. The highly conserved parts of the CYP structures include: the proline-rich cluster, which is close to the N-terminus [30]; the loop following the A helix; the C and I helices; parts of the K helix; and the proton-transfer groove [29,31]. In addition, inserted helices (F′ and G′) between the F and G helices, in which the outer

[1] **CATH** is a hierarchical classification of protein domain structures, which clusters proteins at four major levels: Class (C), Architecture (A), Topology (T) and Homologous superfamily (H). The boundaries and assignments for each protein domain are determined using a combination of automated and manual procedures which include computational techniques, empirical and statistical evidence, literature review and expert analysis (http://www.cathdb.info/).

surface is hydrophobic, are presented in some CYPs; these two helices form interactions with the membrane.

A prosthetic group (heme b), inserted between the I and L helices, is present in the center of all CYPs and is close to the active site. The role of this group is to link the sulfur atom of a cysteine to form a S–Fe bond (cys-pocket). Although CYPs have different orientations and conformations of the heme side chains, the overall porphyrin geometry is basically the same [32]. Heme alters its oxidation state when interacting with various substrates, typically O_2, CO and NO [33]. The highly ordered solvent serves as a direct proton donor to the iron-linked substrates, particularly O_2 [34]. The heme iron takes part in the electron transport for oxygen cleavage and substrate oxidation from different redox partners [35]. The electron transport systems can be divided into two major classes: the adrenal mitochondrial P450 system; and the liver microsomal P450 system [36].

11.5 Diversity of the Substrates of CYPs

CYPs metabolize a wide range of structurally dissimilar substrates, including steroids, fatty acids, vitamin D, cholesterol, carcinogens, retinoic acid and nitrogenous organic bases. The volume, shape and flexibility of the active sites determine the diversity in recognizing substrates, while the access/egress path is a rigid, narrow funnel that controls the regioselectivity of the CYPs.

CYP3A4, as a major member of the P450 superfamily, metabolizes more than 50% of all drugs – a proportion far greater than for any other CYP isoform [37]. This interesting feature of CYP3A4 is due its overall size and the shape of the active site (there is, however, no standard definition for calculating active site volumes, so these cannot be strictly compared). It is generally suggested that the CYP3A4 active site is relatively large, and has been proved to metabolize not only many small molecules but also large substrates such as bromocryptine, erythromycin and cyclosporine [38]. The active site of CYP3A4 also contains multiple substrate-binding sites that could be occupied by small substrates simultaneously [20, 39–41]. The binding of two molecules of a large substrate (such as ketoconazole; $C_{26}H_{28}Cl_2N_4O_4$) to the active site of CYP3A4 serves as a good example of both a large active site and multiple substrate binding (Figure 11.2). Here, the flexibility of the CYP3A4 should be taken into account, since evidence suggests that CYP3A4 is readily denatured to the inactive P420 form under high pressure, this being attributed to the flexibility of the active sites. The results of recent studies have shown that such flexibility also occurred when substrates such as ketoconazole and erythromycin were bound to CYP3A4, with dramatic conformational changes also taking place [6].

11.6 Critical Amino Acids in the Active Sites

Based on the available crystal structures, the homology modeling technique was used to predict many other CYP superfamily protein structures. As a result, many conserved and essential amino acids have been identified in the active sites of these CYPs. It has also been shown that lipophilicity relationships, hydrogen bonding and π–π stacking interactions play important roles in substrate selectivity and binding affinity [12, 42, 43].

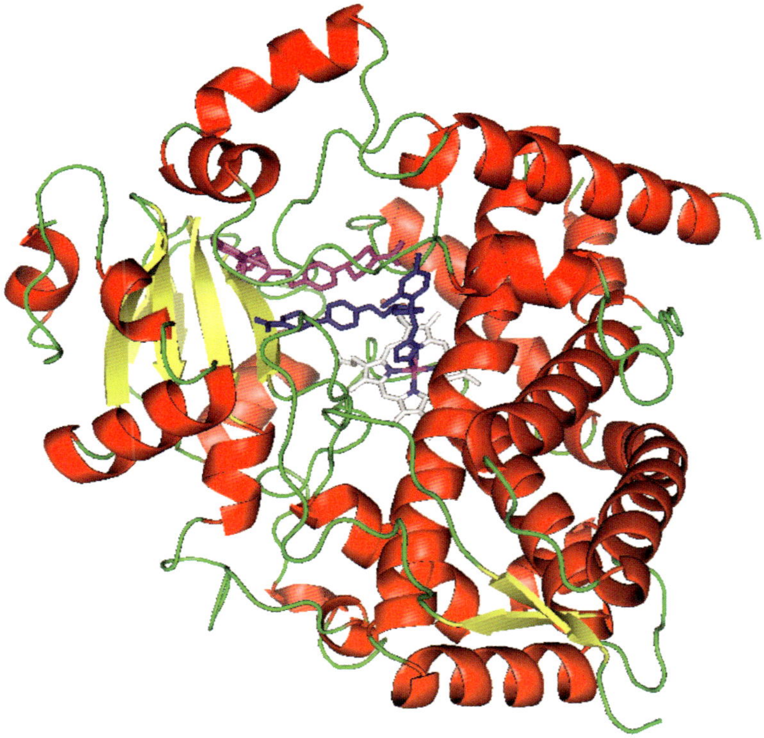

Figure 11.2 *Illustration of CYP3A4 (PDB code: 2V0M) binding with two molecules of ketoconazole, which is shown in stick presentation with colors of blue and magenta. The red, yellow and green colors represent helices, β sheets and loops, respectively, and the molecule-with-stick representation in the center is the heme cofactor*

At this point, we will discuss the critical amino acids of the active site of CYP2D6, which metabolizes about 30% of all known drugs – second only to CYP3A4. However, whilst CYP3A4 metabolizes a diverse range of substrates, CYP2D6 prefers substrates with a planar aromatic ring and a basic nitrogen atom [29]. It has been suggested that Phe120, Asp301, Thr309, Glu216 and Phe483 are the most important residues in the CYP2D6 active sites. An example of metoprolol binding with CYP2D6 (obtained from Autodock studies) is illustrated in Figure 11.3, where Phe120 forms π–π stacking and the Asp301, Glu216 and Ala305 form hydrogen bonds with the substrate. The topological roles of Asp301 and Thr309 have been identified. Mutating this residue into any other amino acid (except glutamate) can have a severe detrimental influence on substrate oxidation [18]. In the T309V mutation, Thr309 was shown to play a pivotal structural role in the active site crevice [44], while docking studies have confirmed that Glu216 is a binding residue which recognizes the ligands [45]. Finally, Phe120 and Phe483 have been identified as two important aromatic residues in the active site, since almost all known drugs metabolized by CYP2D6 have an aromatic ring. In many cases, the substrate has been shown to form π–π stacking interactions with Phe120 and/or Phe483.

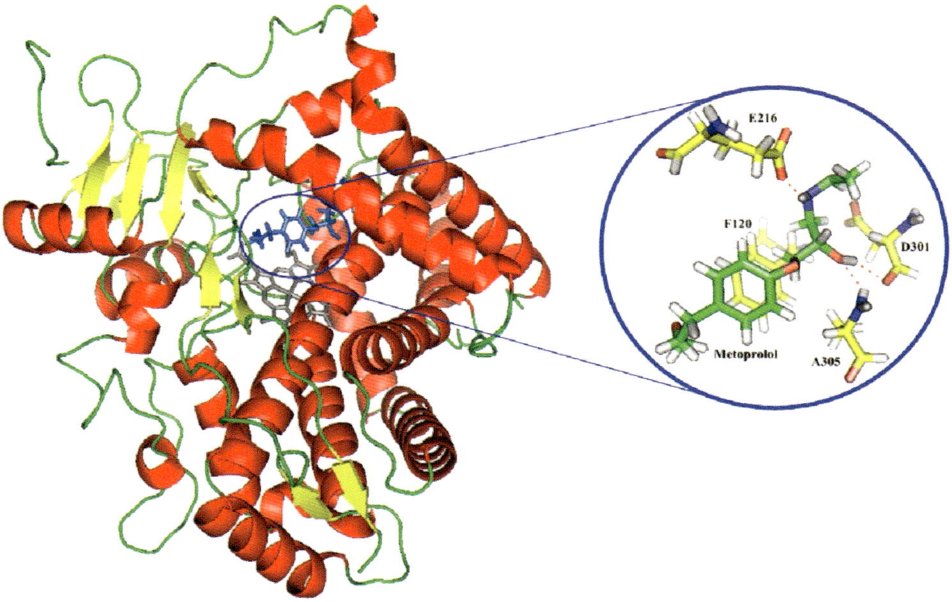

Figure 11.3 Illustration of CYP2D6 (2F9Q), bonds with metoprolol. The enlarged part shows the interaction between metoprolol and CYP2D6 active site amino acids, E216, D301, A305 and F120. This interaction is obtained using Autodock 3.05. The molecules with the stick representation are the ligand and residues close to the ligand, where the red, yellow-green, blue and light-gray colors represent oxygen, carbon, nitrogen and hydrogen, respectively

11.7 SNP Studies of CYP2C19: A Case Study

CYP2C19 is one of the key drug-metabolizing enzymes in the CYP superfamily, having been shown capable of metabolizing proton-pump inhibitors, phenytoin, tricyclic antidepressants, propranolol and benzodiazepines. CYP2C19 is noted for its polymorphisms, from CYP2C19*1 to CYP2C19*21, which induce variability in drug metabolism. Polymorphisms can also lead to adverse drug reactions, to drug–drug interactions, and also constitutes a major factor in drug toxicity.

In order to study the CYP2C19 SNP, two SNP 3-D structures based on the CYP2C19 computational structure have been modeled using structural bioinformatic methods. First, the backbones were the same as the template, but with amino acid residues Trp120 and ILe331 mutated to arginine and valine, respectively. Second, the new structures were optimized by energy minimization. Third, the computational structures obtained were used for further docking and molecular dynamic studies [46].

Molecular docking, using the Metropolis algorithm, was employed to identify the most favorable binding interaction; the Shanghai Molecular Modeling (SIMM) program was used to perform this function. In the docking studies, the ligands 3-cyano-7-ethoxycoumarin (CEC), fluvoxamine, fluvastatin (Lescol™) and ticlopidine, were flexible. The program generates a diversified set of conformations by making random changes to the ligand coordinates. When a new conformation of the ligand was generated, the search for favorable binding configurations was conducted within a specified 3-D docking box, using either simulated annealing or a Tabu search. Both of these methods seek to optimize the purely spatial contacts, as well as the electrostatic interactions. The interaction energy was calculated using the electrostatic and van der Waals potential fields. In all computations, the CHARMM22 force field parameters were used. The computational structures of two SNPs were taken as the receptor, where four ligands – CEC, fluvoxamine, fluvastatin and ticlopidine – were docked. CYP2C19 has a multifaceted and flexible dynamic structural feature which may be reflected by using molecular dynamics (MD) tools. Such MD simulations can be used to solve the classic motion equations for a system formed by target protein (SNPs) and small ligands (CEC, fluvoxamine, fluvastatin and ticlopidine) under specified ensembles.

In the current case, the energy-favorable structures derived by the docking operation mentioned above were further studied with MD simulations triggered by hydrogen bond braking and making of the ligand and receptor interactions. In this way, the final results obtained would provide further conformational searching information in space. The MD simulations were performed at a constant temperature (300 K) and under normal pressure. With the side chains being allowed to move freely, all of the backbone atoms were fixed to maintain the correct protein 3-D structure. In order to preferably represent the motions of heat in and out of the system, some fictitious degrees of freedom were added to the system. This operation could generate a series of conformations in the important phase space area, which can in turn provide configuration and momentum information for each relevant atom, such that the system thermodynamic properties can be calculated.

In the computational 3-D structure of CYP2C19, two binding pockets – named A and B – were considered to be more energy-favorable (Figure 11.4). The binding pocket B is also the active site of wild-type CYP2C19 for drug metabolism [12]. In order to identify the most energy-favorable binding pocket, all four ligands were docked to the two pockets A and B of the two SNPs, respectively. The binding energies obtained by docking CEC, fluvoxamine, fluvastatin and ticlopidine to the SNPs W120R and I331V, are listed in Table 11.2. For SNPI331V, the binding pocket B appeared to be superior, although the

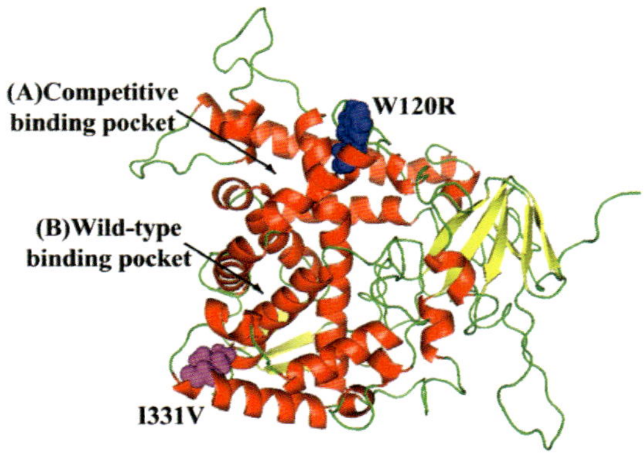

Figure 11.4 *Illustration of a 3-D structure of CYP2C19 which was obtained by homology modeling. The molecules with sphere representation are W120 and I331 with colors of blue and magenta, respectively. The two SNPs generate two binding pockets: A for novel competitive binding pocket; and B for wild-type binding pocket*

Table 11.2 *Binding energies (kcal mol^{-1}) obtained by docking CEC, fluvoxamine, fluvastatin and ticlopidine to CYP2C19 wild-type, SNP W120R pocket A and B, SNP I331V pocket A and B, respectively*

Ligand	E (binding)				
	Wild-type	W120R (A)	W120R (B)	I331V (A)	I331V (B)
CEC	−19.05	−15.0080	−18.2248	−14.9860	−19.1012
Fluvoxamine	−18.09	−19.4766	−19.1819	−15.8665	−22.5504
Fluvastatin	−20.59	−20.3081	−20.0519	−21.3289	−24.9472
Ticlopidine	−19.17	−14.4633	−17.3532	−15.1186	−16.7373

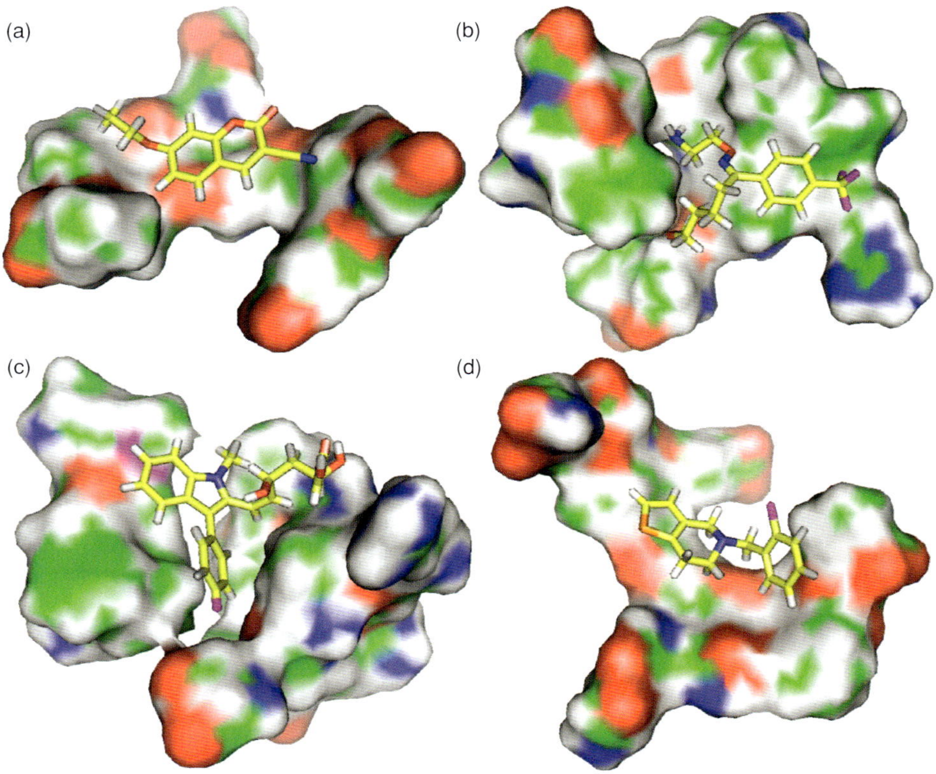

Figure 11.5 *Illustrations showing the surfaces of the binding pockets of SNP W120R for (a) CEC, (b) fluvoxamine, (c) fluvastatin and (d) ticlopidine with stick representation, where the colors light gray, red, blue, orange and pink represent hydrogen, oxygen, nitrogen, sulfur and fluorine, respectively. Carbon atoms are colored yellow within the molecules, and green on the surfaces*

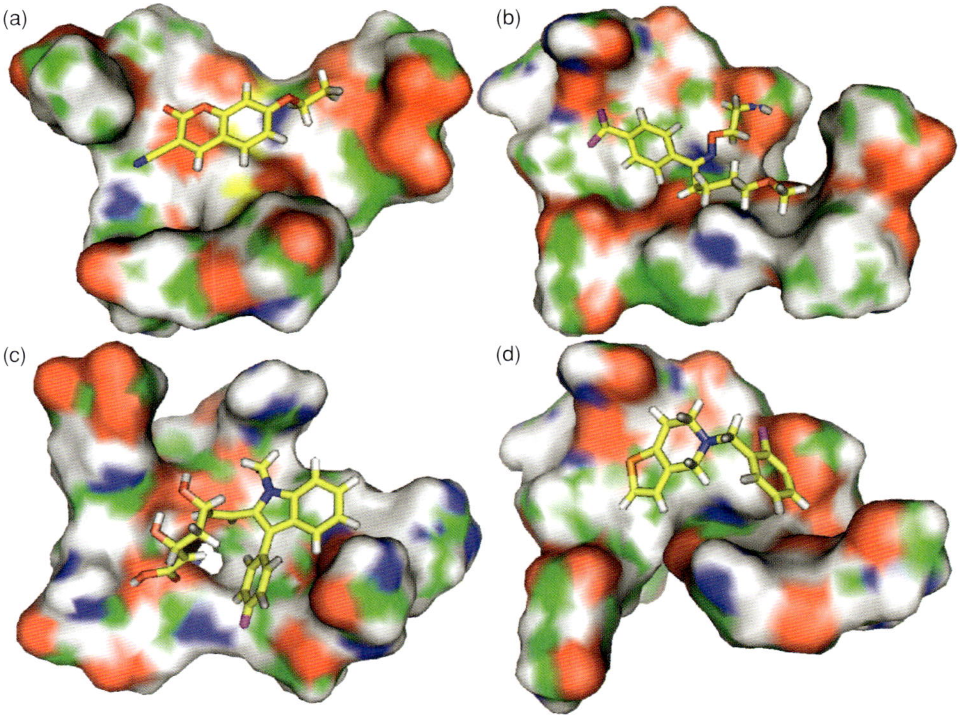

Figure 11.6 Illustrations showing the surfaces of the binding pockets of SNP I331V for (a) CEC, (b) fluvoxamine, (c) fluvastatin and (d) ticlopidine with stick representation, where the colors light gray, red, blue, orange and pink represent hydrogen, oxygen, nitrogen, sulfur and fluorine, respectively. Carbon atoms are colored yellow within the molecules, and green on the surfaces

same situation was not apparent for SNP W120R. For fluvoxamine and fluvastatin, binding pocket A was preferred, which meant that these compounds could not be metabolized by SNP W120R; however, for CEC and ticlopidine the binding pocket B was still preferred.

The surfaces of the two SNPs with the four ligands – (a) CEC; (b) fluvoxamine; (c) fluvastatin; and (d) ticlopidine – are shown in Figures 11.5 and 11.6. The lipophilicity, which can be ascertained from Figures 11.5 and 11.6, is a significant factor for designing orally active drugs due to the complementary lipophilic and hydrophilic interactions between proteins and ligands, and the harmony between lipophilicity and water-solubility. Notably, the latter property is essential for absorption to occur via the intestinal tract.

Whilst it is difficult to understand the hydrophobic effects at the molecular level, remarkable changes in the docking free energies between proteins and ligands might be a key relevant factor. The different lipophilicities in the binding pockets or active-site cavities of SNPs W120R and I331V might also be the main factors behind receptor-binding pockets being identified for CEC, fluvoxamine, fluvastatin and ticlopidine. In addition, for SNP W120R, fluvoxamine and fluvastatin may not be metabolized as they do not bind to the active sites. Nonetheless, all of these findings should prove useful when conducting mutagenesis investigations to identify drug treatments based on the characteristics of individual patients, not only to improve efficacy but also to reduce the extent and severity of adverse reactions.

References

1. Danielson, P.B. (2002) The cytochrome P450 superfamily: biochemistry, evolution and drug metabolism in humans. *Current Drug Metabolism*, **3**, 561–97.
2. Hasler, J.A., Estabrook, R., Murray, M. *et al.* (1999) Human cytochromes P450. *Molecular Aspects of Medicine*, **20**, 12–137.
3. Przybylski, D. and Rost, B. (2004) Improving fold recognition without folds. *Journal of Molecular Biology*, **341**, 255–69.
4. Skolnick, J., Kihara, D. and Zhang, Y. (2004) Development and large scale benchmark testing of the PROSPECTOR_3 threading algorithm. *Proteins: Structure, Function, and Bioinformatics*, **56**, 502–18.
5. Skolnick, J., Zhang, Y., Arakaki, A.K. *et al.* (2003) TOUCHSTONE: a unified approach to protein structure prediction. *Proteins: Structure, Function, and Bioinformatics*, **53**, 411–24.
6. Ekroos, M. and Sjogren, T. (2006) Structural basis for ligand promiscuity in cytochrome P450 3A4. *Proceedings of the National Academy of Sciences of the United States of America*, **103**, 13682–7.
7. Floudas, C.A., Fung, H.K., McAllister, S.R. *et al.* (2006) Advances in protein structure prediction and de novo protein design. *Chemical Engineering Science*, **61**, 966–88.
8. Klepeis, J.L., Floudas, C.A., Morikis, D. *et al.* (2003) Integrated computational and experimental approach for lead optimization and design of compstatin variants with improved activity. *Journal of the American Chemical Society*, **125**, 8422–3.
9. Kim, D.-H., Kim, K.-H., Isin, E.M. *et al.* (2008). Heterologous expression and characterization of wild-type human cytochrome P450 1A2 without conventional N-terminal modification in *Escherichia coli*. *Protein Expression and Purification*, **57**, 188–200.
10. Williams, P.A., Cosme, J., Sridhar, V. *et al.* (2000) Mammalian microsomal cytochrome P450 monooxygenase: structural adaptations for membrane binding and functional diversity. *Molecular Cell*, **5**, 121–31.
11. Lewis, D.F. (2002) Modelling human cytochromes P450 involved in drug metabolism from the CYP2C5 crystallographic template. *Journal of Inorganic Biochemistry*, **91**, 502–14.
12. Wang, J.-F., Wei, D.-Q., Li, L. *et al.* (2007) 3D structure modeling of cytochrome P450 2C19 and its implication for personalized drug design. *Biochemical and Biophysical Research Communications*, **335**, 513–19.
13. Wester, M.R., Johnson, E.F., Marques-Soares, C. *et al.* (2003a) Structure of a substrate complex of mammalian cytochrome P450 2C5 at 2.3 A resolution: evidence for multiple substrate binding modes. *Biochemistry*, **42**, 6370–9.
14. Wester, M.R., Johnson, E.F., Marques-Soares, C. *et al.* (2003b) Structure of mammalian cytochrome P450 2C5 complexed with diclofenac at 2.1 A resolution: evidence for an induced fit model of substrate binding. *Biochemistry*, **42**, 9335–45.
15. Schoch, G.A., Yano, J.K., Wester, M.R. *et al.* (2004) Structure of human microsomal cytochrome P450 2C8. Evidence for a peripheral fatty acid binding site. *Journal of Biological Chemistry*, **279**, 9497–503.
16. Williams, P.A., Cosme, J., Ward, A. *et al.* (2003) Crystal structure of human cytochrome P450 2C9 with bound warfarin. *Nature*, **424**, 464–8.

17. Wester, M.R., Yano, J.K., Schoch, G.A. *et al.* (2004) The structure of human cytochrome P450 2C9 complexed with flurbiprofen at 2.0-A resolution. *Journal of Biological Chemistry*, **279**, 35630–7.
18. Rowland, P., Blaney, F.E., Smyth, M.G. *et al.* (2006) Crystal structure of human cytochrome P450 2D6. *Journal of Biological Chemistry*, **281**, 7614–22.
19. Yano, J.K., Wester, M.R., Schoch, G.A. *et al.* (2004) The structure of human microsomal cytochrome P450 3A4 determined by X-ray crystallography to 2.05-A resolution. *Journal of Biological Chemistry*, **279**, 38091–4.
20. Williams, P.A., Cosme, J., Vinkovic, D.M. *et al.* (2004) Crystal structures of human cytochrome P450 3A4 bound to metyrapone and progesterone. *Science*, **305**, 683–6.
21. Scott, E.E., White, M.A., He, Y.-A. *et al.* (2004) Structure of mammalian cytochrome P450 2B4 complexed with 3-(4-chlorophenyl)imidazole at 1.9 A resolution: insight into the range of P450 conformations and the coordination of redox partner binding. *Journal of Biological Chemistry*, **279**, 27294–301.
22. Zhao, Y.-H., White, M.A., Muralidhara, B.K. *et al.* (2006) Structure of microsomal cytochrome P450 2B4 complexed with the antifungal drug bifonazole: insight into P450 conformational plasticity and membrane interaction. *Journal of Biological Chemistry*, **281**, 5973–81.
23. Zhao, Y.-H., Sun, L., Muralidhara, B.K. *et al.* (2007) Structural and thermodynamic consequences of 1-(4-chlorophenyl)imidazole binding to cytochrome P450 2B4. *Biochemistry*, **46**, 11559–67.
24. Yano, J.K., Hsu, M.-H., Griffin, K.J. *et al.* (2005) Structures of human microsomal cytochrome P450 2A6 complexed with coumarin and methoxsalen. *Nature Structural and Molecular Biology*, **12**, 822–3.
25. Yano, J.K., Deton, T.T., Cerny, M.A. *et al.* (2006) Synthetic inhibitors of cytochrome P-450 2A6: inhibitory activity, difference spectra, mechanism of inhibition, and protein cocrystallization. *Journal of Medicinal Chemistry*, **49**, 6987–7001.
26. Sansen, S., Hsu, M.-H., Stout, C.D. and Johnson, E.F. (2007a) Structural insight into the altered substrate specificity of human cytochrome P450 2A6 mutants. *Archives of Biochemistry and Biophysics*, **464**, 197–206.
27. Sansen, S., Yano, J.K., Reynald, R.L. *et al.* (2007b) Adaptations for the oxidation of polycyclic aromatic hydrocarbons exhibited by the structure of human P450 1A2. *Journal of Biological Chemistry*, **282**, 14348–55.
28. Orengo, C.A., Michie, A.D., Jones, S. *et al.* (1997) CATH: a hierarchic classification of protein domain structures. *Structure*, **5**, 1093–108.
29. Otyepka, M., Skopalik, J., Anzenbacherova, E. and Anzenbacher, P. (2007). What common structural features and variations of mammalian P450s are known to date. *Biochimica et Biophysica Acta*, **1770**, 376–89.
30. Kemper, B. (2004) Structural basis for the role in protein folding of conserved proline-rich regions in cytochromes P450. *Toxicology and Applied Pharmacology*, **199**, 305–15.
31. Mestres, J. (2005) Structure conservation in cytochromes P450. *Proteins: Structure, Function, and Bioinformatics*, **58**, 596–609.
32. Hudecek, J., Anzenbacherova, E., Anzenbacher, P. *et al.* (2000) Structural similarities and differences of the heme pockets of various P450 isoforms as revealed by resonance Raman spectroscopy. *Archives of Biochemistry and Biophysics*, **383**, 70–8.
33. Tsiftsoglou, A.S., Tsamadou, A.I. and Papadopoulou, L.C. (2006) Heme as key regulator of major mammalian cellular functions: Molecular, cellular, and pharmacological aspects. *Pharmacology and Therapeutics*, **111**, 327–45.
34. Poulos, T.L. (2005) Structural biology of heme monooxygenases. *Biochemical and Biophysical Research Communications*, **338**, 337–45.
35. Munro, A.W., Girvan, H.M. and McLean, K.J. (2007) Cytochrome P450: redox partner fusion enzymes. *Biochimica et Biophysica Acta*, **1770**, 345–59.

36. Hannemann, F., Bichet, A., Ewen, K.M. and Bernhardt, R. (2007) Cytochrome P450 systems: biological variations of electron transport chains. *Biochimica et Biophysica Acta*, **1770**, 330–44.
37. Anzenbacher, P. and Anzenbacherová, E. (2001) Cytochromes P450 and metabolism of xenobiotics. *Cellular and Molecular Life Sciences*, **58**, 737–47.
38. Rendic, S. (2002) Summary of information on human CYP enzymes: human P450 metabolism data. *Drug Metabolism Reviews*, **34**, 83–448.
39. Dabrowski, M.J., Schrag, M.L., Wienkers, L.C. and Atkins, W.M. (2002) Pyrene-pyrene complexes at the active site of cytochrome P450 3A4: evidence for a multiple substrate binding site. *Journal of the American Chemical Society*, **124**, 11866–7.
40. Schrag, M.L. and Wienkers, L.C. (2001) Covalent alteration of the CYP3A4 active site: evidence for multiple substrate binding domains. *Archives of Biochemistry and Biophysics*, **391**, 49–55.
41. Scott, E.E. and Halpert, J.R. (2005) Structures of cytochrome P450 3A4. *Trends in Biochemical Sciences*, **30**, 5–7.
42. Kirton, S.B., Murray, C.W., Verdonk, M.L. and Taylor, R.D. (2005) Prediction of binding modes for ligands in the cytochromes P450 and other heme-containing proteins. *Proteins: Structure, Function, and Bioinformatics*, **58**, 836–44.
43. Lewis, D.F., Jacobs, M.N. and Dickins, M. (2004) Compound lipophilicity for substrate binding to human P450s in drug metabolism. *Drug Discovery Today*, **9**, 530–7.
44. Waterschoot, R.A.V., Keizers, P.H., Graaf, C.D. *et al.* (2006) Topological role of cytochrome P450 2D6 active site residues. *Archives of Biochemistry and Biophysics*, **447**, 53–8.
45. Marechal, J.D., Kemp, C.A., Roberts, G.C.K. *et al.* (2008) Insights into drug metabolism by cytochromes P450 from modeling studies of CYP2D6-drug interactions. *British Journal of Pharmacology*, **153**, S82–9.
46. Wang, J.-F., Wei, D.-Q., Chen, C. *et al.* (2008) Molecular modeling of two CYP2C19 SNPs and its implications for personalized drug design. *Protein and Peptide Letters*, **15**, 27–32.

12

Recent Progress of Bioinformatics in Membrane Protein Structural Studies

Hong-Bin Shen[1,2], Jun-Feng Wang[3], Li-Xiu Yao[1], Jie Yang[1] and Kuo-Chen Chou[1,4]

[1]*Institute of Image Processing and Pattern Recognition, Shanghai Jiaotong University, China*
[2]*College of Information Engineering, Southern Yangtze University, China*
[3]*Department of Biological Chemistry and Molecular Pharmacology, Harvard Medical School, Boston, USA*
[4]*Gordon Life Science Institute, USA*

12.1 Introduction

As a 'building block of life', a cell is deemed the most basic structural and functional unit of all living organisms. It is highly organized with many functional units or organelles according to cellular anatomy, with most such units being 'enveloped' by one or more membranes, which form the structural basis for many important biological functions. Although the lipid bilayer is the basic structure of membranes, most of the specific functions of the cell membrane are performed by the membrane proteins [1, 2]. For example, it is through membrane proteins that molecules can be transported into and out of cells by such methods as ion pumps, channel proteins and carrier proteins; that various chemical messages such as nerve impulses and hormone activity can be passed between cells; that parts of the cytoskeleton can be attached to the cell membrane in order to provide shape; that cells can be attached to an extracellular matrix in grouping cells together to form tissues; and that metabolic processes and the body's defense mechanisms can be completed.

Membrane-embedded α-helical, polytopic proteins constitute the majority of ion channels, transporters and receptors in living organisms. This class of proteins, which accounts

Automation in Proteomics and Genomics: An Engineering Case-Based Approach Edited by Gil Alterovitz, Roseann Benson and Marco Ramoni

for approximately 40% of all membrane proteins, is infamously difficult to target for high-resolution structural studies. Due to the intrinsic structural plasticity associated with many of these proteins, the chance of obtaining crystals suitable for X-ray or electron diffraction studies is small. Although helical membrane proteins pose a higher degree of experimental difficulty, their conformation is, in a number of ways, more predictable than that of water-soluble proteins. For example, the transmembrane helices must span the hydrophobic layer of membrane or membrane-mimetic detergent micelles, whereas the amphipathic helices or loops are either associated with the head group region or exposed to bulk solvent. These conditions effectively reduce the search problem in the three-dimensional (3-D) conformational space to that in a much more restricted pseudo-two-dimensional (2-D) space.

The knowledge gap between known membrane protein sequences and their structures cries out for automated efforts, as automated efforts that can augment our membrane protein structures knowledge rapidly will expedite closing the knowledge gap. For example, automation robots have been successfully applied in automating the crystallization of membrane proteins during the past few years [3,4]. Indeed, during the past decade automated computational intelligence algorithms have also been widely studied in membrane protein research, such as membrane protein type prediction [5–11], membrane transmembrane segment prediction [12–23], and so on. In this chapter, we review the recent progress of bioinformatics researches in membrane protein structural studies.

12.2 Automated Membrane Protein Type Prediction

Membrane proteins comprise different types, the function of a membrane protein being closely correlated with the type to which it belongs. For instance, transmembrane proteins can function on both sides of a membrane or transport molecules across it, whereas proteins that function on only one side of the lipid bilayer are often associated exclusively with either the lipid monolayer or a protein domain on that side. Therefore, information about membrane protein type may offer important clues towards determining the function of an uncharacterized membrane protein. Furthermore, owing to the fluid nature of their infrastructure, membrane proteins can move around the cell membrane to where their function is required. Knowing the type of a membrane protein can provide insight into this kind of motion, which is indispensable for studying the biological process at the cellular level from a dynamic point of view [24]. Therefore, the pace at which the function of uncharacterized membrane proteins could be determined, and their action processes understood, would be clearly expedited if knowledge of their type could be timely ascertained. Notably, in recent times the number of sequences entering into databanks has rapidly increased. For example, in 1986 the number of total protein sequence entries in SWISS-PROT was only 3939 but, according to the version 52.4 released on 1 May 2007 at http://www.ebi.ac.uk/swissprot/, this number has now leapt to 265 950. In other words, the number of the entries now is more than 67 times the number listed in 1986! When combining the explosion of protein sequences entering into databanks with the fact that membrane proteins are encoded by 20–35% of genes, but represent less than 1% of known protein structures to date [25], it becomes clear that it would be highly desirable to develop a sequence-based automated method for the fast and effective identification of newly-found proteins according to the following two questions. (1) Is it a membrane protein? (2) If it is, to which type does it belong?

During the past few years, although a wide variety of predictive methods have been proposed in this area [8–11, 26], most of these have exhibited the following problems that need to be addressed:

1. They were developed based on a prerequisite that the query protein was already known that belonged to the membrane proteins, without any effort being made to identify whether the query protein was a membrane protein or a nonmembrane protein. To make the case logically more reasonable and practically more useful, such a procedure is indispensable.
2. None of the methods was based on a benchmark dataset with a clear data-culling operation to avoid redundancy and homologous bias. Hence, the reported success rates therein might be overestimated.
3. Only five membrane types were covered; with the development of the Swiss-Prot database, more types should be included to increase the scope of practical application.
4. None of these methods has provided a Web server for public use and, consequently, their practical application value is quite limited. In view of this, Chou and Shen recently developed a new membrane protein-type prediction tool called MemType-2L [5]; this is an online prediction tool and can be freely accessed through http://chou.med.harvard.edu/bioinf/MemType or http://www.csbio.sjtu.edu.cn/bioinf/MemType/.

Protein sequences in the benchmark training dataset of the MemType-2L predictor were collected from the Swiss-Prot database at http://www.ebi.ac.uk/swissprot/ (version 51.0, released on 6 October 2006). In order to collect as much desired information as possible, while ensuring optimal quality for the benchmark dataset, the data were screened strictly according to the following criteria and order.

1. Sequences annotated with 'fragment' were excluded; also, sequences with less than 50 amino acid residues were excluded because they may possibly be fragments.
2. Sequences annotated with ambiguous or uncertain terms, such as 'potential', 'probable', 'probably', 'maybe' or 'by similarity', were removed for further consideration.
3. For the sequences left after the preceding two screen procedures, those annotated with 'membrane protein' were stored in the membrane protein reservoir R_{mem}; while the rest were stored in the nonmembrane protein reservoir $R_{non\text{-}mem}$.
4. Eight different membrane protein types (Figure 12.1) were found in R_{mem}; to reduce the homology bias, a redundancy cut-off was operated by an in-house program to winnow sequences down to those which have $\geq$80% sequence identity to any other in a same membrane type.
5. A similar cut-off procedure was operated for the sequences in $R_{non\text{-}mem}$ from the data obtained after such a redundancy-reducing cut-off procedure. Sequences were randomly picked to form the benchmark dataset for nonmembrane proteins.

MemType-2L is a two-layer predictor (Figure 12.2): the first layer prediction engine is to identify whether a query protein is membrane or not; the second layer is to identify its

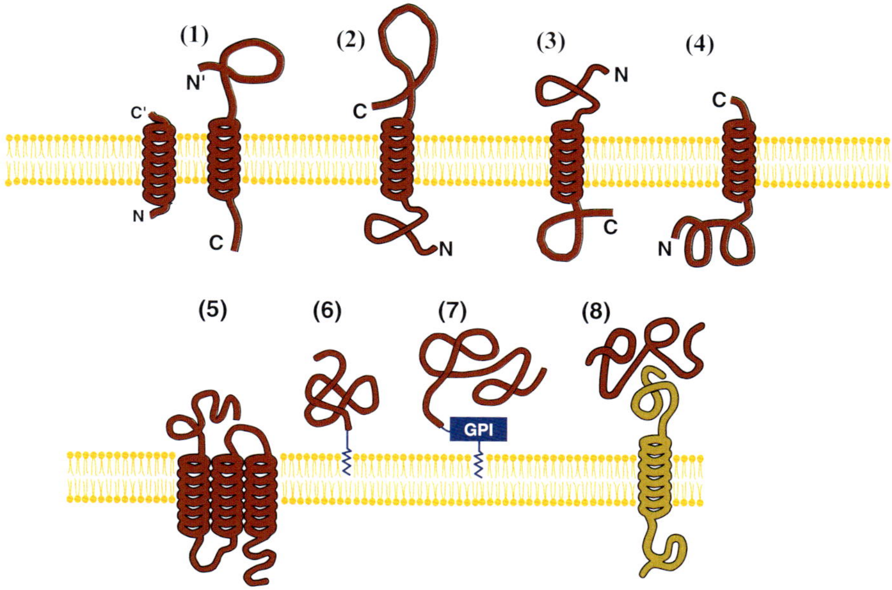

Figure 12.1 *Schematic illustration to show the eight types of membrane proteins. (1) type I transmembrane; (2) type II; (3) type III; (4) type IV; (5) multipass transmembrane; (6) lipid-chain-anchored membrane; (7) GPI-anchored membrane; (8) peripheral membrane. As shown in the figure, types I, II, III and IV are all single-pass transmembrane proteins (see Ref. [24] for a detailed description of differences in these proteins)*

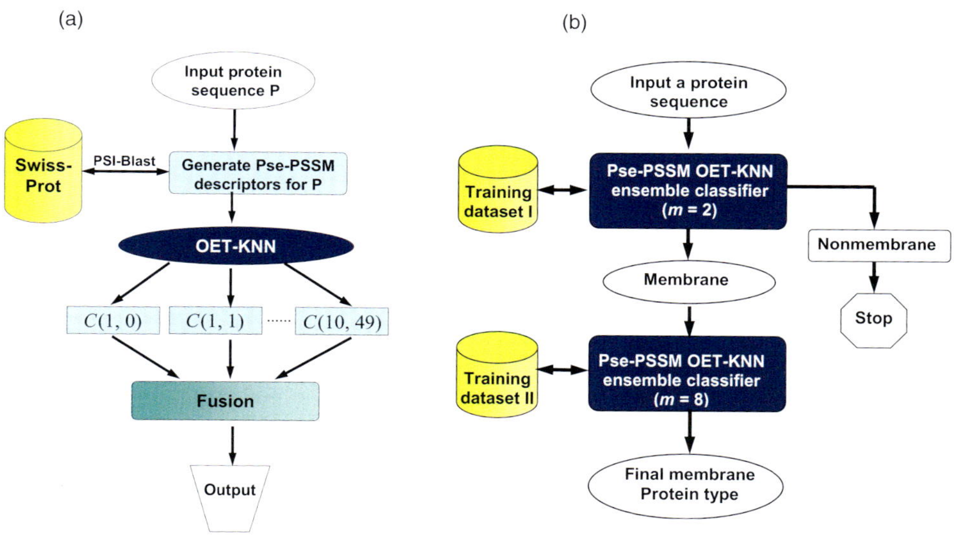

Figure 12.2 *Flowchart showing (a) the Pse-PSSM OET-KNN ensemble classifier and (b) the two-layer engine of MemType-2L*

type if the outcome from the first layer turns out to be positive. Compared with the existing predictors which cover only five to six membrane protein types, MemType-2L can cover eight types. Experimental results on the benchmark dataset show that:

- The overall jackknife success rate by the current MemType-2L in discriminating membrane and nonmembrane proteins is 92.7%, which is about 13–16% higher than those by the least Euclidean algorithm [27] and ProtLoc [28] based on the conventional amino acid composition.
- The overall jackknife success rate by MemType-2L in identifying the membrane protein type is 85.0%, which is about 33% higher than those by the other methods.
- The overall independent dataset test success rate is 91.6%, which is about 30–54% higher than those by the other methods. All these indicate that MemType-2L is indeed very useful in identifying membrane proteins and their types. The high success rates obtained by MemType-2L is that: (i) it takes into account the evolution information by representing the protein samples with the Pse-PSSM (pseudo-position-specific scoring matrix) vectors derived from the results generated by PSI-BLAST; and (ii) it operates by fusing many powerful individual OET-KNN (optimized evidence theoretic K-nearest neighbor) classifiers so as to minimize both the information-missing problem and the overfitting problem.

12.3 Predicting 2-D Structures

12.3.1 Automated Transmembrane Helix Prediction

Two overall transmembrane (TM) fold types have been observed in membrane proteins, namely α-helix bundle and β-barrel, indicating that the TM segments can be either TM helix (TMH) or TM β-strand [29, 30]. Until now, β-barrel membrane proteins have been observed only in Gram-negative bacterial outer membrane proteins and their relatives [31].

Membrane-embedded α-helical, polytopic proteins constitute the majority of ion channels, transporters and receptors in living organisms. These classes of proteins, which account for approximately 40% of all membrane proteins, are difficult targets for high-resolution structural studies. Although experimentally determined structures of integral membrane proteins have been increasing rapidly in recent years, they only sum to less than 1% of the structures in the Protein Data Bank (PDB). Usually, the first analysis that researchers perform when studying a helical membrane protein, whether it is for functional or structural characterization, is a prediction of the transmembrane helix (TMH) from the protein amino acid sequence. Knowledge of the TMH is very useful in an initial elucidation of the overall topology of the protein, as well as in the rational design of protein constructs for structural studies.

Computational tools for TMH prediction are widely available. In general, residues of TMH are mostly hydrophobic; hence, earlier TMH prediction programs (such as TOP-PRED [32]) compute sequence hydrophobicity from amino acid hydrophobicity scales assigned by biophysical and chemical measurements [13, 33, 34], and predict TMH propensity based on the average hydrophobicity score of a sliding prediction window of N successive residues along the sequence. Later predictors have used more statistics-based,

Table 12.1 TMHs predictors that widely used in the literature, and their web sites

TMH predictor	Web site address
THUMBU [30]	http://sparks.informatics.iupui.edu/Softwares-Services_files/thumbup.htm
SOSUI [41]	http://bp.nuap.nagoya-u.ac.jp/sosui/
DAS-TMfilter [19]	http://mendel.imp.ac.at/sat/DAS/DAS.html
TOP-PRED [32]	http://bioweb.pasteur.fr/seqanal/interfaces/toppred.html
TMHMM [21]	http://www.cbs.dtu.dk/services/TMHMM/
Phobius [18]	http://phobius.cgb.ki.se/
PHDhtm [22]	http://rostlab.org/predictprotein/submit_adv.html
Split4 [37]	http://split.pmfst.hr/split/4/
TMAP [38]	http://bioinfo.limbo.ifm.liu.se/tmap/index.html
MEMSAT3 [36]	http://bioinf.cs.ucl.ac.uk/psipred/

machine learning techniques. For example, PHDhtm [22] is based on neural networks, while TMHMM [21] and Phobius [18] are based on the hidden Markov model. The available TMH predictors are used routinely in membrane protein characterization and, in most cases, are sufficiently reliable in providing qualitative information about the number of TMHs in a membrane protein [35]. The TMHs predictors that are most widely used in the literature, together with their web addresses, are listed in Table 12.1.

Generally speaking, in order to cross the membrane, the TMH requires at least 15 amino acids [36–38]. However, as more high-resolution structures of helical membrane proteins become available, it was discovered that TMH has a wide length distribution. About 5% of the TMHs in the known structures are very short (<15 residues) and only span the membrane partially. These helices are known as the 'half-TMHs' (see an example in the structure of the glycerol-conducting channel [39]). Very long TMHs (>40 residues) have also been found in the membrane proteins, such as the metalloenzyme protein [40]. Figure 12.3 shows the TMH length distribution in 70 known high-resolution membrane protein structures as shown in Table 12.2, where it is clear that some 'half-TMHs' are around ten amino acids long, although there are also very long TMHs in the structures (>40 residues). None of the existing TMH predictors perform satisfactorily in detecting TMHs of irregular lengths. For example, TOP-PRED [32] predicts all of the TMHs to be 21 residues long, TMHMM [21] cannot predict TMHs shorter than 16 residues or longer than 35 residues, and SOSUI [41] cannot predict TMHs longer than 25 residues. Hence, there is a great opportunity for improving the sensitivity and accuracy of predicting TMHs.

12.3.2 Automated Methods for Predicting N-Terminal Signal Peptide

A signal peptide is a short sequence chain, which functions as an 'address tag' that directs nascent proteins to their proper cellular and extracellular locations, and also controls the entry of virtually all proteins to the secretory pathway, both in eukaryotes and prokaryotes [42]. If the signal peptide for a nascent protein were to be changed, the protein could end up in the wrong cellular location, resulting in a variety of weird diseases. All secreted proteins, as well as many transmembrane proteins, are synthesized with N-terminal signal

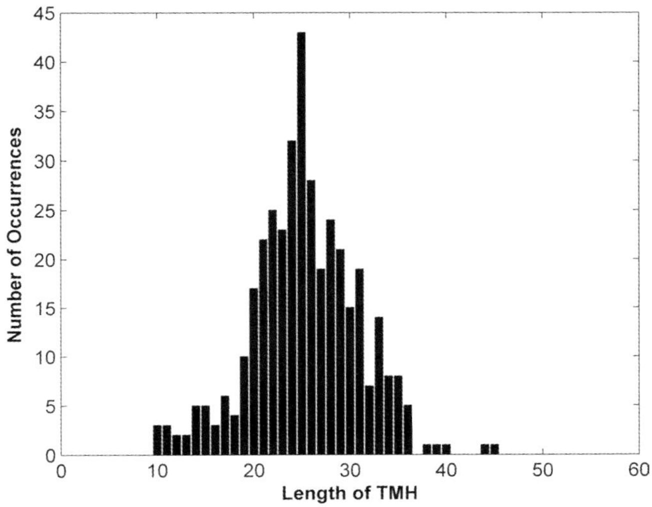

Figure 12.3 *TMH length distribution of 70 known high-resolution membrane protein structures*

peptides (Figure 12.4). According to statistics, approximately 20% of membrane proteins have N-terminal signal peptides and, since the signal peptide also has a 'hydrophobic core' of 7–15 residues in the h-region [43], they are often falsely predicted as the TMH by most of the current predictors. In fact, most of the TMH prediction servers do not have modules to detect N-terminal signal peptides [29]. Signal peptides are usually N-terminal extensions of 3–60 amino acids length, although they can also be located within a protein or at its C-terminal end [44]. The correct prediction of N-terminal signal peptides is very helpful for improving the sensitivity of predicting membrane topology [45, 46]. It should be stressed that, on the one hand, a TMH predictor could falsely predict the signal peptide as TMH yet, on the other hand, a signal peptide predictor could also falsely consider the TMH as a signal peptide. Hence, in order to discriminate effectively the N-terminal signal peptide

Table 12.2 *The PDB accession codes of the 70 membrane proteins*

1AP9_A	1AR1_A	1AT9_A	1BCC_C	1EHK_A	1EYS_L
1EYS_M	1EYS_H	1FX8_A	1IH5_A	1IWG_A	1JGJ_A
1KQG_B	1KQG_C	1L7V_A	1LGH_A	1LGH_B	1NEK_C
1NEK_D	1NKZ_A	1OCC_D	1OCC_G	1OCC_J	1OCC_K
1OCC_L	1OCC_M	1OED_A	1OED_B	1OED_C	1OED_E
1OKC_A	1PRC_M	1PSS_L	1PSS_M	1PV7_A	1PW4_A
1Q90_D	1QHJ_A	1QLB_C	1RC2_B	1RHZ_A	1RWT_A
1SOR_A	1U7G_A	1UAZ_A	1VF5_C	1VF5_D	1VGO_A
1XIO_A	1XQF_A	1YCE_A	1ZCD_A	2A65_A	2AHZ_A
2B2J_A	2B5F_A	2BBJ_A	2BL2_A	2BRD_A	2H8A_A
2HI7_B	2IRV_A	2IUB_A	2J7A_C	2JO1_A	2NQ2_A
2ONK_C	2PNO_A	2Q7M_A	2QTS_A		

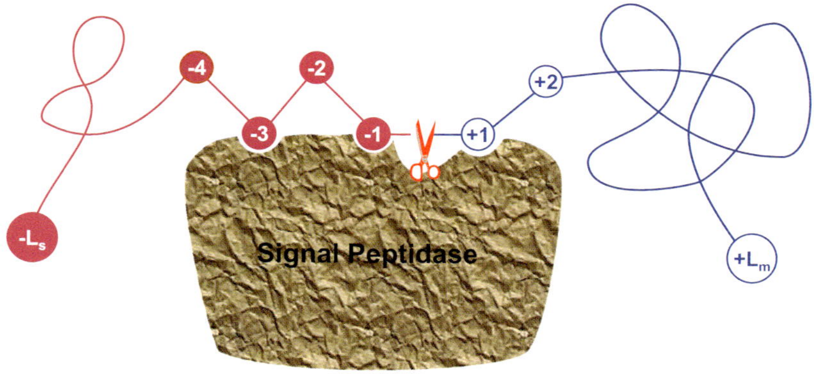

Figure 12.4 *A schematic diagram showing the signal sequence of a protein and how it is cleaved by the signal peptidase. An amino acid in the signal part is depicted as a red circle with a white number to indicate its sequential position, while that in the mature protein is depicted as an open circle with a blue number. The signal sequence contains L_s residues and mature protein L_m residues. The cleavage site is at the position $(-1, +1)$; that is, between the last residue of the signal sequence and the first residue of the mature protein*

from the TMHs, many efforts are underway to automate a reliable process to predict the signal peptide [46–56]. A brief introduction for most of these methods can be found in several reviews [43, 57].

Recently, Shen and Chou [58] proposed another signal peptide predictor called Signal-3L, and which has three layers. The target of the first-layer is to identify a query protein as secretory or nonsecretory with the powerful OET-KNN (optimized evidence-theoretic K nearest neighbor) classifier in a PseAAC (pseudo amino acid composition) space. If the protein is identified as secretory, the process will be automatically continued by entering into the second layer, where a set of candidates for its signal peptide cleavage site are to be selected with a subsite-coupled discriminator, or $\{-3, -1, +1\}$ coupling model (Figure 12.5), by sliding a scaled window along the protein sequence. The third layer is to finally determine the unique cleavage site by fusing the global sequence alignment outcome for each of the selected candidates through a voting system. Figure 12.6 illustrates the working flowchart of Signal-3L (which is freely available as a web server at http://chou.med.harvard.edu/bioinf/Signal-3L/ or http://www.csbio.sjtu.edu.cn/bioinf/Signal-3L/).

PrediSi [54] and SignalP [53] are two popular web-server predictors developed for identifying the signal peptide and its cleavage site. Compared to PrediSi, Signal-3L [58] can achieve 5–18% higher success rates on the benchmark datasets by the jackknife test. Because SignalP is a predictor with a built-in training dataset covering only three different organisms, in order to compare it with the current predictor Signal-3L, Shen and Chou used both SignalP and Signal-3L to deal with the proteins, the signal peptides of which have been experimentally verified. The results showed that many protein signal peptides predicted incorrectly by SignalP were successfully corrected by Signal-3L. It was also shown that some of the results predicted by SignalP 3.0-NN and SignalP 3.0-HMM – two important signal peptide predictors in the SignalP package – were often inconsistent. For

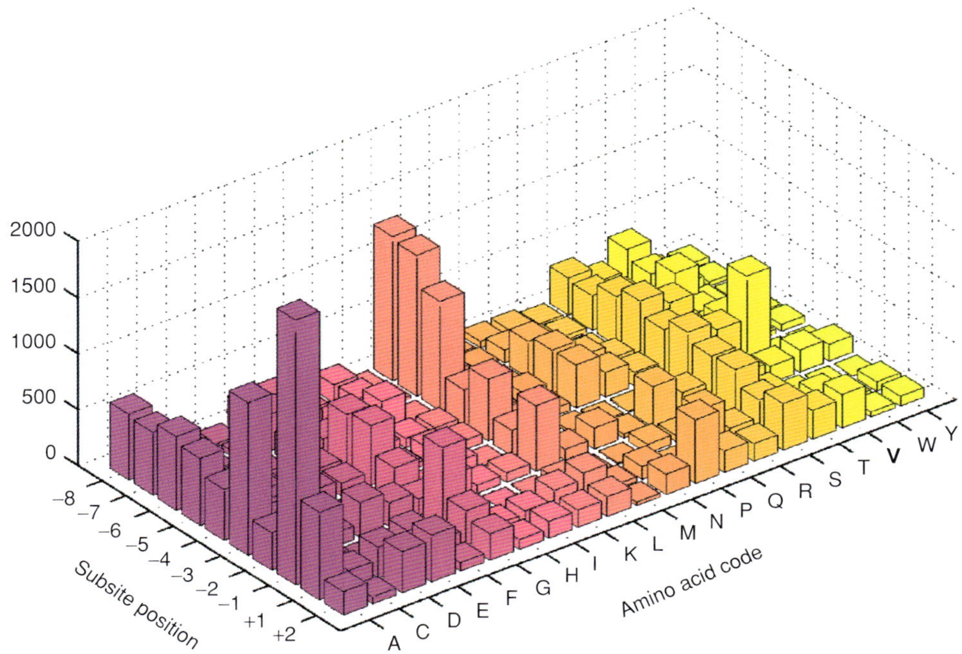

Figure 12.5 A 3-D histogram showing the frequency of the 20 native amino acids that occur at the subsites proximal to the cleavage site. As shown, the occurrence frequencies of Ala at subsites −3, −1 and +1 are overwhelming in comparison with the other 19 amino acids, suggesting a high selectivity of Ala at the three key subsites. The data used to derive this figure are from the 4184 secretory proteins [53]

example, the signal peptide of FZD3_HUMAN predicted by SignalP 3.0-NN was 1–17, but that by SignalP 3.0-HMM was 1–22. This type of inconsistency might cause confusion if no experiment results were timely available. Nevertheless, the predicted result by Signal-3L supported the latter, and was fully consistent with the experimental observation. However, for a different protein, such as IBP7_HUMAN, Signal-3L supported the result obtained by SignalP 3.0-NN rather than SignalP 3.0-HMM, which also was fully consistent with the experimental observation.

The above results and discussion indicate that the Signal-3L is a powerful tool for predicting signal peptides, and can at least play an important complementary role to SignalP and PrediSi, which are widely used in the relevant areas. It is also expected that, by combining both the powerful TMHs predictors and the signal peptide predictors, more reliable and robust membrane protein secondary structure predictors can be developed.

12.4 Predicting 3-D Structures

As for the prediction of membrane protein 3-D structures, the methods can be generally classified into two classes, namely *homology modeling* and *ab initio* and *de novo methods*.

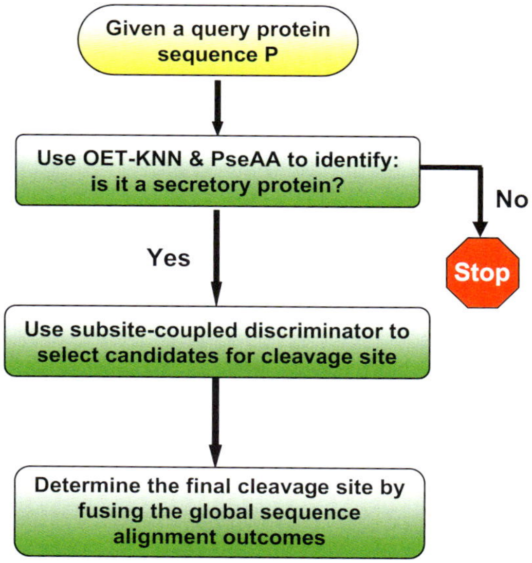

Figure 12.6 *Flowchart showing how the three-layer predictor is able to identify a query protein as secretory or nonsecretory, selecting the candidates of its signal peptide cleavage site if the protein is secretory, and determining the final cleavage site*

12.4.1 Homology Modeling

Homology modeling attempts to construct the 3-D model of a given membrane protein sequence according to the high-resolution 3-D structure of a similar protein sequence that has been determined experimentally [59–62]. The general steps of homology modeling are as follows:

- Identify homologous proteins and determine the extent of their sequence similarity with one another and the unknown.
- Align the sequences.
- Identify structurally conserved and structurally variable regions.
- Generate coordinates for core (structurally conserved) residues of the unknown structure from those of the known structure(s).
- Generate conformations for the loops (structurally variable) in the unknown structure.
- Build the side-chain conformations.
- Refine and evaluate the unknown structure [63–65].

The performance of homology modeling is heavily dependent on the 'homology similarity' between the target sequence and the template sequence. Past research projects have revealed that, if the identity between the two sequences is below 30%, then the accuracy of the obtained model will decrease dramatically [66–68]. Although homology modeling has been successfully applied to predicting the 3-D structures of globular proteins [66, 67, 69, 70], it is still a very challenging problem in membrane protein structure prediction for the following reasons:

- Few high-resolution membrane protein structures are available in the database currently. According to Stephen White's membrane protein database at http://blanco.biomol.uci.edu/Membrane_Proteins_xtal.html, there are only 154 unique high-resolution membrane protein structures that have been solved and deposited into the protein data bank. This accounts for approximately 1% of all the protein structures solved to date, although membrane proteins account for approximately 20–35% of all the proteins.
- The loop regions of integral membrane proteins exhibit remarkable structural variability, which makes modeling in these regions unreliable. Although it is very difficult to apply homology modeling techniques to predict the whole-membrane protein 3-D structures currently, research results still show much progress in this field. For example, less than 2 Å of the RMSD (root mean-square deviation) between the modeled and template structures have been obtained in the transmembrane regions when the corresponding sequence identity is >30% [71].

12.4.2 *Ab Initio* or *De Novo* Methods

Ab initio or *de novo* methods are very useful when minimal experimental information and homology structures are available for target membrane protein sequences. The methods include prediction of the 2-D structures, such as TMH [21], and prediction of the helix-helix interactions [72–75]. Although *ab initio* or *de novo* methods and software have been successfully applied in the prediction of globular protein 3-D structures, the success in membrane proteins has been quite limited until recently. The reasons for this are twofold. First, the membrane proteins and globular proteins belong to different environments, and it is difficult to directly apply those methods that are suitable for the globular proteins to membrane proteins. Second, the size of membrane proteins are larger than the globular proteins that have been correctly predicted by using *ab initio* methods. In view of this, there is still a long way to go to develop effective *ab initio* or *de novo* methods to directly predict membrane protein 3-D structures [76].

12.5 Conclusions and Future Directions

Membrane proteins are encoded by 20–35% of genes, and approximately 40% of such membranes are polytopic in nature. Yet, solved membrane protein structures represent less than 1% of known protein structures to date, which makes the investigation of membrane protein structures a most challenging problem. Although, helical membrane proteins pose a higher degree of experimental difficulty, their conformation is, in many ways, more predictable than that of water-soluble proteins. For example, the transmembrane helices must span the hydrophobic layer of membrane or membrane-mimetic detergent micelles, whereas the amphipathic helices or loops are either associated with the head group region or exposed to bulk solvent. These conditions effectively reduce the search problem in a 3-D conformational space to that in a much more restricted pseudo-2-D space. During the past few decades, a large number of bioinformatics methods and tools have been developed and successfully applied to predict membrane protein structures and functional features. At the same time, an increasing amount of evidence has shown that the prediction of membrane protein structures is more difficult than once was imagined, as the resolved high-resolution membrane protein structures have exhibited significant structural differences. For example, current TMHs predictors cannot accurately predict half-TMHs and very long TMHs. Some new structural features have also been observed in membrane proteins, an example being the large number of helices found at the membrane-water interface [77]. Another intriguing observation, made from known membrane protein structures, was that there are often homologous domains with opposite or parallel membrane orientations, leading to proteins with a quasi-twofold axis in the plane of the membrane [78]. All of these situations call for the development of more automated bioinformatics research methods to help further our understanding of membrane proteins and their structures. It is believed that, as an increasing number of high-resolution membrane protein structures are resolved by structural biologists, the resultant expansion of membrane protein knowledge will enable the prediction of membrane protein structures to be much more successful.

Acknowledgments

The authors thank James Chou, Kirill Oxenoid and Matthew Call for useful discussions. These studies were supported by the National Natural Science Foundation of China (Grant No. 60704047), Science and Technology Commission of Shanghai Municipality (Grant No. 08ZR1410600, 08JC1410600) and sponsored by Shanghai Pujiang Program.

References

1. Alberts, B., Bray, D., Lewis, J. *et al.* (1994) *Molecular Biology of the Cell, Chap. 1*, 3rd edn, Garland Publishing, New York and London.
2. Lodish, H., Baltimore, D., Berk, A. *et al.* (1995) *Molecular Cell Biology, Chap. 3*, 3rd edn, Scientific American Books, New York.
3. Cherezov, V., Peddi, A., Muthusubramaniam, L. *et al.* (2004) A robotic system for crystallizing membrane and soluble proteins in lipidic mesophases. *Acta Crystallographica D Biological Crystallography*, **60**(Pt 10), 1795–807.
4. Barnard, T.J., Wally, J.L. and Buchanan, S.K. (2007) Crystallization of integral membrane proteins. *Current Protocols in Protein Science*, **Chapter 17**, Unit 17. 9.
5. Chou, K.C. and Shen, H.B. (2007) MemType-2L: a web server for predicting membrane proteins and their types by incorporating evolution information through Pse-PSSM. *Biochemical and Biophysical Research Communications*, **360**(2), 339–45.
6. Chou, K.C. and Elrod, D.W. (1999) Prediction of membrane protein types and subcellular locations. *Proteins: Structure, Function, and Genetics*, **34**(1), 137–53.
7. Feng, Z.P. and Zhang, C.T. (2000) Prediction of membrane protein types based on the hydrophobic index of amino acids. *Journal of Protein Chemistry*, **19**, 269–75.
8. Shen, H.B., Yang, J. and Chou, K.C. (2006) Fuzzy KNN for predicting membrane protein types from pseudoamino acid composition. *Journal of Theoretical Biology*, **240**(1), 9–13.
9. Wang, M., Yang, J., Xu, Z.J. and Chou, K.C. (2005) SLLE for predicting membrane protein types. *Journal of Theoretical Biology*, **232**(1), 7–15.
10. Wang, M., Yang, J., Liu, G.P. *et al.* (2004) Weighted-support vector machines for predicting membrane protein types based on pseudoamino acid composition. *Protein Engineering Design and Selection*, **17**(6), 509–16.

11. Cai, Y.D., Liu, X.J. and Chou, K.C. (2001) Artificial neural network model for predicting membrane protein types. *Journal of Biomolecular Structure and Dynamics*, **18**(4), 607–10.
12. Claros, M.G., Brunak, S. and von Heijne, G. (1997) Prediction of N-terminal protein sorting signals. *Current Opinion in Structural Biology*, **7**, 394–8.
13. Kyte, J. and Doolittle, R.F. (1982) A simple method for displaying the hydropathic character of a protein. *Journal of Molecular Biology*, **157**(1), 105–32.
14. Jones, D.T. (2007) Improving the accuracy of transmembrane protein topology prediction using evolutionary information. *Bioinformatics*, **23**(5), 538–44.
15. Cuthbertson, J.M., Doyle, D.A. and Sansom, M.S. (2005) Transmembrane helix prediction: a comparative evaluation and analysis. *Protein Engineering Design and Selection*, **18**(6), 295–308.
16. Yuan, Z., Mattick, J.S. and Teasdale, R.D. (2004) SVMtm: support vector machines to predict transmembrane segments. *Journal of Computational Chemistry*, **25**(5), 632–6.
17. Harris, M.A., Clark, J., Ireland, A. *et al.* (2004) The Gene Ontology (GO) database and informatics resource. *Nucleic Acids Research*, **32** (Database issue), D258–61.
18. Kall, L., Krogh, A. and Sonnhammer, E.L. (2004) A combined transmembrane topology and signal peptide prediction method. *Journal of Molecular Biology*, **338**(5), 1027–36.
19. Cserzo, M., Eisenhaber, F., Eisenhaber, B. and Simon, I. (2004) TM or not TM: transmembrane protein prediction with low false positive rate using DAS-TMfilter. *Bioinformatics*, **20**(1), 136–7.
20. Chamberlain, A.K., Lee, Y., Kim, S. and Bowie, J.U. (2004) Snorkeling preferences foster an amino acid composition bias in transmembrane helices. *Journal of Molecular Biology*, **339**(2), 471–9.
21. Krogh, A., Larsson, B., von Heijne, G. and Sonnhammer, E.L. (2001) Predicting transmembrane protein topology with a hidden Markov model: application to complete genomes. *Journal of Molecular Biology*, **305**(3), 567–80.
22. Rost, B., Fariselli, P. and Casadio, R. (1996) Topology prediction for helical transmembrane proteins at 86% accuracy. *Protein Science*, **5**(8), 1704–18.
23. Gromiha, M.M., Ahmad, S. and Suwa, M. (2004) Neural network-based prediction of transmembrane beta-strand segments in outer membrane proteins. *Journal of Computational Chemistry*, **25**(5), 762–7.
24. Spiess, M. (1995) Heads or tails–what determines the orientation of proteins in the membrane. *FEBS Letters*, **369**(1), 76–9.
25. Douglas, S.M., Chou, J.J. and Shih, W.M. (2007) DNA-nanotube-induced alignment of membrane proteins for NMR structure determination. *Proceedings of the National Academy of Sciences Online (US)*, **104**(16), 6644–8.
26. Shen, H. and Chou, K.C. (2005) Using optimized evidence-theoretic K-nearest neighbor classifier and pseudoamino acid composition to predict membrane protein types. *Biochemical and Biophysical Research Communications*, **334**(1), 288–92.
27. Nakashima, H., Nishikawa, K. and Ooi, T. (1986) The folding type of a protein is relevant to the amino acid composition. *Journal of Biochemistry*, **99**, 152–62.
28. Cedano, J., Aloy, P., P'erez-Pons, J.A. and Querol, E. (1997) Relation between amino acid composition and cellular location of proteins. *Journal of Molecular Biology*, **266**, 594–600.
29. Punta, M., Forrest, L.R., Bigelow, H. *et al.* (2007) Membrane protein prediction methods. *Methods*, **41**(4), 460–74.
30. Zhou, H. and Zhou, Y. (2003) Predicting the topology of transmembrane helical proteins using mean burial propensity and a hidden-Markov-model-based method. *Protein Science*, **12**(7), 1547–55.
31. Koebnik, R., Locher, K.P. and Van Gelder, P. (2000) Structure and function of bacterial outer membrane proteins: barrels in a nutshell. *Molecular Microbiology*, **37**(2), 239–53.
32. Claros, M.G. and von Heijne, G. (1994) TopPred II: an improved software for membrane protein structure predictions. *Computer Applications in the Biosciences*, **10**(6), 685–6.

33. Chamberlain, A.K., Lee, Y., Kim, S. and Bowie, J.U. (2004) Snorkeling preferences foster an amino acid composition bias in transmembrane helices. *Journal of Molecular Biology*, **339**(2), 471–9.
34. Wimley, W.C. and White, S.H. (1996) Experimentally determined hydrophobicity scale for proteins at membrane interfaces. *Nature Structural Biology*, **3**(10), 842–8.
35. White, S.H. (2004) The progress of membrane protein structure determination. *Protein Science*, **13**(7), 1948–9.
36. Cao, B., Porollo, A., Adamczak, R. *et al.* (2006) Enhanced recognition of protein transmembrane domains with prediction-based structural profiles. *Bioinformatics*, **22**(3), 303–9.
37. Juretic, D., Zoranic, L. and Zucic, D. (2002) Basic charge clusters and predictions of membrane protein topology. *Journal of Chemical Information and Computer Sciences*, **42**(3), 620–32.
38. Persson, B. and Argos, P. (1997) Prediction of membrane protein topology utilizing multiple sequence alignments. *Journal of Protein Chemistry*, **16**(5), 453–7.
39. Fu, D., Libson, A., Miercke, L.J. *et al.* (2000) Structure of a glycerol-conducting channel and the basis for its selectivity. *Science*, **290**(5491), 481–6.
40. Lieberman, R.L. and Rosenzweig, A.C. (2005) Crystal structure of a membrane-bound metalloenzyme that catalyses the biological oxidation of methane. *Nature*, **434**(7030), 177–82.
41. Hirokawa, T., Boon-Chieng, S. and Mitaku, S. (1998) SOSUI: classification and secondary structure prediction system for membrane proteins. *Bioinformatics*, **14**(4), 378–9.
42. Gierasch, L.M. (1989) Signal sequences. *Biochemistry*, **28**(3), 923–30.
43. Chou, K.C. (2002) Prediction of protein signal sequences. *Current Protein and Peptide Science*, **3**(6), 615–22.
44. Kutay, U., Ahnert-Hilger, G., Hartmann, E. *et al.* (1995) Transport route for synaptobrevin via a novel pathway of insertion into the endoplasmic reticulum membrane. *The EMBO Journal*, **14**(2), 217–23.
45. Agnihothram, S.S., York, J., Trahey, M. and Nunberg, J.H. (2007) Bitopic membrane topology of the stable signal peptide in the tripartite Junin virus GP-C envelope glycoprotein complex. *Journal of Virology*, **81**(8), 4331–7.
46. Nielsen, H., Engelbrecht, J., Brunak, S. and von Heijne, G. (1997) Identification of prokaryotic and eukaryotic signal peptides and prediction of their cleavage sites. *Protein Engineering Design and Selection*, **10**(1), 1–6.
47. McGeoch, D.J. (1985) On the predictive recognition of signal peptide sequences. *Virus Research*, **3**(3), 271–86.
48. von Heijne, G. (1986) A new method for predicting signal sequence cleavage sites. *Nucleic Acids Research*, **14**(11), 4683–90.
49. Folz, R.J. and Gordon, J.I. (1987) Computer-assisted predictions of signal peptidase processing sites. *Biochemical and Biophysical Research Communications*, **146**(2), 870–7.
50. Ladunga, I., Czako, F., Csabai, I. and Geszti, T. (1991) Improving signal peptide prediction accuracy by simulated neural network. *Computer Applications in the Biosciences*, **7**(4), 485–7.
51. Emanuelsson, O., Nielsen, H. and von Heijne, G. (1999) ChloroP, a neural network-based method for predicting chloroplast transit peptides and their cleavage sites. *Protein Science*, **8**(5), 978–84.
52. Chou, K.C. (2001) Using subsite coupling to predict signal peptides. *Protein Engineering Design and Selection*, **14**(2), 75–9.
53. Bendtsen, J.D., Nielsen, H., von Heijne, G. and Brunak, S. (2004) Improved prediction of signal peptides: SignalP 3.0. *Journal of Molecular Biology*, **340**(4), 783–95.
54. Hiller, K., Grote, A., Scheer, M. *et al.* (2004) PrediSi: prediction of signal peptides and their cleavage positions. *Nucleic Acids Research*, **32** (Web Server issue), W375–9.

55. Liu, H., Yang, J., Liu, D.Q. *et al.* (2007) Using a new alignment kernel function to identify secretory proteins. *Protein and Peptide Letters*, **14**(2), 203–8.
56. Chou, K.C. and Shen, H.B. (2007) Signal-CF: a subsite-coupled and window-fusing approach for predicting signal peptides. *Biochemical and Biophysical Research Communications*, **357**(3), 633–40.
57. Nielsen, H., Brunak, S. and von Heijne, G. (1999) Machine learning approaches for the prediction of signal peptides and other protein sorting signals. *Protein Engineering Design and Selection*, **12**(1), 3–9.
58. Shen, H.B. and Chou, K.C. (2007) Signal-3L: A 3-layer approach for predicting signal peptides. *Biochemical and Biophysical Research Communications*, **363**(2), 297–303.
59. Meyer, B. and Kuever, J. (2008) Homology modeling of dissimilatory APS reductases (AprBA) of sulfur-oxidizing and sulfate-reducing prokaryotes. *PLoS ONE*, **3**(1), e1514.
60. Li, D., Tang, H.Y. and Speicher, D.W. (2008) A structural model of the erythrocyte spectrin heterodimer initiation site determined using homology modeling and chemical cross-linking. *Journal of Biological Chemistry*, **283**(3), 1553–62.
61. Li, M. and Wang, B. (2007) Homology modeling and examination of the effect of the D92E mutation on the H5N1 nonstructural protein NS1 effector domain. *Journal of Molecular Modeling*, **13**(12), 1237–44.
62. Fernandez-Fuentes, N., Rai, B.K., Madrid-Aliste, C.J. *et al.* (2007) Comparative protein structure modeling by combining multiple templates and optimizing sequence-to-structure alignments. *Bioinformatics*, **23**(19), 2558–65.
63. Leach, A.R. (1996) *Molecular Modelling: Principles and Applications*, Vol. **xvi**, Longman, Harlow, England, p. 595.
64. Hinchliffe, A. (2003) *Molecular Modelling for Beginners*, Vol. **xviii**, John Wiley & Sons, Inc., Chichester, West Sussex, England; Hoboken, NJ, p. 410.
65. Ciobanu, G. and Rozenberg, G. (2004) *Modelling in Molecular Biology*. Natural computing series, Vol. **x**, Springer, Berlin; New York, p. 304.
66. Kairys, V., Gilson, M.K. and Fernandes, M.X. (2006) Using protein homology models for structure-based studies: approaches to model refinement. *Scientific World Journal*, **6**, 1542–54.
67. Rockey, W.M. and Elcock, A.H. (2006) Structure selection for protein kinase docking and virtual screening: homology models or crystal structures? *Current Protein and Peptide Science*, **7**(5), 437–57.
68. Dunbrack, R.L. Jr. (2006) Sequence comparison and protein structure prediction. *Current Opinion in Structural Biology*, **16**(3), 374–84.
69. Chaney, M.O., Webster, S.D., Kuo, Y.M. and Roher, A.E. (1998) Molecular modeling of the Abeta1-42 peptide from Alzheimer's disease. *Protein Engineering Design and Selection*, **11**(9), 761–7.
70. Srinivasan, S., March, C.J. and Sudarsanam, S. (1993) An automated method for modeling proteins on known templates using distance geometry. *Protein Science*, **2**(2), 277–89.
71. Forrest, L.R., Tang, C.L. and Honig, B. (2006) On the accuracy of homology modeling and sequence alignment methods applied to membrane proteins. *Biophysical Journal*, **91**(2), 508–17.
72. Prodohl, A., Weber, M., Dreher, C. and Schneider, D. (2007) A mutational study of transmembrane helix-helix interactions. *Biochimie*, **89**(11), 1433–7.
73. Guharoy, M. and Chakrabarti, P. (2007) Secondary structure based analysis and classification of biological interfaces: identification of binding motifs in protein-protein interactions. *Bioinformatics*, **23**(15), 1909–18.
74. Sal-Man, N., Gerber, D., Bloch, I. and Shai, Y. (2007) Specificity in transmembrane helix-helix interactions mediated by aromatic residues. *Journal of Biological Chemistry*, **282**(27), 19753–61.

75. Liang, Y., Fotiadis, D., Filipek, S. *et al.* (2003) Organization of the G protein-coupled receptors rhodopsin and opsin in native membranes. *Journal of Biological Chemistry*, **278**(24), 21655–62.
76. Fleishman, S.J. and Ben-Tal, N. (2006) Progress in structure prediction of alpha-helical membrane proteins. *Current Opinion in Structural Biology*, **16**(4), 496–504.
77. Granseth, E., von Heijne, G. and Elofsson, A. (2005) A study of the membrane-water interface region of membrane proteins. *Journal of Molecular Biology*, **346**(1), 377–85.
78. Rapp, M., Seppala, S., Granseth, E. and von Heijne, G. (2007) Emulating membrane protein evolution by rational design. *Science*, **315**(5816), 1282–4.

13

Trends in Automation for Genomics and Proteomics

Gil Alterovitz[1,2,3,4], Roseann Benson[4], Marco Ramoni[1,2,4] and Dmitriy Sonkin[4]

[1]*Harvard/MIT Health Science and Technology Division, Massachusetts Institute of Technology, Cambridge, MA, USA*
[2]*Children's Hospital Informatics Program, Harvard Medical School, Boston, MA, USA*
[3]*Department of Electrical Engineering and Computer Science, Massachusetts Institute of Technology, Cambridge, MA, USA*
[4]*Harvard Partners Center for Genetics and Genomics, Harvard Medical School, Boston, MA, USA*

Initially motivated by the Human Genome Project, tremendous effort was expended to develop cost-efficient, high-throughput DNA sequencing instruments capable of decoding any genome of interest. In order for DNA sequencers to become more amenable to automation, extensive modification of the original chemistry was required, an avoidance of radiolabeling was introduced (and indeed proved to be a key innovation), while major investments were made in robotics and parallelized sample preparation. Hence, the next-generation sequencing technologies included cyclic array sequencing, pyrosequencing of emulsion polymerase chain reaction (PCR) features, and the sequencing of emulsion PCR features by ligation, of bridge PCR and single molecule features by synthesis, of dideoxy sequencing-based microfluidics, and of sequencing by hybridization, by mass spectrometry, by exonuclease digestion, by nanopore threading and by scanning probes.

As expected, next-generation proteomics and genomics instruments are continuing to increase the amount of data generated – a situation which makes the evaluation of new methods for pipelining laboratory automation [1] paramount. A well-conceived information technology strategy, including improved integration with laboratory information

Automation in Proteomics and Genomics: An Engineering Case-Based Approach Edited by Gil Alterovitz, Roseann Benson and Marco Ramoni

management systems (LIMS), is becoming an integral part of this approach. Unfortunately, the supply of an appropriate application-programming interface (API) for integration with LIMS is often, at best, an afterthought. For example, whilst next-generation sequencers provide integrated mechanisms to copy run results to network storage, many of them omit a convenient API to use a LIMS-generated run name and sample unique identifiers. As a result, instrument integration with LIMS becomes all the more challenging.

There is a great deal of anticipation regarding the development of a sub-$1000 sequencer that will allow human genome sequencing for sequence lengths over 1000 base pairs. It is hoped that these sequencers will also include a mechanism for integration with LIMS. Indeed, the establishment of a solid and widely used open-source LIMS framework could help manufacturers meet the LIMS/instrument integration challenge by making integration point and ready-to-use modules [2] widely available.

In practice, the need to integrate and perform complex analyses on heterogeneous data sources results in ad hoc connections between databases and software tools by writing small scripts, cutting and pasting queries, and basic manual labor. As a result, the recent years have witnessed a surge in the development of new software tools focused on automating and simplifying these tasks. These bioinformatics resource tools tend to fall into four categories; semantic mapping; interoperation of heterogeneous bioinformatics databases; automated workflow analyses; and programs that integrate the data with bioinformatics software.

Open-source software solutions have been a boon to the industry. An example is the ability to satisfy the demand for using existing computational tools in the identification of uncharacterized proteins subcellular locations based on their sequences without the need for detailed mathematical understanding. The details of a web-server package called 'Cell-PLoc' are provided in Chapter 5. One disadvantage of open-source software has been the lack of usable user interfaces for freely available code, although good progress has recently been made in this area [3]. However, as has been the case with clinical trials, the necessary financial resources are, on occasion, available only through commercial enterprises.

A number of hybrid solutions exist where software initially developed in academia has been commercialized but remains available – free of charge – to educational institutions and nonprofit organizations. It is conceivable that instrument manufacturers could support the widespread availability of bioinformatics software as a viable business model, since such software – especially for new types of arrays – might increase the new product adoption rate. For example, gene expression arrays manufacturers could provide complementary integrated data analysis tools, with such tools being based on open-source solutions and providing fundamental types of analysis.

Systems biology frameworks for exchanging data between independently developed software tools and databases to enable interactive exploration of systems biology data, such as Gaggle [4], are very exciting developments in the bioinformatics field. However, many laboratories – and especially the smaller ones – will, potentially, have difficulty in taking advantage of these new tools due to constraints of human resources, as applied bioinformaticians are scarce even in large institutional settings. The applied bioinformatician, rather than solely developing new algorithms, will work alongside the biologist to grapple with a diverse range of bioinformatics tasks. In order to function effectively in this role, the applied bioinformatician must possess a solid understanding of bioinformatics algorithms and resources, while remaining aware of current developments in the field. He or she must also be knowledgeable of biological research in order to understand the biologist's goals and requirements. Bioinformatics, as a discipline, is a burgeoning field that has had to acclimatize to rapid changes in research priorities and, as bioinformatics academic programs mature, an overall increase in the quality of bioinformatics education can be expected. Yet, old-fashioned, two-way communication also appears to foster optimal cooperation between biologists and bioinformaticians. To that end, new advances in automation systems are being directed towards facilitating biologist–bioinformatician interaction within the automated laboratory pipeline [5] so that as much information as possible can be extracted from well-designed and well-executed experiments.

Automation in fields outside genomics and proteomics enables relevant biological data to be generated. For example, one biopharmaceutical company is currently detecting synergistic effects by using dose-specific combinations of various FDA (Food and Drug Administration)-approved drugs. Here, testing is conducted in an automated manner using cell-based assays. The benefits of this research include not only the potential of finding useful combinations of drugs but also the provision of insights into affected pathways and crosstalk between pathways. Another application is to introduce *network concepts* into computational pharmacology studies, including drug target identification and drug discovery, both of which are aimed at improving our understanding of drug actions through a variety of biochemical networks. Other ongoing studies include automated, system-wide searches for disease biomarkers in peripheral tissues, the ultimate aim being that diagnoses can be effected using less invasive procedures [6].

Functional genomics, including DNA microarrays to monitor gene expression over time, has led to the development of experimental techniques that allow interactions between genes to be elucidated on a large scale. In that respect, information obtained from DNA microarrays is of great interest, because these are high-capacity systems for monitoring the expression of many genes in parallel. Moreover, they also provide the opportunity to measure simultaneously the expression of several thousand genes within a cell. Thus, microarray gene expression does indeed represent a popular and powerful technique. There is also a possibility that interpretation of the intensities of such arrays could be improved with a better understanding of on-chip hybridization kinetics [7]. Although the effects of sequence complementarities and base composition are well known, they do not completely explain differences in intensities between probes for the same gene, and this is especially evident in exon level-type arrays. Taken together, these studies provide interesting insides on other potential influences on hybridization signals. Indeed, if it were possible to better adjust hybridization signals from microarrays, this would not only help future experiments but also help in the acquisition of better information from a significant number of microarrays data accumulated in public databases.

Gene set enrichment analysis (GSEA) [8] permits gene expression changes to be explored, not only on an individual gene level but also on a number of related genes, known as gene sets. Examples of possible gene sets include genes from the same pathway or related pathways, genes from a response to particular substance, genes related to a particular disease, and genes from the same chromosomal region. GSEA allows the identification of modest – but related – changes that are difficult to detect by investigating differentially expressed genes on an individual basis. As GSEA depends on the availability and accuracy of gene sets, the expert curation of gene sets is essential, and the Broad Institute's free, publicly available GSEA implementation and gene set database represents a highly valuable resource for this type of analysis. GSEA and other tools can also be used to determine clusters of significant biological categories that are responsible for certain gene expression activities. These programs often rely on standard categorization schemes – namely ontologies – that hierarchically categorize biological information and assign genes to categories based on experimental or computationally determined information. The results thus depend on the organization and content of these ontologies. Today, efforts are ongoing to determine the optimal design and utilization of such ontologies for automated, optimal inference [9, 10].

The results of recent studies have shown that some phenotypes cannot be explained by one or two genes, but rather by a complex interplay of many genes [11]. Currently, new automated network-based methods, from Bayesian networks to random walks, are being used to further elucidate the relationships between genes, as well as using existing relationships to determine those phenotypes that are associated with genes and their associated proteins [12–15].

Pathways-based analyses, in general, represent a very important technique. Today, several commercial options are available, as well as open-source tools such as Cytoscope [16] and VisANT [17], all of which allow the visualization of pathways, as well as the overlying of other data. This type of functionality is especially useful in cancer research, where the ability to easily overlay pathways with gene expression, DNA mutations and epigenetic data from the same samples represents a welcome addition to pathway tools. The ability to determine easily which cancer-related mutations are known in general on genes of a particular pathway(s) could also focus research by correlating the proposed effect of mutations with observed gene expression in pathway(s). Most biological network visualization tools implement a variant of the initial force-directed layout algorithms, and use either animation or resource-constrained incremental calculations to strike a balance between optimal equilibrium and the timely layout of the network. Whilst these tools complement each other in terms of their respective performance, user query capability, declarative query capability, flexibility and ease of integration of network biology data management, the trend of future tools is to enable not only 'dynamic' or 'integrative' aspects of the visual networks but also 'data-driven' and knowledge discovery-oriented tasks performed with user query scripts. Towards this goal, all visualization software tools require significant further development.

Based on the tools developed during the past few decades, and on those currently under development, a detailed molecular understanding of most biological processes is within reach, and the design of such processes is no longer considered 'science fiction fodder'. The ability to design and write DNA blueprints freely from scratch will provide the opportunity to create novel biological systems and to revolutionize biomedical research. Today, synthetic biology is an emerging automation area where the goal is to synthesize sequences, proteins and even entire cells in an automated manner [18, 19]. Synthetic biology currently offers the ability to study cellular regulation and behavior using *de novo* networks, while future applications of synthetic systems will surely extend to the fields of medicine and biotechnology. Unfortunately, the overall development of DNA synthesis technology has, so far, lagged behind that of DNA sequencing and, consequently, has been unable to meet current and future demands from synthetic biology. Yet, the development of new frameworks for regulatory costs, trade-offs and the energy consumption of network structures represents a major challenge that might eventually lead to the construction of viable minimal cells. In 2003, a functional bacteriophage genome was synthesized in two weeks, and attempts at the *de novo* synthesis of bacterial genomes are currently under way. With further falls in price and increases in throughput, however, the time will surely come when *de novo* DNA writing will become a routine and standard method for molecular biology and bioengineering. The capability for *de novo* DNA writing will offer the freedom to obtain any DNA molecule with convenience, and will undoubtedly transform biomedical research.

A foundation for *de novo* design is the understanding of the sequence–structure–function relationship of biological molecules, focusing mostly proteins and RNAs. In this respect, small interfering RNA (siRNA) has become a very effective research tool, as it may be used to selectively suppress the expression of proteins, one at a time, while providing valuable information on gene function. Some siRNA experiments have been conducted using cell lines which allow a higher degree of automation. It may also be possible to use a combination of well-characterized siRNAs to affect multiple genes in a pathway, in particular parts of a pathway, or even multiple pathways. Such experiments may provide a better understanding of individual pathways, of pathway redundancies and of pathways crosstalk.

The incorporation of next-generation cell-free metagenomics capabilities to screen for unusual new polymerases should result in a self-reinforcing cycle of discovery and evolved polymerases that promise significant changes in the biochemical capabilities of existing enzymes. A knowledge of metabolic pathways is indispensable to understand a living system at the level of molecular networks. Moreover, an automated method, or a complementary tool, for the rapid prediction of network relationships of enzymes and substrates/products in a living system, would surely expedite this understanding.

Today, the deciphering of the structure and organization of gene regulatory networks remains in its infancy. However, in order to better understand the biology of the systems under investigation, the trend is clearly towards the aggregation of multiple sources of biological information and, to that end, it is challenging to choose appropriate experimental data and computational approaches. The modeling of a static and dynamic regulatory network remains an open-ended problem that has yet to be solved; therefore, the choice of a procedure that agrees with the initial biological question is imperative.

Clearly, this is a fascinating time in which to be involved in biological research. A comprehensive understanding will benefit from an integrated approach that simultaneously incorporates the individual and contextual properties of all constituents. Progress in research requires successful collaboration between by biologists, engineers, bioinformaticians, biophysicists, mathematicians and physicians in this exciting and ever-evolving field.

References

1. Alterovitz, G., Afkhami, E., Barillari, J. and Ramoni, M. (2006) *Proteomics*, in *Encyclopedia of Biomedical Engineering* (ed. M. Akay), John Wiley & Sons, Inc., New York.
2. Alterovitz, G., Liu, J., Chow, J. and Ramoni, M.F. (2006) Automation, parallelism, and robotics for proteomics. *Proteomics*, **6**(14), 4016–22.
3. Alterovitz, G., Jiwaji, A. and Ramoni, M.F. (2008) Automated programming for bioinformatics algorithm deployment. *Bioinformatics*, **24**(3), 450–1.

4. Shannon, P.T., Reiss, D.J., Bonneau, R. and Baliga, N.S. (2006) The Gaggle: an open-source software system for integrating bioinformatics software and data sources. *BMC Bioinformatics*, **7**, 176.
5. Alterovitz, G., Aivado, M., Spentzos, D. *et al.* (2004) Analysis and robot pipelined automation for SELDI-TOF mass spectrometry. *Conference Proceedings of the IEEE Engineering in Medicine and Biology Society*, **4**, 3068–71.
6. Alterovitz, G., Xiang, M., Liu, J. *et al.* (2008) System-wide peripheral biomarker discovery using information theory. *Pacific Symposium on Biocomputing*, **13**, 231–42.
7. Khomyakova, E., Livshits, M.A., Steinhauser, M.-C. *et al.* (2008) On-chip hybridization kinetics for optimization of gene expression experiments. *Biotechniques*, **44**(1), 109–17.
8. Subramanian, A., Livshits, M.A., Steinhauser, M.-C. *et al.* (2005) Gene set enrichment analysis: a knowledge-based approach for interpreting genome-wide expression profiles. *Proceedings of the National Academy of Sciences of the United States of America*, **102**(43), 15545–50.
9. Dolan, M.E., Ni, L., Camon, E. and Blake, J.A. (2005) A procedure for assessing GO annotation consistency. *Bioinformatics*, **21**(Suppl. 1), i136–43.
10. Alterovitz, G., Xiang, M., Mohan, M. and Ramoni, M.F. (2007) GO PaD: the gene ontology partition database. *Nucleic Acids Research*, **35** (Database issue), D322–7.
11. Sebastiani, P., Ramoni, M.F., Nolan, V. *et al.* (2005) Genetic dissection and prognostic modeling of overt stroke in sickle cell anemia. *Nature Genetics*, **37**(4), 435–40.
12. Jeong, H., Mason, S.P., Barabási, A.-L. and Oltvai, Z.N. (2001) Lethality and centrality in protein networks. *Nature*, **411**(6833), 41–2.
13. Alterovitz, G., Muralidhar, V. and Ramoni, M.F. (2006) Gene lethality detection and characterization via topological analysis of regulatory networks. *IEEE Transactions on Circuits and Systems I*, **53**(11), 2438–43.
14. Alterovitz, G. and Ramoni, M.F. (eds) (2007) *Systems Bioinformatics: An Engineering Case-Based Approach*, Artech House, Boston, MA.
15. Alterovitz, G. (2006) Bayesian methods in proteomics, in *Bayes Boot Camp*, Harvard Medical School, Boston, MA.
16. Shannon, P., Markiel, A., Ozier, O. *et al.* (2003) Cytoscape: a software environment for integrated models of biomolecular interaction networks. *Genome Research*, **13**(11), 2498–504.
17. Hu, Z., Ng, D.M., Yamada, T. *et al.* (2007) VisANT 3.0: new modules for pathway visualization, editing, prediction and construction. *Nucleic Acids Research*, **35** (Web Server issue), W625–32.
18. Wood, C. and Alterovitz, G. (2007) Fundamentals of design for synthetic biology, in *Systems Bioinformatics: An Engineering Case-Based Approach* (eds G. Alterovitz and M.F. Ramoni), Artech House, New York.
19. Forster, A.C. and Church, G.M. (2006) Towards synthesis of a minimal cell. *Molecular Systems Biology*, **2**, 45.

Index

Page references in bold refer to tables. Those in italics refer to figures.

Automation in Proteomics and Genomics: An Engineering Case-Based Approach Edited by Gil Alterovitz, Roseann Benson and Marco Ramoni